EXTRAORDINAI
ROBERT KANIGEL'S

THE MAN WHO KNEW INFINITY

"ENLIGHTENING. . . . a magic, tragic ugly-duckling fable. . . . Ramanujan's remarkable story comes through. . . ."

—*The New York Times*

"The most luminous expression ever of . . . genius interacting with genius . . . I've seen nothing to compare with it."

—Hugh Kenner, *BYTE*

"ENTHRALLING . . . one of the best scientific biographies I've ever seen."

—Dr. John Gribbin,
author of *In Search of Shrödinger's Cat*

"COMPELLING . . . a work of arduous research and rare insight . . . Kanigel deserves high praise."

—*Booklist*

". . . a REMARKABLE book. . . . a model of the biographer's art: Kanigel has taken a man, a social context and a specialist field and made each accessible and convincing. He has done so with a rare combination of skills—encyclopedic thoroughness, meticulous research, genuine sympathy for his subjects and first-rate writing of exceptional lucidity and verve. THOUGHTFUL, COMPASSIONATE AND CLEAR, *THE MAN WHO KNEW INFINITY* IS A MASTERPIECE. . . . BREATHTAKING."

—*The Washingon Post Book World*

A BOOK-OF-THE-MONTH-CLUB
FEATURED SELECTION

FINALIST FOR THE *LOS ANGELES TIMES BOOK AWARD*

"[A] SUPERBLY CRAFTED biography. . . . Kanigel succeed[s] in giving a taste of Ramanujan the mathematician, but his exceptional triumph is in the telling of this wonderful human story. . . . a pleasure to read . . . *THE MAN WHO KNEW INFINITY* is a thoughtful and deeply moving account of a signal life."

—*Science*

"A simple story VIVIDLY TOLD. . . . Kanigel excels in descriptions that will appeal to both the lay and scholarly reader."

—*San Francisco Chronicle*

"PERSPICACIOUS, INFORMED, IMAGINATIVE, *THE MAN WHO KNEW INFINITY* is . . . the best mathematical biography I have ever read."

—*The New York Review of Books*

"[An] extremely well-researched and well-written biography."

—*Library Journal*

"Mr. Kanigel has a wonderful gift. . . . The drama of Ramanujan's 'Spring' and 'Autumn' comes through magnificently."

—Freeman Dyson,
author of *Disturbing the Universe*

"MOVING AND ASTONISHING."

—*Publishers Weekly*

"*THE MAN WHO KNEW INFINITY* is a story at least as compelling as Brian Epstein's discovery of the Beatles . . . [a] richly detailed road map to strange, wondrous foreign cultures. . . . Kanigel expertly intertwines the details of Ramanujan's odd, doomed life with his soaring professional accomplishments."

—*Los Angeles Times Book Review*

"BRILLIANTLY REALIZED. . . . [the] fascinating story of a difficult but astoundingly fruitful cross-cultural collaboration."

—*Kirkus Reviews*

"*THE MAN WHO KNEW INFINITY* is an accessible look at an almost romantic episode in the enormously rich intellectual world of mathematics . . . Robert Kanigel also gives a real sense of Ramanujan's creative compulsion which, like Mozart's, contained the seeds of both success and tragedy."

—*Baltimore Evening Sun*

". . . more fascinating than a novel . . . a verbal portrait, A VIRTUAL MASTERPIECE, complete with vibrant scenes from all the places graced by the presence of Ramanujan. . . . ENCHANTING."

—*Lexington Herald-Leader* (Kentucky)

"*THE MAN WHO KNEW INFINITY* tells of the plight of unrecognized genius. . . . this story of romance with mathematics makes for lively reading . . . with a heartbreaking end."

—*Christian Science Monitor*

"SPLENDID . . . One of Robert Kanigel's achievements in *THE MAN WHO KNEW INFINITY* is to make the math magic . . . accessible. . . . a very human story. . . . EXCITING."

—*San Diego Union*

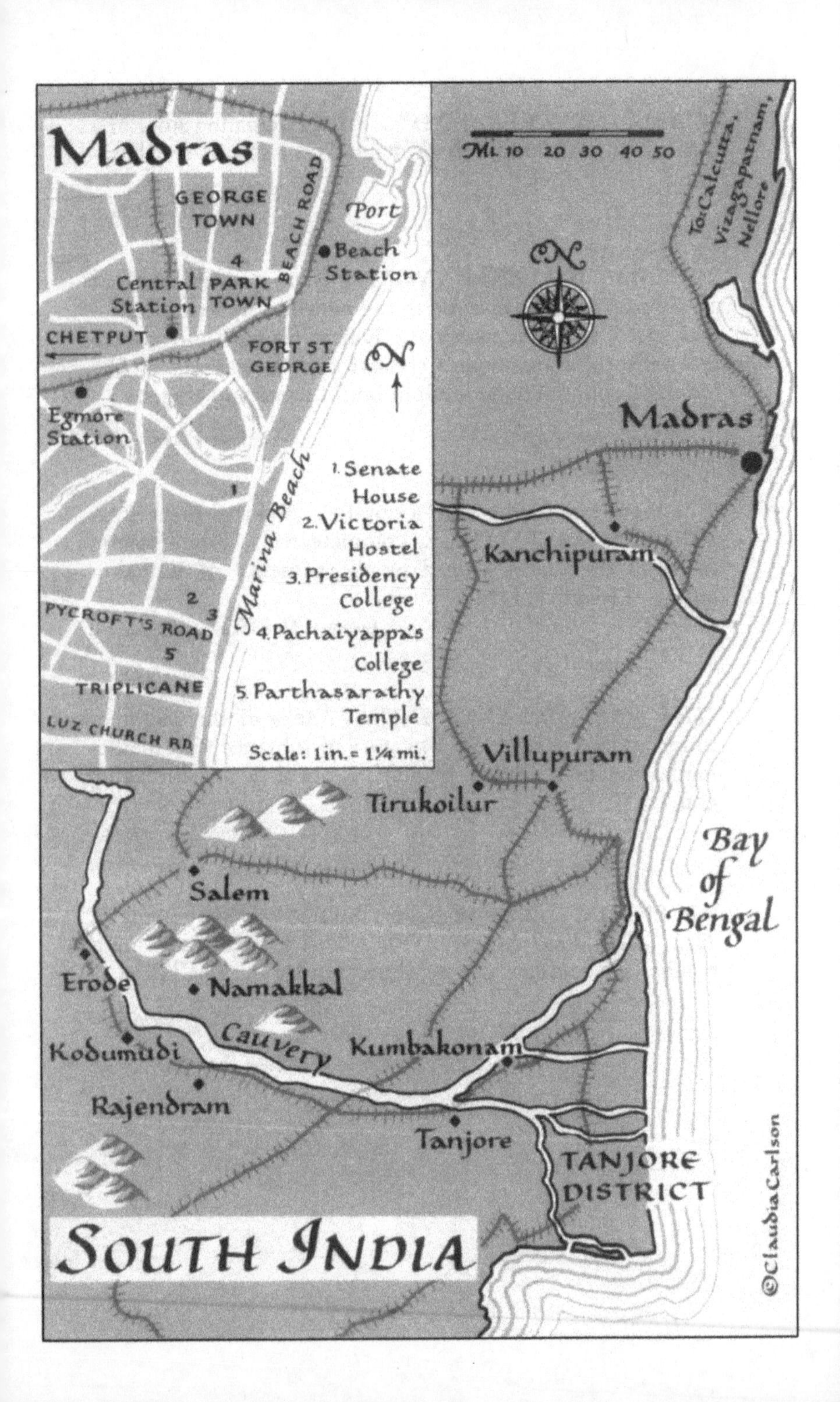
Madras
GEORGE TOWN
BEACH ROAD
Port
Beach Station
4
Central Station
PARK TOWN
CHETPUT
FORT ST GEORGE
Egmore Station
Marina Beach
1
1. Senate House
2. Victoria Hostel
3. Presidency College
4. Pachaiyappa's College
5. Parthasarathy Temple
2
3
PYCROFT'S ROAD
5
TRIPLICANE
LUZ CHURCH RD
Scale: 1in. = 1¼ mi.
Mi. 10 20 30 40 50
To: Calcutta, Vizagapatnam, Nellore
N
Madras
Kanchipuram
Villupuram
Tirukoilur
Salem
Bay of Bengal
Erode
Namakkal
Cauvery
Kodumudi
Kumbakonam
Rajendram
Tanjore
TANJORE DISTRICT
SOUTH INDIA
©Claudia Carlson

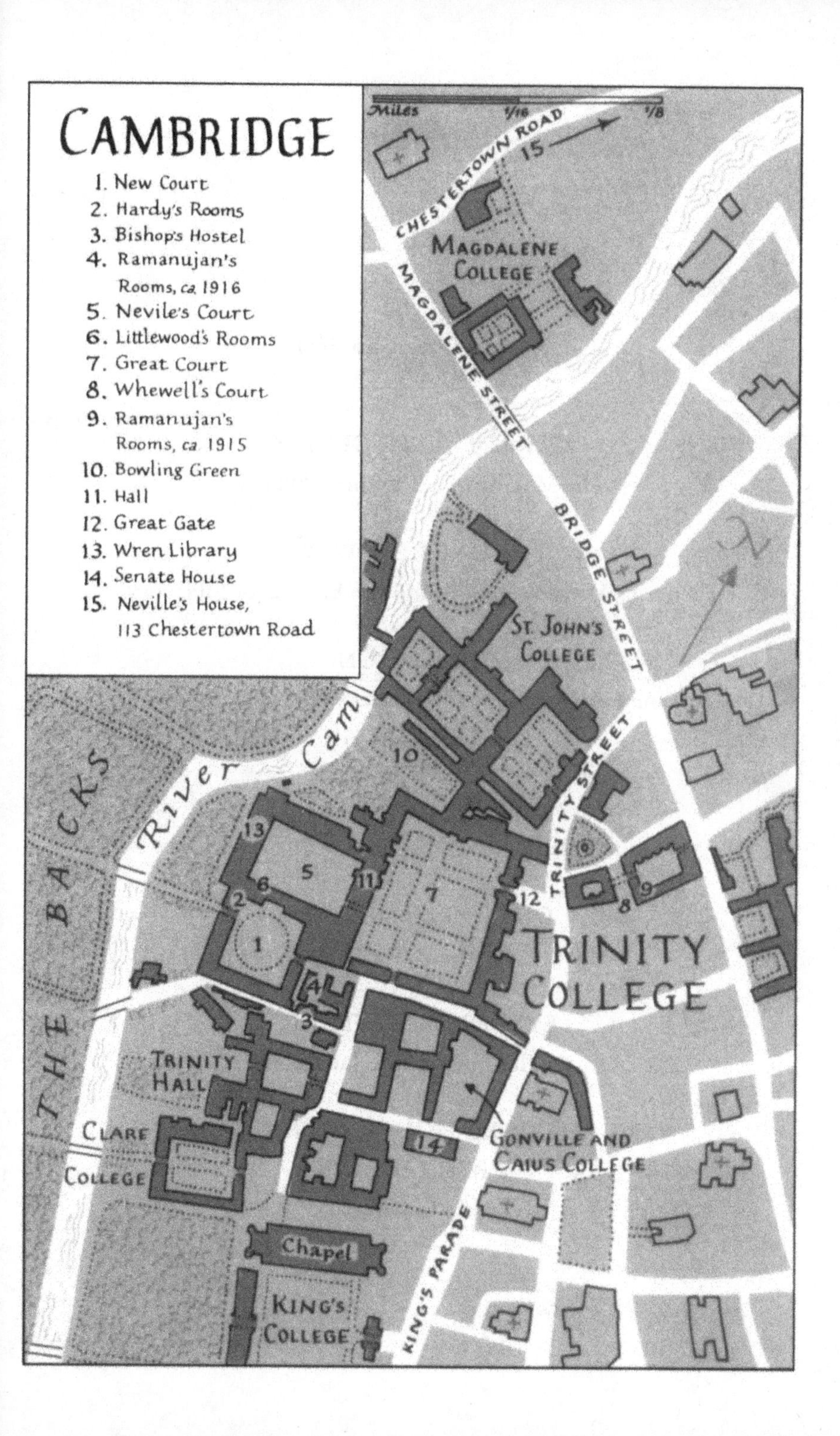
CAMBRIDGE
1. New Court
2. Hardy's Rooms
3. Bishop's Hostel
4. Ramanujan's Rooms, ca. 1916
5. Nevile's Court
6. Littlewood's Rooms
7. Great Court
8. Whewell's Court
9. Ramanujan's Rooms, ca. 1915
10. Bowling Green
11. Hall
12. Great Gate
13. Wren Library
14. Senate House
15. Neville's House, 113 Chestertown Road
Miles
1/16
1/8
CHESTERTOWN ROAD
MAGDALENE COLLEGE
MAGDALENE STREET
BRIDGE STREET
ST. JOHN'S COLLEGE
TRINITY STREET
River Cam
THE BACKS
TRINITY COLLEGE
TRINITY HALL
CLARE COLLEGE
GONVILLE AND CAIUS COLLEGE
Chapel
KING'S COLLEGE
KING'S PARADE

Also by Robert Kanigel

APPRENTICE TO GENIUS:
The Making of a Scientific Dynasty

THE ONE BEST WAY:
Frederick Winslow Taylor and the Enigma of Efficiency

HIGH SEASON:
How One French Riviera Town Has Seduced Travelers for Two Thousand Years

FAUX REAL:
Genuine Leather and 200 Years of Inspired Fakes

ON AN IRISH ISLAND:
The Lost World of the Great Blasket

EYES ON THE STREET:
The Life of Jane Jacobs

HEARING HOMER'S SONG:
The Brief Life and Big Idea of Milman Parry

THE MAN WHO KNEW INFINITY

A Life of the Genius Ramanujan

ROBERT KANIGEL

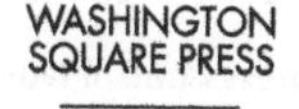

ATRIA

New York London Toronto Sydney New Delhi

For Mom and Dad
with love and thanks

The author gratefully acknowledges permission from the following sources to reprint material in their control: The Master and Fellows of Trinity College, Cambridge, for S. Ramanujan and G. H. Hardy documents residing in Trinity Library; St. Catherine's School magazine, 1933, for G. E. Hardy poem; Reading University Archives for E. H. Neville manuscript; Freeman Dyson for his letter to C. P. Snow; Cambridge University Press for material from *A Mathemetician's Apology* by G. H. Hardy, copyright 1940; The London Mathematical Society for G. H. Hardy documents residing in Trinity Library; the Syndics of Cambridge University Library for S. Ramanujan documents residing in Cambridge University Library. Author's failure to obtain a necessary permission for the use of any other copyrighted material included in this work is inadvertent and will be corrected in future printings following notification in writing to the Publisher of such omission accompanied by appropriate documentation.

ATRIA

An Imprint of Simon & Schuster, Inc.
1230 Avenue of the Americas
New York, NY 10020

This Washington Square Press/Atria Paperback edition April 2016

WASHINGTON SQUARE PRESS / **ATRIA** PAPERBACK and colophons are trademarks of Simon & Schuster, Inc.

Manufactured in the United States of America

11 13 15 17 19 20 18 16 14 12

Library of Congress Control Number: 90049788

ISBN 978-0-684-19259-8
ISBN 978-1-4767-6349-1 (pbk)
ISBN 978-1-4391-2186-3 (ebook)

Contents

Prologue

One day in the summer of 1913, a twenty-year-old Bengali from an old and prosperous Calcutta family stood in the chapel of King's College in the medieval university town of Cambridge, England. A glorious, grandly proportioned place, more cathedral than chapel, it was the work of three kings of England going back to 1446. Light streamed in through stained glass panels ranging across the south wall. Great fluted columns reached heavenward, flaring out into the massive splayed vault of the roof.

Prasantha Chandra Mahalanobis was smitten. Scarcely off the boat from India and planning to study in London, he had come up on the train for the day to sightsee. But now, having missed the last train back to London and staying with friends, he couldn't stop talking about the chapel and its splendors, how moved he'd been, how . . .

Perhaps, proposed a friend, he should forget London and come to King's instead. That was all Mahalanobis needed to hear. The next day he met with the provost, and soon, to his astonishment and delight, he was a student at King's College, Cambridge.

He had been at Cambridge for about six months when his mathematics tutor asked him, "Have you met your wonderful countryman Ramanujan?"

He had not yet met him, but he had heard of him. Ramanujan was a self-taught mathematical prodigy from a town outside Madras, in South India, a thousand miles from the sophisticated Calcutta that Mahalanobis knew best, a world as different from his own as Mahalanobis's was from England. The South, as educated North Indians were wont to see it, was backward and superstitious, scarcely brushed by the enlightened rationality of Bombay and Calcutta. And yet, somehow, out of such a place, from a poor family, came a mathematician so alive with genius that the English had practically hand-delivered him to Cambridge, there to share his gifts with the scholars of Trinity College and learn whatever they could teach him.

Among the colleges of Cambridge University, Trinity was the largest, with the most lustrous heritage, home to kings, poets, geniuses. Isaac Newton himself had studied there; since 1755, his marble likeness, holding the prism he'd used to explore the polychromatic nature of light, stood in its chapel. Lord Byron had gone to Trinity. So had Tennyson, Thackeray, and Fitzgerald. So had the historian Macaulay, and the physicist Rutherford, and the philosopher Bertrand Russell. So had five English prime ministers.

And now, Ramanujan was at Trinity, too.

Soon Mahalanobis did meet him, and the two became friends; on Sunday mornings, after breakfast, they'd go for long walks, talk about life, philosophy, mathematics. Later, looking back, Mahalanobis would date the flowering of their friendship to one day in the fall following Ramanujan's arrival. He'd gone to see him at his place in Whewell's Court, a three-story stone warren of rooms built around a grassy quadrangle laced with arched Gothic windows and pierced at intervals by staircases leading to rooms. One such portal led to Ramanujan's small suite, on the ground floor, a step or two off the court.

It had turned cold in Cambridge and as Mahalanobis came in, he saw Ramanujan, with his fleshy, pockmarked face, sitting huddled by the fire. Here was the pride of India, the man whom the English had moved heaven and earth to bring to Cambridge. But well-laid plans had gone awry. It was the shameful year of 1914, and Europe had gone to war. The graceful arched cloisters of Nevile's Court, Sir Christopher Wren's eternal stamp on Trinity, had become an open-air hospital. Thousands had already left for the front. Cambridge was deserted. And cold.

Are you warm at night? asked Mahalanobis, seeing Ramanujan beside the fire. No, replied the mathematician from always-warm Madras, he slept with his overcoat on, wrapped in a shawl. Figuring his friend hadn't enough blankets, Mahalanobis stepped back into the little sleeping alcove on the other side of the fireplace from the sitting room. The bedspread was loose, as if Ramanujan had just gotten up. Yet the blankets lay perfectly undisturbed, tucked neatly under the mattress.

Yes, Ramanujan had enough blankets; he just didn't know what to *do* with them. Gently, patiently, Mahalanobis showed him how you peeled them back, made a little hollow for yourself, slipped inside . . .

* * *

For five years, walled off from India by the war, Ramanujan would remain in strange, cold, distant England, fashioning, through twenty-one major papers, an enduring mathematical legacy. Then, he would go home to India, to a hero's welcome, and die.

"Srinivasa Ramanujan," an Englishman would later say of him, "was a mathematician so great that his name transcends jealousies, the one superlatively great mathematician whom India has produced in the last thousand years." His leaps of intuition confound mathematicians even today, seven decades after his death. His papers are still plumbed for their secrets. His theorems are being applied in areas—polymer chemistry, computers, even (it has recently been suggested) cancer—scarcely imaginable during his lifetime. And always the nagging question: *What might have been*, had he been discovered a few years earlier, or lived a few years longer?

Ramanujan was a simple man. His needs were simple. So were his manners, his humor. He was no idiot savant; he was intelligent in realms outside mathematics, persistent, hardworking, and even, in his own way, charming. But by the lights of Cambridge or, for that matter, of Calcutta or Bombay, he was supremely narrow and naive. Something so small as Mahalanobis's lesson in the art of blanketing could leave him "extremely touched." He was shamed by the most insignificant slight. His letters, outside their mathematical content, are barren of grace or subtlety.

In this book I propose to tell Ramanujan's story, the story of an inscrutable intellect and a simple heart.

It is a story of the clash of cultures between India and the West—between the world of Sarangapani Sannidhi Street in Kumbakonam in South India, where Ramanujan grew up, and the glittering world of Cambridge; between the pristine proofs of the Western mathematical tradition and the mysterious powers of intuition with which Ramanujan dazzled East and West alike.

It is a story of one man and his stubborn faith in his own abilities. But it is not a story that concludes, *Genius will out*—though Ramanujan's, in the main, did. Because so nearly did events turn out otherwise that we need no imagination to see how the least bit less persistence, or the least bit less luck, might have consigned him to obscurity. In a way, then, this is also a story about social and educational *systems*, and about how they matter, and how they can sometimes nurture talent and sometimes crush it. How many Ramanujans, his life begs us to ask, dwell in India today, unknown and unrecognized? And how many in America and Britain, locked away in racial or economic ghettos, scarcely aware of worlds outside their own?

This is a story, too, about what you do with genius once you find it. Ramanujan was brought to Cambridge by an English mathematician of aristocratic mien and peerless academic credentials, G. H. Hardy, to whom he had written for help. Hardy saw that Ramanujan was a rare flower, one not apt to tolerate being stuffed methodically full of all the mathematical knowledge he'd never acquired in India. "I was afraid," he wrote, "that if I insisted unduly on matters which Ramanujan found irksome, I might destroy his confidence and break the spell of his inspiration."

Ramanujan was a man who grew up praying to stone deities; who for most of his life took counsel from a family goddess, declaring it was she to whom his mathematical insights were owed; whose theorems would, at intellectually backbreaking cost, be proved true—yet leave mathematicians baffled that anyone could divine them in the first place. This is also a book, then, about an uncommon and individual mind, and what its quirks may suggest about creativity, intuition, and intelligence.

Like most books, this one started with an idea. Sadly, it was not mine, but that of Barbara Grossman, then senior editor at Crown Publishers, now publisher at Scribners. Barbara first encountered the name of Ramanujan in late 1987, a time when magazines and newspapers in the United States, India, and Britain were full of articles marking the hundredth anniversary of his birth. Like Mahalanobis in King's College Chapel, Barbara was smitten. First, with the sheer romance of his life—the *story* in it. But also with how today, years after his death and long into the computer age, some of his theorems were, as she put it later, being "snatched back from history."

"Ramanujan who?" I said when my agent, Vicky Bijur, told me of Barbara's interest in a biography of him. Though skeptical, I did some preliminary research into his life, as recorded by his Indian biographers. And the more I learned, the more I, too, came under Ramanujan's spell. His was a rags-to-intellectual-riches story. Parts of it, wrote an English mathematician, B. M. Wilson, in the 1930s, "might be lifted almost unchanged by a scenario-writer for the talkies." My doubts fell away. My excitement mounted at the prospect of delving into the life of this strange genius.

Early on, I viewed a documentary about Ramanujan's life by the

British filmmaker Christopher Sykes. Released by the BBC the previous year as *Letters from an Indian Clerk,* Sykes's film superbly distilled, into a single hour, something of the romance of Ramanujan's life. But watching it, I grew beguiled by G. H. Hardy, too. Hardy, it turned out, was the *third* English mathematician to whom Ramanujan had appealed; the other two declined to help. And Hardy did not just recognize Ramanujan's gifts; he went to great lengths to bring him to England, school him in the mathematics he had missed, and bring him to the attention of the world.

Why Hardy?

Was it sheer mathematical acumen? Probably not; the other two mathematicians were equally distinguished. There must have been other, less purely intellectual traits demanded of him—a special openness, perhaps, a willingness to disrupt his life and stake his reputation on someone he'd never seen.

Hardy, I learned, was a bizarre and fascinating character—a cricket aficionado, a masterful prose stylist, a man blessed with gorgeous good looks who to his own eyes was so repulsively ugly he couldn't look at himself in the mirror. And this enfant terrible of English mathematics was, at the time he heard from Ramanujan, working a revolution on his field that would be felt for generations to come.

One is, of course, moved to praise Hardy's ability to see genius in the tattered garb in which it was clothed, and to agree that the world was enriched as a result. But, it struck me, *Hardy* was enriched, too. His whole life was shaped by his time with Ramanujan, which he called "the one romantic incident in my life." The story of Ramanujan, then, is a story about *two* men, and what they meant to each other.

I must say one thing more, though in doing so I risk alienating a few readers. To Americans, India and Cambridge are, indeed, foreign countries. And as L. P. Hartley has written, the past is alien territory, too: "They do things differently there." Thus, the years around the turn of the century when our story begins, a time when India was still British and Victoria still queen, represent a second foreign country to explore. Now, to these two worlds remote in place and time, I must add a third—the mathematics that Ramanujan and Hardy did, alone and together, as their life's work.

It is tempting to concentrate exclusively on the exotic and flavorful elements of Ramanujan's life and skip the mathematics altogether. Indeed, virtually all who have taken Ramanujan as their subject have severed his work from his life. Biographies as do exist either ignore the mathematics, or banish it to the back of the book. Similarly, scholarly papers devoted to Ramanujan's mathematics normally limit to a few paragraphs their attention to his life.

And yet, can we understand Ramanujan's life without some appreciation for the mathematics that he lived for and loved? Which is to say, can we understand an artist without gaining a feel for his art? A philosopher without some glimpse into what he believed?

Mathematics, I am mindful, presents a special problem to the general reader (and writer). Art, at least, you can *see*. Philosophy and literature, too, have the advantage that, however recondite, they can at least be rendered into English. Mathematics, however, is mired in a language of symbols foreign to most of us, explores regions of the infinitesimally small and the infinitely large that elude words, much less understanding. So specialized is mathematics today, I am told, that most mathematical papers appearing in most mathematics journals are indecipherable even to most mathematicians. Pennsylvania State's George Andrews, who rediscovered a long-forgotten Ramanujan manuscript at Trinity College, says it took someone already expert in the narrow area of mathematics with which it dealt to recognize it—that merely being a professional mathematician with a Ph.D. would not have sufficed.

What hope, then, has the general reader faced with Ramanujan's work?

Little, certainly, if we set as the task to follow one of Ramanujan's proofs through twenty pages of hieroglyphics in a mathematics journal—*especially* in the case of Ramanujan, who routinely telescoped a dozen steps into two, leaving his reader to find the connections. But to come away with some flavor of his work, the paths by which he got there, its historical roots? These pose no insuperable problem—certainly no more than following a philosophical argument, or a challenging literary exegesis.

In one sense, at least, Ramanujan's mathematics is *more* accessible than some other fields; much of it comes under the heading of number theory, which seeks out properties of, and patterns among, the ordinary numbers with which we deal every day; and 8s, 19s, and 376s are surely more familiar than quarks, quasars, and phosphocreatine. While the mathematical tools Ramanujan used were subtle and powerful, the problems to which he applied them were often surprisingly easy to formulate.

In a note at the end of the book I mention by name many who have helped me along the way. But here I wish to especially thank the people of South

India as a whole for making my time there so personally satisfying.

I spent five weeks in the South, traveling to places that figured in Ramanujan's life. I rode trains and buses, toured temples, ate with my hands off banana leaves. I was butted in the behind by a cow on the streets of Kumbakonam, shared a room with a lizard in Kodumudi. I saw Erode, the town where Ramanujan was born. I toured the house in which he grew up, participated in opening exercises at his alma mater, wandered through the grounds of the temple in Namakkal to which he came at a turning point in his life, and saw the room in which he died in Madras. The people of South India took me into their homes. They bestowed upon me every kindness. Auto rickshaw and bicycle rickshaw drivers often went to extraordinary lengths to get me to out-of-the-way places, forever struggling to understand what to them was my atrocious pronunciation of South Indian place-names. They stared at the whiteness of my skin but always treated me with gentleness and goodwill.

Spend any length of time among the people of South India and it is hard not to come away with a heightened sense of spirituality, a deepened respect for hidden realms, that implicitly questions Western values and ways of life. In the West, all through the centuries, artists have sought to give expression to religious feeling, creating Bach fugues and Gothic cathedrals in thanks and tribute to their gods. In South India today, such religious feeling hangs heavy in the air, and to discern a spiritual resonance in Ramanujan's mathematics seems more natural by far than it does in the secular West.

Ramanujan's champion, Hardy, was a confirmed atheist. Yet when he died, one mourner spoke of his

> profound conviction that the truths of mathematics described a bright and clear universe, exquisite and beautiful in its structure, in comparison with which the physical world was turbid and confused. It was this which made his friends . . . think that in his attitude to mathematics there was something which, being essentially spiritual, was near to religion.

The same, but more emphatically, goes for Ramanujan, who all his life believed in the Hindu gods and made the landscape of the Infinite, in realms both mathematical and spiritual, his home. "An equation for me has no meaning," he once said, "unless it expresses a thought of God."

Baltimore
May 1990

CHAPTER ONE

In the Temple's Coolness

[1887 to 1903]

1. DAKSHIN GANGE

He heard it all his life—the slow, measured *thwap . . . thwap . . . thwap* . . . of wet clothes being pounded clean on rocks jutting up from the waters of the Cauvery River. Born almost within sight of the river, Ramanujan heard it even as an infant. Growing up, he heard it as he fetched water from the Cauvery, or bathed in it, or played on its sandy banks after school. Later, back in India after years abroad, fevered, sick, and close to death, he would hear that rhythmic slapping sound once more.

The Cauvery was a familiar, recurring constant of Ramanujan's life. At some places along its length, palm trees, their trunks heavy with fruit, leaned over the river at rakish angles. At others, leafy trees formed a canopy of green over it, their gnarled, knotted roots snaking along the riverbank. During the monsoon, its waters might rise ten, fifteen, twenty feet, sometimes drowning cattle allowed to graze too long beside it. Come the dry season, the torrent became a memory, the riverbanks wide sandy beaches, and the Cauvery itself but a feeble trickle tracing the deepest channels of the riverbed.

But always it was there. Drawing its waters from the Coorg Mountains five hundred miles to the west, branching and rebranching across the peninsula, its flow channeled by dams and canals some of which went back fifteen hundred years, the Cauvery painted the surrounding countryside an intense, unforgettable green. And that single fact, more than any other, made Ramanujan's world what it was.

Kumbakonam, his hometown, flanked by the Cauvery and one of its tributaries, lay in the heartland of staunchly traditional South India, 160 miles south of Madras, in the district then known as Tanjore. Half the district's thirty-seven hundred square miles, an area the size of the state of Delaware, was watered directly by the river, which fell gently, three feet per mile, to the sea, spreading its rich alluvial soil across the delta.

The Cauvery conferred almost unalloyed blessing. Even back in 1853, when it flooded, covering the delta with water and causing immense damage, few lives were lost. More typically, the great river made the surrounding land immune to year-to-year variation in the monsoon, upon whose caprices most of the rest of India hung. In 1877, in the wake of two straight years of failed monsoons, South India had been visited by drought, leaving thousands dead. But Tanjore District, nourished by the unfailing Cauvery, had been scarcely touched; indeed, the rise in grain prices accompanying the famine had brought the delta unprecedented prosperity.

No wonder that the Cauvery, like the Ganges a thousand miles north, was one of India's sacred rivers. India's legendary puranas told of a mortal known as Kavera-muni who adopted one of Brahma's daughters. In filial devotion to him, she turned herself into a river whose water would purify from all sin. Even the holy Ganges, it was said, periodically joined the Cauvery through some hidden underground link, so as to purge itself of pollution borne of sinners bathing in its waters.

Dakshin Gange, the Cauvery was called—the Ganges of the South. And it made the delta the most densely populated and richest region in all of South India. The whole edifice of the region's life, its wealth as well as the rich spiritual and intellectual lives its wealth encouraged, all depended on its waters. The Cauvery was a place for spiritual cleansing; for agricultural surfeit; for drawing water and bathing each morning; for cattle, led into its shallow waters by men in white dhotis and turbans, to drink; and always, for women, standing knee-deep in its waters, to let their snaking ribbons of cotton or silk drift out behind them into the gentle current, then gather them up into sodden clumps of cloth and slap them slowly, relentlessly, against the water-worn rocks.

2. SARANGAPANI SANNIDHI STREET

In September 1887, two months before her child was due to be born, a nineteen-year-old Kumbakonam girl named Komalatammal traveled to Erode, her parental home, 150 miles upriver, to prepare for the birth of the child she carried. That a woman returned to her native home for the birth of her first child was a tradition so widely observed that officials charged with monitoring vital statistics made a point of allowing for it.

Erode, a county seat home to about fifteen thousand people, was located at the confluence of the Cauvery and one of its tributaries, the Bhavani, about 250 miles southwest of Madras. At Erode—the word means "wet skull," recalling a Hindu legend in which an enraged Siva tears off one of Brahma's five heads—the Cauvery is broad, its stream bed littered with great slabs of protruding rock. Not far from the river, in "the fort," as the town's original trading area was known, was the little house, on Teppukulam Street, that belonged to Komalatammal's father.

It was here that a son was born to her and her husband Srinivasa, just after sunset on the ninth day of the Indian month of Margasirsha—or Thursday, December 22, 1887. On his eleventh day of life, again in accordance with tradition, the child was formally named, and a year almost to the day after his birth, Srinivasa Ramanujan Iyengar and his mother returned to Kumbakonam, where he would spend most of the next twenty years of his life.

"Srinivasa"—its initial syllable pronounced *shri*—was just his father's name, automatically bestowed and rarely used; indeed, on formal documents, and when he signed his name, it usually atrophied into an initial "S." "Iyengar," meanwhile, was a caste name, referring to the particular branch of South Indian Brahmins to which he and his family belonged. Thus, with one name that of his father and another that of his caste, only "Ramanujan" was his alone. As he would later explain to a Westerner, "I have no proper surname." His mother often called him Chinnaswami, or "little lord." But otherwise he was, simply, Ramanujan.

He got the name, by some accounts, because the Vaishnavite saint Ramanuja, who lived around A.D. 1100 and whose theological doctrines injected new spiritual vitality into a withered Hinduism, was also born on a Thursday and shared with him other astrological likenesses.

"Ramanujan"—pronounced Rah-MAH-na-jun, with only light stress on the second syllable, and the last syllable sometimes closer to jum—means younger brother (*anuja*) of Rama, that model of Indian manhood whose story has been handed down from generation to generation through the Ramayana, India's national epic.

Ramanujan's mother, Komalatammal, sang *bhajans*, or devotional songs, at a nearby temple. Half the proceeds from her group's performances went to the temple, the other half to the singers. With her husband earning only about twenty rupees per month, the five or ten she earned this way mattered; never would she miss a rehearsal.

Yet now, in December 1889, she *was* missing them, four or five in a row. So one day, the head of the singing group showed up at Komalatammal's house to investigate.

There she found, piled near the front door, leaves of the margosa tree; someone, it was plain to her, had smallpox. Stepping inside, she saw a small, dark figure lying atop a bed of margosa leaves. His mother, chanting all the while, dipped the leaves in water laced with ground turmeric, and gently scoured two-year-old Ramanujan's pox-ridden body—both to relieve the infernal itching and, South Indian herbalists believed, subdue the fever.

Ramanujan would bear the scars of his childhood smallpox all his life. But he recovered, and in that was fortunate. For in Tanjore District, around the time he was growing up, a bad year for smallpox meant four thousand deaths. Fewer than one person in five was vaccinated. A cholera epidemic when Ramanujan was ten killed fifteen thousand people. Three or four children in every ten died before they'd lived a year.

Ramanujan's family was a case study in the damning statistics. When he was a year and a half, his mother bore a son, Sadagopan. Three months later, Sadagopan was dead.

When Ramanujan was almost four, in November 1891, a girl was born. By the following February, she, too, was dead.

When Ramanujan was six and a half, his mother gave birth to yet another child, Seshan—who also died before the year was out.

Much later, two brothers did survive—Lakshmi Narasimhan, born in 1898, when Ramanujan was ten, and Tirunarayanan, born when he was seventeen. But the death of his infant brothers and sister during those early years meant that he grew up with the solicitous regard and central position of an only child.

After the death of his paternal grandfather, who had suffered from leprosy, Ramanujan, seven at the time, broke out in a bad case of itching and boils. But this was not the first hint of a temperament inclined to extreme and unexpected reactions to stress. Indeed, Ramanujan was a sensitive, stubborn, and—if a word more often reserved for adults in their prime can be applied to a little boy—eccentric child. While yet an infant back in Erode, he wouldn't eat except at the temple. Later, in Kumbakonam, he'd take all the brass and copper vessels in the house and line them up from one wall to the other. If he didn't get what he wanted to eat, he was known to roll in the mud in frustration.

For Ramanujan's first three years, he scarcely spoke. Perhaps, it is tempting to think, because he simply didn't choose to; he was an enormously self-willed child. It was common in those days for a young wife to shuttle back and forth between her husband's house and that of her parents, and Komalatammal, worried by her son's muteness, took Ramanujan to see her father, then living in Kanchipuram, near Madras. There, at the urging of an elderly friend of her father's, Ramanujan began the ritual practice of *Akshara Abhyasam*: his hand, held and guided by his grandfather, was made to trace out Tamil characters in a thick bed of rice spread across the floor, as each character was spoken aloud.

Soon fears of Ramanujan's dumbness were dispelled and he began to learn the 12 vowels, 18 consonants, and 216 combined consonant-vowel forms of the Tamil alphabet. On October 1, 1892, the traditional opening day of school, known as *Vijayathasami*, he was enrolled, to the accompaniment of ancient Vedic chants, in the local *pial* school. A *pial* is the little porch in front of most South Indian houses; a *pial* school was just a teacher meeting there with half a dozen or so pupils.

But five-year-old Ramanujan, disliking the teacher, bristled at attending. Even as a child, he was so self-directed that, it was fair to say, unless he was ready to do something on his own, in his own time, he was scarcely capable of doing it at all; school for him often meant not keys to knowledge but shackles to throw off.

Quiet and contemplative, Ramanujan was fond of asking questions like, Who was the first man in the world? Or, How far is it between clouds? He liked to be by himself, a tendency abetted by parents who, when friends called, discouraged him from going out to play; so he'd talk to them from the window overlooking the street. He lacked all interest in sports. And in a world where obesity was virtually unknown, where bones protruded from humans and animals alike, he was, first as a child and then for most of his life, fat. He used to say—whether as boast, joke, or lament remains unclear—that if he got into a fight with another boy he had only to fall on him to crush him to pieces.

For about two years, Ramanujan was shuffled between schools. Beginning in March 1894, while still at his mother's parents' house in Kanchipuram, he briefly attended a school in which the language of instruction was not his native Tamil but the related but distinct Telugu. There, sometimes punished by having to sit with his arms folded in front of him and one finger turned up to his lips in silence, he would at times stalk out of class in a huff.

In a dispute over a loan, his grandfather quit his job and left Kanchipuram. Ramanujan and his mother returned to Kumbakonam, where he enrolled in the Kangayan Primary School. But when his other grandfather died, Ramanujan was bounced back to his maternal grandparents, who by now were in Madras. There he so fiercely fought attending school that the family enlisted a local constable to scare him back to class.

By mid-1895, after an unhappy six months in Madras, Ramanujan was once more back in Kumbakonam.

Kumbakonam was flanked by the Cauvery and the Arasalar, its tributary. Most streets ran parallel to these rivers or else marched straight down to their banks, perpendicular to the first set, making for a surprisingly regular grid system. And there, near the middle of this compact grid, on Sarangapani Sannidhi Street, a dirt road about thirty feet wide with squat little buildings close packed on either side, was Ramanujan's house.

The one-story structure, thatched with palm leaves, stood back about ten feet from the street, insulated, as it were, by its two-tiered, covered *pial*: it was a step or two up from the dusty (or muddy) street, another few up to the little porch. The stucco house faced the street with a twelve-foot-wide wall broken by a window to the left and a door to the right. A bystander in the street, peering through the open door and into the gloom of the interior, could sight all the way through to the back, where his gaze would be arrested by a splash of sunlight from the open rear court. The rest of the house, meanwhile, was offset to the left, behind the front window. Here was the main living area and, behind it, a small kitchen, redolent with years of cooking smells.

South India was not always hot; but it was never cold. At a latitude of about eleven degrees north. Kumbakonam lay comfortably within the tropics; even on a January winter's night, the thermometer dropped, on average, only to seventy degrees. And that climatological fact established an architectural fact, for it gave South Indian homes a kind of permeability; their interiors always savored a little of the outside (a feeling familiar to Americans with screened-in porches). Windows there were, but these were merely cutouts in the wall, perhaps with bars or shutters, never space-sealing panes of glass that left you conscious of being on one side or another. In the middle of most houses was a small courtyard, the *muttam*, open to the sky—like a skylight but again without the glass—that brought rain into the center of the house, where it was funneled to a drain that led back outside. In Ramanujan's house, smells from outside wafted inside. Lizards crawled, mosquitos flew unimpeded. The soft South Indian air, fragrant with roses, with incense, with cow dung burned as a fuel, wafted over everything.

Just outside the door lay Kumbakonam, an ancient capital of the Chola Empire. The Cholas had reached their zenith around A.D. 1000, when Europe wallowed in the Dark Ages, and had ruled, along with northern Ceylon, most of what, during Ramanujan's day, was known as the Madras Presidency (which, with those of Bombay and Calcutta, constituted the chief administrative and political units of British-ruled India). The dozen or so major temples dating from this period made Kumbakonam a magnet to pilgrims from throughout South India.

Every twelve years, around February or March, they came for the Mahammakham festival, commemorating a legendary post-Deluge event in which the seeds of creation, drifting upon the waters in a sacred pot—or *kumba,* source of the town's name—was pierced by the god Siva's arrow. The nectar thus freed, it was said, had collected in the Mahammakham "tank," the outdoor pool for ritual bathing that was part of every temple. At such times—as in 1897, when Ramanujan was nine—three-quarters of a million pilgrims might descend on the town. And the great tank, surrounded by picturesque mandapams, or halls, and covering an expanse of twenty acres, would be so filled with pilgrims that its water level was said to rise several inches.

When not in use, temple tanks could seem anything but spiritually uplifting. Open, stone-lined reservoirs, sometimes stocked with fish, frequently green with algae, they often served as breeding grounds for malarial mosquitos. Situated on low ground between two rivers, Kumbakonam was notorious for its bad water, its mosquitos, and its filarial elephantiasis, a mosquito-borne disease that left its victims with grotesquely deformed limbs, sometimes with scrota the size of basketballs. When Ramanujan was six, the town completed a drainage system. But this was designed to carry off only surface water, not sewage, and most of the town's health problems continued unabated.

Kumbakonam, a day's train ride from Madras, which was almost 200 miles north and the nearest real metropolis, had a seventy-two-bed hospital. It had four police stations, two lower secondary English schools, three conducting classes in Tamil, a high school of excellent reputation, and a college. Indeed, with a population during Ramanujan's day of more than fifty thousand, it was no mere village, but a major town, sixth largest in the Madras Presidency.

Just outside town and all through the surrounding districts ranged some of the richest cropland in all of India. Two-thirds of the population—including whole castes given over to agricultural labor alone, like the Paraiyans and the Pallans—worked the land for a few annas a day. Silt carried down to the delta by the Cauvery made the use of expensive manure as fertilizer unnecessary. Narrow strips of land beside the river, annually submerged in the silt-laden water of monsoon-borne floods, were especially valued and used to raise bamboo, tobacco, or banana. Meanwhile, most of the rest of the delta's arable cropland, more than three-quarters of it, was given over to rice.

In many parts of South India, the land was, for much of the year, a bleak brown. But here, midst the rice fields of the Cauvery, the landscape suddenly thickened with lush greenery in a rich palette of shades and textures. Farmers nursed delicate infant rice seedlings in small, specially watered plots whose rich velvety green stood out against neighboring fields. When, after thirty or forty days, the plants were healthy and strong, laborers individually scooped them out with their root pods and transplanted them to large, flooded fields; these made for a softer green. There the plants grew until a yellower hue signaled they were ready for harvest.

Almost every square foot of the delta was under the plow, and had been since time immemorial. Cattle and sheep found little room in which to graze; the land was just too valuable. Forests were few, just isolated coconut, banyan, or fig trees and, toward the coast, palmyra palms and Alexandrian laurel. The 342 square miles of the *taluk,* or county, of which Kumbakonam was chief town, comprised just about that same number of villages. Most were little more than tiny inhabited islands midst a sea of waving crops—a couple of dozen thatched-roof huts and a few hundred inhabitants, half-hidden by coconut palms, sitting on cramped little sites a few feet above the neighboring rice fields.

And yet, whatever their debt to the land, Ramanujan's family was not *of* the land. They were townspeople. They were poor, but they were *urban* poor; they inhabited not just the ground on which they lived but a wider world of the mind and spirit. The Cauvery freed the town from undue preoccupation with the day's weather and the season's crops, bestowing upon it a measure of wealth. And Ramanujan's family was among the many who, indirectly, lived off its bounty.

Like the American city of Des Moines, with its similar relationship to the corn-rich countryside of Iowa, Kumbakonam was more cosmopolitan than its surroundings, was a center for the work of eye, hand, and brain, which needs a degree of leisure to pursue. A census taken around the time Ramanujan was growing up found it had a higher proportion of professionals than anywhere in the presidency but Madras itself. The crafts were especially strong. One specialty was fine metalwork; Kumbakonam craftsmen, six hundred of them, it was estimated, kept European markets stocked with deities of the Hindu pantheon executed in copper, silver, and brass.

Another specialty was silk saris, the product of two thousand small looms manned by three thousand people. No place in South India was better known for its fine silk saris, dazzling in bright colors, embroidered in silver stripes, fringed with gold, than Kumbakonam and neighboring Tanjore. Saris woven in Kumbakonam could cost as much as a hundred rupees—a year's income to many poor families.

Bountiful harvests made the delta home to many wealthy farmers, and the marriage of one of their daughters might mean the purchase of a dozen or more saris. Before the wedding, the whole family would troop into town, make their selections, only later to be billed for what they took away; the merchants were happy to extend credit to such well-heeled customers. Otherwise, though, it was normally the husbands who did the buying, worried lest their wives, as one Kumbakonam sari weaver and shop owner told an English visitor around this time, spent too much.

Srinivasa Iyengar, twenty-four at his son Ramanujan's birth and about five years older than his wife, was a clerk in one such shop, just as his own father, Kupuswamy, had been. Normally, such a clerk remained one all his life—waiting on customers, taking orders, performing routine paperwork, perhaps traveling to nearby villages to collect bills. Occasionally a clerk might be taken into the business or would go off to start his own. But that required some special drive or entrepreneurial temperament. Apparently Srinivasa was good at appraising fabrics, a skill upon which his employer relied; but beyond that, whatever it took to step to a better job he could never muster.

Clerks like Srinivasa reported to work at eight or so in the morning and didn't get home till long after dark (which, so close to the equator, varied little across the year from about 6:00 P.M.). Sometimes they would return home at midday for lunch, though more likely their wives packed food in small metal cannisters for them to eat at the shop. Because certain months were deemed unpropitious, weddings would often stack up in months reckoned as lucky, making business quite seasonal and leaving clerks to sit idle for long periods. At such times, Srinivasa might well have been found asleep in the shop in the middle of the day.

Day after day, year after year, he was at the shop, largely absent from Ramanujan's early life. Indian society generally left the father little role to play at home, casting him as an aloof, physically undemonstrative, even unwelcoming figure whose relationship with his children was largely formal. Srinivasa was almost invisible, his name largely absent from family accounts. "Very quiet," a boyhood friend of Ramanujan called him. Someone else would resort to the word "weightless." But even had he been otherwise, he could scarcely have competed with Komalatammal as an influence on their son.

Years later, while away in England, and with at least one letter to his father confined to reminders to keep up the house and not let the gutter run over, Ramanujan wrote his mother about the titanic struggle unleashed in Europe with the onset of the Great War, down to details of the number of men fighting, the width of the battle fronts, the use of airplanes in combat, and the contribution of Indian rajas to the British war effort.

He must have known such an account would interest her. He and his mother understood each other. They talked the same language, enjoyed one another's intelligent company, shared the same intensity of feeling. When he was young, the two of them dueled at Goats and Tigers, played with pebbles, on a grid resembling a perspective view of railroad tracks receding to the horizon, crossed by other tracks perpendicular to them. Three "tigers" sought to kill fifteen "goats" by jumping them, as in checkers, while the goats tried to encircle the tigers, immobilizing them. The game demanded logic, strategy, and fierce, chesslike concentration. The two of them reveled in it.

Komalatammal, whom Ramanujan resembled physically, was, in the words of one account, "a shrewd and cultured lady." Her family could claim Sanskrit scholars in its line, scholars upon whom local kings had bestowed gifts. She was the daughter of Narayana Iyengar, well known in Erode as amin, an official in the district court charged with calling witnesses, taking court notes, and conferring with lawyers. When Ramanujan was about four, her father offended some higher-up and lost his job. It was then that he and his wife, Rangammal, moved to Kanchipuram, the temple city near Madras. There he managed a choultry, a temple annex where marriages are held and pilgrims put up.

A picture of Komalatammal survives, probably taken in her forties or fifties. It shows a woman whose corpulence even nine yards of sari cannot hide. Only her hands, resting lightly over the arms of her chair, suggest ease. The whole rest of her body conveys raw intensity: head cocked to one side, eyes alive, almost glaring, mouth set, leaning a little forward in the chair, only the balls of her bare feet touching the floor, poised as if ready to spring. The overall impression is one of great personal force only barely contained within her body.

She was an intense, even obsessive woman, never shy about thrusting her powerful personality onto objects of her interest. And her primary object all the years he was growing up was her son, Chinnaswami. In India, strong ties between son and mother are legendary; close indeed must have been the relationship between Ramanujan and his mother that even his Indian biographers invariably saw fit to comment upon it.

Komalatammal fed him his yogurt and rice, his spicy, pickled fruits and vegetables, his lentil soup. She combed his hair and coiled it into the traditional tuft, sometimes placing in it a flower. She tied his dhoti (or, as it was known in Tamil, *veshti*), the long piece of cloth wrapped around the waist and pulled up between the legs that all but the most Westernized men wore. She applied the *namam,* the powdery caste mark, to his forehead. She walked him to school; before going, Ramanujan would touch her feet in the traditional Indian sign of respect and secure her blessing. She monitored his friends and his time, made his decisions. Later, when Ramanujan didn't get the treatment at school she thought he deserved, she stormed into the principal's office and protested. And when she decided he ought to marry, she found him a wife and arranged for the wedding—all without bothering to consult her husband.

She poured prodigious energy into her spiritual life. In Hindu families, the women were apt to be more pious, and more scrupulous about observing tradition, than the men. So it had been in her own family; her mother was said to have gone into hypnotic trances that placed her in communion with the gods. And so it was in Ramanujan's family. Komalatammal was fiercely devout, held prayer meetings at her home, sang at the temple, pursued astrology and palmistry. Always, the name of their family deity, the Goddess Namagiri of Namakkal, was on her lips. "An exceptionally gifted lady with psychic powers and a remarkable imagination" was how one friend of the family described her. She had "'a remarkable repertoire of mythological stories and used to tell me stories from [the] ancient Mahabharata and Ramayana to [the] later Vikramaditya legends." Any pause in the telling was cause for yet another murmured appeal to Namagiri.

From his mother, Ramanujan absorbed tradition, mastered the doctrines of caste, learned the puranas. He learned to sing religious songs, to attend pujas, or devotions, at the temple, to eat the right foods and forswear the wrong ones—learned, in short, what he must do, and what he must never do, in order to be a good Brahmin boy.

3. A BRAHMIN BOYHOOD

For thousands of years Brahmins had been the learned men, teachers, and interpreters of Hindu life. Brahmins with heads so shaved in front that they looked prematurely bald, prominent caste marks of dried, colored paste upon their foreheads, locks of hair in the back like little ponytails, and thin, white, knotted threads worn diagonally across their bare chests, were an everyday sight on the streets of Kumbakonam and within its temples. Kumbakonam was a bedrock of Brahminism, the traditional Hinduism associated with its highest, priestly caste.

Four percent of the South Indian population, Brahmins were to most Hindus objects of veneration and respect; in pre-British India, at least, wealthy patrons acquired religious merit and washed away sins by giving them land, houses, gold. Brahmins were the temple priests, the astrologers, the gurus, the pandits specializing in sacred law and Vedic exegesis, indispensable at every wedding and funeral, occupying the most exalted niche in the Indian caste system.

Books about India by British writers around this time seemed to delight in regaling their readers with the horrors of the caste system—of men and women punished for sins committed in past lives by being consigned in this one to low and pitiable stations. There were four castes, these accounts recorded: Brahmins, at the top of the heap; Kshatriyas, or warriors; Vaisyas, or merchants and traders; and Sudras, or menials. A fifth group, the untouchables, lay properly outside the caste system. The first three castes were entitled to wear the sacred thread that affirmed them "twice-born." The Sudras could not, but could enter the temples. The untouchables could not even do that. Nor could they draw water from the village well. Nor, traditionally, could even their shadows cross the path of a Brahmin without his having to undergo a purification ritual.

Even this rudimentary breakdown, based on caste law first set down in the *Institutes of Manu,* a Sanskrit work dating to the third century, didn't quite apply in South India; for the Kshatriyas went largely unrepresented in the South. But more, it omitted the reality of India's several thousand self-governing subcastes, each with rules as to who could eat with whom, and whom one could marry. Most were originally, and often still, rooted in occupational categories. Thus, there were castes of agricultural workers, barbers, weavers, carpenters. It was these subcastes, or *jatis,* to which one really belonged. One simply *was* a Vanniar, or a Chettiar. Or, as in Ramanujan's case, a Vaishnavite Brahmin; his very name, Iyengar, labeled him one.

From the Hindu pantheon of Brahma, Vishnu, and Siva, Vaishnavites—about one Brahmin in four—singled out Vishnu as object of special devotion. Further theological nuances—for example, over just how much human effort was needed to secure divine grace—lay in the split between its Tengalai and Vedagalai, or northern and southern, branches. Such distinctions were not unlike those marking off, first, Christians from Jews, then Protestants from Roman Catholics, and finally, Lutherans from Methodists. And like their Western counterparts, the differences were often as much matters of style, tone, ritual emphasis, and historical accident as theological doctrine.

All Hindus believed in reincarnation and karma, heard the same tellings of the great Indian epics, shared certain sensibilities, values, and beliefs. But Vaishnavite Brahmins, as a rule, simply did not marry Shaivite Brahmins, those devoted to Siva. Each group had its own temples, shrines, and centers of religious teaching. Ramanujan wore a caste mark on his forehead—the *namam,* a broad white "U" intersected by a red vertical slash—wholly distinct from the three white horizontal stripes worn by Shaivites.

Caste barriers rose highest at mealtime. A Brahmin ate only with other Brahmins, could be served only by other Brahmins. In the cities, restaurants and hotels employing Brahmin chefs prominently advertised that fact. A Brahmin away from home went to elaborate pains to verify the source of food he ate. Brahmin families on pilgrimages to distant shrines would pull over to the side of the road to eat what they'd brought with them rather than chance food prepared by who-knew-whom.

Most often, it was a Brahmin male's wife who prepared and served his meals. But he never ate with her—another example of heathen ways the English cited as repugnant to proper Christians: women prepared the meals of the men and children of the household, serving them from vessels of silver, copper, and brass (not china, which was deemed insufficiently clean), and hovering over them during mealtime to dispense fresh helpings. The men would eat, largely oblivious to them, then rise together at meal's end. Only then, once having cleaned up, would the women retreat to the kitchen and eat whatever remained.

Ramanujan ate while seated on the floor, from a round metal tray or, more often, banana leaves set before him and later discarded, like paper plates. He ate with his hands. This did not mean using bread to scoop or sop up food. The staple food up North was wheat, that of the South rice; bread played little role in its diet. So Ramanujan ate precisely as every Western toddler learns *not* to eat—with his fingers.

Into the center of the banana leaf would be ladled a helping of rice. Toward the periphery of the leaf—about the size of a place mat in a Western household and still green and fresh, with a thick, muscular rib running down the middle—would go dollops of sharply pickled fruits or vegetables, like mangos, onions, or oranges; spiced fruit chutneys; sambhar, a thick lentil soup stocked with potatoes; and yogurt. Sometimes just a few selections, sometimes, for a festive meal, as many as a dozen. With the fingers of his right hand (and only his right hand), Ramanujan would mix rice with one or several other foods. Then, with four fingers and thumb formed into a pincer, he'd shape some of the loose mixture into a pasty ball and plop it onto his tongue.

South Indian cuisine was tasty and nutritious, if not always subtle. It was never bland; the curried dishes were sharp and spicy, the others almost maddeningly sweet. Rice and yogurt, beyond their nutritive value, softened and blunted the bite of the spices themselves. Coconuts and bananas (or actually plantains, a shorter, stubbier variety, tasting much the same) were the main fruits, along with mango and guava.

That Ramanujan never ate meat, then, was no act of painful self-denial. Like virtually all Brahmins, he was a strict vegetarian. And yet to say meat was "prohibited" to him subtly misses the point. It scarcely needed to be prohibited, and for the same simple, invisible reason an orthodox Jew or Muslim needn't be told not to eat pork: *you just didn't do it.* Others ate meat; *he* didn't. He would have gagged at the thought. Some of his friends even avoided ingredients, like beetroot, that gave food a reddish cast reminiscent of blood.

Ramanujan absorbed such dos and don'ts of Brahmin life as naturally as he learned to walk and talk. "As the child learned to accept responsibility for its own bodily cleanliness, it was also taught the importance of avoiding the invisible pollution conferred by the touch of members of the lowest castes," is how one scholar, G. Morris Carstairs, would later depict the Indian socializing process at work. "The mother or grandmother would call him in and make him bathe and change his clothes if this should happen, until his repugnance for a low caste person's touch became as involuntary as his disgust for the smell and touch of feces."

Every morning a Hindu male underwent an elaborate cleansing ritual. He defecated, using his left hand only to clean himself with water. Then he bathed, preferably in a holy river like the Cauvery, but always paying special heed to ears, eyes, and nostrils. In drinking, he never brought a cup to his lips but rather spilled water from it into his mouth. After a meal, he got up, left the eating area, and ceremoniously poured water over his hands. For all the dirt and lack of modern sanitary facilities which so bothered English visitors, there was a fastidiousness about Hindu life that no one observed more scrupulously than orthodox Brahmins.

Though sometimes scorned as haughty, Brahmins felt pride that, in their own estimation, even the poorest among them were cleaner and purer than others; that the least educated Brahmin knew some Sanskrit, the ancient language of Hinduism's sacred texts; that normally they were accorded deference and respect by others; that educationally and professionally, they excelled. All this contributed to a sense almost universal among them—and nothing suggests Ramanujan failed to share it—that Brahmins were, in a real sense, *chosen.*

4. OFF-SCALE

Among Brahmins, traditionally, a sanyasi, or itinerant beggar who gave up worldly interests for spiritual, was not deemed a failure. An ascetic streak ran through Brahmin culture. As Sanskrit scholar Daniel Ingalls has written in an essay, "The Brahmin Tradition," "Asceticism and mysticism have been, for many centuries now, to the respectable Indian classes what art has been for the last century and a half to the bourgeoisie of Western Europe"—something to which, whether aspiring to it themselves or not, they at least gave lip service, and respected.

This tradition lifted an eyebrow toward any too-fevered a rush toward worldly success, lauded a life rich in mind and spirit, bereft though it might be of physical comfort. Even wealthy Brahmin families often kept homes that, both by Western standards and those of other well-off Indians, were conspicuous by their simplicity and spartan grace, with bare floors, the meanest of furniture. "Simple living and high thinking," is how one South Indian Brahmin would, years later, characterize the tradition.

But in the years Ramanujan was growing up, things were changing. Brahmins were still the priests and gurus, the logicians and poets, the Sanskrit scholars and sanyasis of Hindu life. But now the old contemplative bent was taking new form; the spiritual was being transmuted into the secular. Like Jews in Europe and America at about the same time (with whom South Indian Brahmins would, almost a century later, routinely compare themselves), they were becoming professionals.

The census following Ramanujan's birth noted that of South India's six hundred thousand male Brahmins, some 15 percent—an extraordinarily high number—held positions in the civil service, the learned professions, and minor professional fields. They already dominated the ranks of the college educated, and within a generation, by 1914, of 650 graduates of the University of Madras no fewer than 452 would be Brahmins—more than ten times their proportion of the population. The old middle class of traders and barristers had traditionally been drawn from their own distinct castes. But the British had helped build a new middle class of brokers, agents, teachers, civil servants, journalists, writers, and government clerks. And these positions Brahmins now began to fill.

In Brahminically steeped Kumbakonam, one in five adult males could read and write, more than anywhere else in South India with the possible exception of Tanjore, the district seat, and Madras itself. Kumbakonam Brahmins had a taste for philosophical and intellectual inquiry, a delight in mental exercise, that led one English observer to pronounce them "proverbial for ability and subtlety." Ramanujan's parents, when not mired in outright poverty, clung to the nethermost reaches of the middle class and were illiterate in English, though not in their native Tamil; his friends, however, mostly came from better-off families and were bound for positions as lawyers, engineers, and government officials.

In doing so, they trod career paths with one thing in common: the way was always marked in English.

Ramanujan's native language was Tamil, one of a family of Dravidian languages that includes Malayalam, Canarese, and the musical-sounding Telugu. European scholars acclaimed Tamil for its clear-cut logic; "a language made by lawyers and grammarians," someone once called it. Spoken from just north of Madras within a broad, kidney-shaped region west to the Nilgiri Hills and south to Cape Comorin at the tip of the subcontinent, as well as in northern Ceylon, Tamil represented no out-of-the-way linguistic outpost. It had its own rich literature, distinct from the Hindi of the north, going back to the fifth century B.C., boasted a verse form reminiscent of ancient Greek, and was spoken by almost twenty million people.

But in the early 1900s, as now, English was ascendant in India. It was the language of the country's rulers. It fueled the machinery of government. It was the lingua franca to which Indians, who spoke more than a dozen distinct languages, turned when they did not otherwise understand one another. Among Indians as a whole, to be sure, the proportion who spoke English was small. Even among relatively well-educated Tamil Brahmin males, only about 11 percent were (in 1911) literate in it. So, those who did speak and read it were, in obedience to the law of supply and demand, propelled onto the fast track. As a clerk, even a smattering of it got you an extra few rupees' pay. It was the ticket of admission to the professions.

While a pupil at Kangayan Primary School, Ramanujan studied English from an early age, and in November 1897, just shy of ten, he passed his primary examinations—in English, Tamil, arithmetic, and geography—scoring first in the district. The following January, he enrolled in the English language high school, Town High.

Town High School had its origins in 1864 in two houses on Big Street, a main thoroughfare near the heart of town. When, some years later, the local college dropped its lower classes, a group of public-spirited citizens rushed to fill the vacant academic niche from below, through an expanded Town High. They would tear down the old buildings, erect a new one on the existing site . . . No, pronounced Thambuswami Mudaliar, a magnificently mustachioed eminence on the school's managing committee, better to start afresh. And for the school's new campus, he offered seven prime acres then harboring a banana orchard. There, he personally supervised construction of the first buildings.

Today, Town High's cluster of handsome white buildings occupies an oasis of tropical charm insulated from the noisy street out front by a sandy field shaded by tall margosa trees. At the time Ramanujan attended, however, the first block of classrooms, with its roof of densely layered red clay tiles and porch overhangs of palm leaf thatching, had gone up just a few years before. Its classrooms were laid end-to-end, making for a building one room wide, with windows on both sides to catch any hint of breeze.

The windows would have caught any adolescent clamor, too, but there was probably little to carry. Years later an alumnus would recall the long coats and turbans of the teachers and the respect they commanded among the students. Headmaster during Ramanujan's time, and for twenty-two years in all, was S. Krishnaswami Iyer, a severe-faced man partial to impromptu strolls between classes. The tapping of his walking stick would alert both teachers and students to his coming. Sometimes he'd step into a class, take over from the teacher, question students, and teach the rest of the class—with enough flair, it seems, that when he taught Grey's "Eton College" one student imagined little Town High as Eton, the irrigation ditch crossing the campus as the Thames.

The school, which stood about a five-minute walk from Ramanujan's house, drew the cream of Kumbakonam youth and launched them into college and career. Alumni would later recall it with genuine fondness. And it nourished Ramanujan for six years, bringing him as close as he'd ever come to a satisfying academic experience.

Even allowing for the retrospective halo that sees in every schoolboy exploit of the famous a harbinger of future greatness, it's plain that Ramanujan's gifts became apparent early. Ramanujan entered Town High's first form at the age of ten, corresponding to about an American seventh grade. And already while he was in the second form, his classmates were coming to him for help with mathematics problems.

Soon, certainly by the third form, he was challenging his teachers. One day, the math teacher pointed out that anything divided by itself was one: Divide three fruits among three people, he was saying, and each would get one. Divide a thousand fruits among a thousand people, and each would get one. So Ramanujan piped up: "But is zero divided by zero also one? If no fruits are divided among no one, will each still get one?"

Ramanujan's family, always strapped for cash, often took in boarders. Around the time he was eleven, there were two of them, Brahmin boys, one from the neighboring district of Trichinopoly, one from Tirunelveli far to the south, studying at the nearby Government College. Noticing Ramanujan's interest in mathematics, they fed it with whatever they knew. Within months he had exhausted their knowledge and was pestering them for math texts from the college library. Among those they brought to him was an 1893 English textbook popular in South Indian colleges and English preparatory schools, S. L. Loney's *Trigonometry*, which actually ranged into more advanced realms. By the time Ramanujan was thirteen, he had mastered it.

Ramanujan learned from an older boy how to solve cubic equations. He came to understand trigonometric functions not as the ratios of the sides in a right triangle, as usually taught in school, but as far more sophisticated concepts involving infinite series. He'd rattle off the numerical values of π and e, "transcendental" numbers appearing frequently in higher mathematics, to any number of decimal places. He'd take exams and finish in half the allotted time. Classmates two years ahead would hand him problems they thought difficult, only to watch him solve them at a glance.

Occasionally, his powers were put to good use. Some twelve hundred students attended the school and each had to be assigned to classrooms, and to the school's three dozen or so teachers, while satisfying any special circumstances peculiar to particular students. At Town High, the senior math teacher, Ganapathi Subbier, was regularly shackled with the maddening job—and he would give it to Ramanujan.

By the time he was fourteen and in the fourth form, some of his classmates had begun to write Ramanujan off as someone off in the clouds with whom they could scarcely hope to communicate. "We, including teachers, rarely understood him," remembered one of his contemporaries half a century later. Some of his teachers may already have felt uncomfortable in the face of his powers. But most of the school apparently stood in something like respectful awe of him, whether they knew what he was talking about or not.

He became something of a minor celebrity. All through his school years, he walked off with merit certificates and volumes of English poetry as scholastic prizes. Finally, at a ceremony in 1904, when Ramanujan was being awarded the K. Ranganatha Rao prize for mathematics, headmaster Krishnaswami Iyer introduced him to the audience as a student who, were it possible, deserved higher than the maximum possible marks.

An A-plus, or 100 percent, wouldn't do to rate him. Ramanujan, he was saying, was off-scale.

Still, during most of his time in school, Ramanujan's life remained in rough balance. At graduation, he was his mother's son, motivated and successful in school, getting set to enroll the following year, with a scholarship, in the Government College at the other end of town, looking ahead to academic achievement, a career, marriage . . .

But soon, very soon, that uneasy balance would be destroyed, and Ramanujan would be led out into a new, mentally unsettling realm of intellectual passion and fierce, unbending intensity that would rule the rest of his life.

For beside the reasoned, rational side of Ramanujan lay an intuitive, even irrational streak that most of his Western friends later could never understand—but with which he was at ease, and to which he happily surrendered himself.

5. THE GODDESS OF NAMAKKAL

It would take a few minutes for his eyes to adjust to the shadows. There, in the Sarangapani temple's outer hall, it seemed gloomy after the bright sun outside. What light there was swept in from the side, softly modeling the intricate sculpted shapes, the lions and geometrically cut stone, of the hall's closely spaced columns.

Away further from the light, nestled among the columns, were areas favored by bats for nesting. Sometimes Ramanujan could hear the quick, nervous swatting of their wings. Or even see them hanging from the ceiling, chirping away, then abruptly fluttering into flight.

Unlike Western churches which, architecturally, drew you higher and higher, here the devout were pulled, as it were, inner and inner. Within the high stone walls of the temple complex stood a broad court, open to the sky and, within that, the roofed columned area. In further yet, you came to the great chariot, its enormous wheels, several feet in diameter, drawn by sculpted horses and elephants. Within the building-within-a-building that was the chariot stood, in a dark stone cell where a lamp burned night and day, the sanctum sanctorum, the primary deity himself—the great god Vishnu, rising up from his slumber beside the many-headed serpent representing Eternity.

Always the temple stirred with little bright devotional fires, the chanting of mantras, the smell of incense in small shrines and dark niches devoted to secondary deities. The closer one approached to the central *shrine itself*, the darker it grew—more mysterious, more intimately scaled, progressively smaller, tighter, closer. What from the noisy street beyond the temple walls might have seemed a fit site for great public spectacles, here, inside, within stone grottos blackened by centuries of ritual fire presided over by bare-chested Brahmin priests, was a place for one man and his gods.

From the outside, the gopuram, or entrance tower, of this great temple built by Nayak kings sometime before A.D. 1350 was a massive twelve-story trapezoid of intricately sculpted figures, 90 feet across at its base and rising 146 into the sky. It was so high you could scarcely discern the images at the top, much less the facial expressions upon which their sculptors had lavished attention. There were figures clothed and naked, figures sitting and standing, with human shapes and animal, realistic and utterly fantastic. There were figures dancing, on horseback, making love, strumming instruments—a full panoply of human activity, densely realized in stone.

To Ramanujan, growing up within sight of the temple, these were not neutral images. Each represented legends onto which, since his earliest childhood, layers of imagery and significance had been heaped up—scenes and stories he had heard at his mother's knee, stories from the great epics, the Ramayana and the Mahabharata, stories meant to edify, or amuse. Every Hindu child learned of mischievous little Krishna—a child now, not yet a god—coming upon a group of women bathing, stealing their saris, and escaping up a tree with them, the women frantically imploring him for their return. Here, Ramanujan had only to lift his gaze to the wall of the gopuram to see Krishna perched in the legendary tree.

All his life, for festivals, or devotions, or just to pass the time, with his family or by himself, Ramanujan came to the temple. He'd grown up virtually in its shadow. Stepping out of his little house, he had but to turn his head to see, at the head of the street, close enough that he could make out the larger figures, the great gopuram. Indeed, the very street on which he lived bore the temple's name. It was Sarangapani Sannidhi Street; *sannidhi* meant entrance or procession way.

There was no special premium on silence within the temple; it was natural for Ramanujan to strike up conversations there. But the prevailing feeling was that of quiet and calm, a stone oasis of serenity, while outside all India clamored with boisterous life.

Here, to the sheltered columned coolness, Ramanujan would come. Here, away from the family, protected from the high hot sun outside, he would sometimes fall asleep in the middle of the day, his notebook, with its pages of mathematical scrawl, tucked beneath his arm, the stone slabs of the floor around him blanketed with equations inscribed in chalk.

More than a dozen major temples studded the town and nearby villages, some devoted to Lord Siva, some to Lord Vishnu. Each had its prominent gopurams, its columned halls, its dark inner sanctums, its tanks, or large, ritual purifying pools. The town fairly exuded spirituality. That once every twelve years the great Mahammakham tank received water from the Ganges—from which geography books showed it hopelessly remote—was, in Kumbakonam, a truth stated not with apology to secular sensibilities, or qualifiers like "tradition has it," or "according to legend," but simply, baldly, as fact.

It was a world in which the spiritual, the mystical, and the metaphysical weren't consigned to the fringes of life, but lay near its center. Ramanujan had but to step outside his house, wander along the street, or loll about the temple, to find someone eager to listen to a monologue on the traits of this or that deity, or the mystic qualities of the number 7, or man's duties as set forth in the *Bhagavad-Gita*.

Not that practical matters were dismissed in the high-caste Brahmin world in which Ramanujan grew up; money, comfort, and security had their place. But so did Vishnu and his incarnations, and what they meant, and how they might be propitiated, and upcoming festivals, and the proper form for devotions. These were not mere distractions or diversions from the business of everyday life. They were integral to it, as central to most South Indians as afternoon tea and cricket were to upper-class Englishmen, or free enterprise and their automobiles to Americans.

Years later, after he was dead, some of his Western friends who thought they knew him would say that Ramanujan was not really religious, that his mind was indistinguishable from any brilliant Westerner's, that he was a Hindu only by mechanical observance, or for form's sake alone.

They were wrong.

All the years he was growing up, he lived the life of a traditional Hindu Brahmin. He wore the *kutumi,* the topknot. His forehead was shaved. He was rigidly vegetarian. He frequented local temples. He participated in ceremonies and rituals at home. He traveled all over South India for pilgrimages. He regularly invoked the name of his family deity, the goddess Namagiri of Namakkal, and based his actions on what he took to be her wishes. He attributed to the gods his ability to navigate through the shoals of mathematical texts written in foreign languages. He could recite from the Vedas, the Upanishads, and other Hindu scriptures. He had a penchant for interpreting dreams, a taste for occult phenomena, and a mystical bent upon which his Indian friends unfailingly commented.

Once a year during the years he was growing up in Kumbakonam, he would set out along the road heading east past the railroad station. Outside of town, the mud houses with their thatched roofs hugging the side of the road thinned out. He could see bullocks tied to stakes beside the road, goats wandering in and out of houses, little roadside shrines, trails leading off the road and into the flat, green countryside. About four miles from Kumbakonam, he'd reach a broad looping curve in the road where the town of Thirunageswaram began, and where the ancient Uppiliapan Koil temple stood. Here Ramanujan came every year, at the time of the full moon, in the month of *Sravana* (around August) to renew his sacred thread.

When he was five years old, participating in a time-honored ceremony of fire and chanting that typically lasted four days, Ramanujan had been invested with the sacred thread—three intertwined strands of cotton thread draped across the bare chest, from the left shoulder diagonally down to the right hip, like a bandolier. The *upanayanam* ceremony solemnized his "twice-born" status as a Brahmin; the first birth, said the ancient lawgiver Manu, is from the mother, the second from the taking of the sacred thread. Thenceforth, he could read the sacred Vedas and perform the rites of his caste. And each year during *Sravanam*, midst food offerings and sacred fires and worship, he renewed it in the company of other Brahmins at the Uppiliapan temple.

One time, a friend recalled later, he and Ramanujan walked through the moonlight the six miles to the nearby town of Nachiarkovil, site of a Vishnu temple, to witness a religious festival. All the while, Ramanujan recited passages from the Vedas and the Shastras, ancient Sanskrit tomes, and gave running commentaries on their meaning.

Another time, when he was twenty-one, he showed up at the house of a teacher, got drawn into conversation, and soon was expatiating on the ties he saw between God, zero, and infinity—keeping everyone spellbound till two in the morning. It was that way often for Ramanujan. Losing himself in philosophical and mystical monologues, he'd make bizarre, fanciful leaps of the imagination that his friends did not understand but found fascinating anyway. So absorbed would they become that later all they could recall was the penetrating set of his eyes.

"Immensely devout," R. Radhakrishna Iyer, a classmate of his, would later term him. "A true mystic . . . intensely religious," recalled R. Srinivasan, a former professor of mathematics. Toward the end of his life, influenced by the West, Ramanujan may have edged toward more secular, narrowly rational values. But that came much later. And growing up midst the dense and ubiquitous spirituality of South India, he could scarcely have come away untouched by it—even if only in rebellion.

Ramanujan never did rebel. He did not deny the unseen realm of spirit, nor even hold it at arm's length; rather, he embraced it. His was not a life set in tension with the South India from which he came, but rather one resonating to its rhythms.

South India was a world apart. All across India's northern plains, the centuries had brought invasion, war, turmoil, and change. Around 1500 B.C. light-skinned Aryans swept in through mountain passes from the north. For eight centuries, Buddhism competed with traditional Brahminism, before at last being overpowered by it. Beginning in the tenth century, it was the Muslims who invaded, ultimately establishing their own Moghul Empire. One empire gave way to another, the races mingled, religions competed, men fought.

And yet by all this, the South remained largely untouched, safe behind its shield of mountains, rivers, and miles.

North of what would become the modern city of Bombay, stretching across the western edge of the subcontinent at roughly the latitude of the Tropic of Cancer, loomed the Vindhya mountains, a broken chain of rugged hills rising as high as three thousand feet and reaching inland almost seven hundred miles. At their base lay further obstacles to movement south into the tapering Indian peninsula—the Narbada and Tapti rivers, flowing west into the Arabian Sea, and the Mahanadi River, flowing into the Bay of Bengal on the east. These, together with sheer distance, exhausted most invaders before they reached very far south.

Thus, spared both the fury of the North and the fresh cultural winds forever sweeping through it, the South remained a place unto itself, remarkably "pure." No part of India was more homogenous. Racially, the South was populated mostly by indigenous Dravidian peoples with black, often curly hair, broad noses, and skin almost as dark as native Africans;

even the Brahmins, thought to be derived from Aryan stock, were not so light-skinned as those seen up North. Linguistically, North and South were divided, too. Tamil and the other Dravidian languages bore few ties to Hindi and the other Sanskrit-based languages of the North. Religiously, the South was more purely Hindu than any other part of India; nine in ten of those in Ramanujan's Tanjore District, for example, were Hindu, only about 5 percent Muslim. So special and distinct was the South in the minds of its inhabitants that in writing overseas they were apt to make "South India" part of the return address. On the political map, no such place existed; yet it expressed a profound cultural truth.

In South India an undiluted spirituality had had a chance to blossom. If the North was like Europe during the Enlightenment, the South was, religiously, still rooted in the Middle Ages. If Bombay was known for commerce, and Calcutta for politics, Madras was the most single-mindedly religious. It was a place where there was less, as it were, to distract you—just rice fields, temples, and hidden gods.

Here, in this setting, with the secular world held at bay, within a traditional culture always willing to see mystical and magical forces at work, Ramanujan's belief in the unseen workings of gods and goddesses, his supreme comfort with a mental universe tied together by invisible threads, came as naturally as breath itself.

All through South India, every village of a few dozen huts had a shrine to Mariamma or Iyenar, Seliamma or Angalamma—gods and goddesses whose origins went back to the very dawn of agricultural communities. These deities represented powers which villagers hoped to propitiate, like smallpox, cholera, and cattle plague. Most were reckoned as female. A few were recent, incorporating the spirits of murder victims or women who had died in childbirth. In 1904, some boys thought they heard trumpets coming from an anthill, and soon the deity of the anthill was attracting thousands of people from nearby villages, who would lie "prostrate on their faces, rapt in adoration."

Grama devata, or village gods, these deities were called, and they had virtually nothing to do with the formal Brahminic Hinduism a student of comparative religion might learn about in college. The villagers might give lip service to Vishnu and Siva, the two pillars of orthodox worship. But at time of pestilence or famine, they were apt to turn back to their little shrines—perhaps a brick building three or four feet high, or a small enclosure with a few rude stones in the middle—where guardianship of their village lay.

Mere idol worship? No more than a primitive, aboriginal animism? So some critics of Hinduism argued. And to the extent that these Dravidian gods were part of Hinduism, one could argue, the critics weren't far off.

But the Hinduism of which Kumbakonam was such a stronghold, and in which Ramanujan was steeped, was a world apart from all this. In Tanjore District, one English observer would note, "Brahminical Hinduism is here a living reality and not the neglected cult, shouldered out by the worship of aboriginal godlings, demons and devils which it so often is in other districts."

The great temples of the South fairly shouted out the difference. Temples in Kumbakonam, in Kanchipuram, in Tanjore, Madurai, and Rameswaram, were, as one authority put it, as superior to more famous ones in the North, say, "as Westminster Abbey and St. Paul's are to the other churches of London." One at Rameswaram, to which Ramanujan and his father, mother, and baby brother went on a pilgrimage in 1901, built over a hundred-year span during the seventeenth century on an island off the coast opposite Ceylon, was 1000 feet long and 650 feet across, built with gopurams 100 feet high on each face, with almost 4000 feet of corridors rich in extravagantly sculpted detail.

A Western observer to such a temple might still be brought up short by the bewildering variety of deities he'd find there—sculpted figures, statues large and small, in wood and stone, sometimes garlanded with flowers, even dressed in rude clothing. But in mainstream Hinduism, these could all be seen as part of a grand edifice of belief vastly more sophisticated than the religion of the villages.

The three chief deities in the Hindu pantheon, Brahma, Siva, and Vishnu, were traditionally represented as, respectively, the universe's creative, destructive, and preserving forces. In practice, however, Brahma, once having fashioned the world, was seen as cold and aloof, and tended to be ignored. So the two great branches of Brahminic Hinduism became Shaivism and Vaishnavism.

Shaivism had a kind of demonic streak, a fierceness, a malignity, a raw sexual energy embodied in the stylized phallic symbol known as a lingam that was the centerpiece of every Shaivite temple. Think of sweeping change, of cataclysmic destruction, and you invoked Lord Siva.

Vaishnavism, befitting its identification with the conserving god Vishnu, had more placid connotations. One contemporary English account likened it to the Spirit of Man—a distinctly gentler idea. Figuring largely in Vaishnavism were Rama and Krishna, heroes of Indian legend, and two of the incarnations, or avatars, in which Vishnu appears.

In Hindu lore, each of the three primal gods appeared in many forms. Siva could be Parmeswara. Vishnu could be Narasimha or Venkatarama. They had consorts and relatives, each of whom themselves had, over the centuries, become the objects of worship, the centers of their own cults. Vishnu, for example, was worshipped in the form of his consort Lakshmi, and as the monkey god, Hanuman. Each was endowed with distinct personalities; each gained its own adherents.

Some worshippers, certainly, construed those stone figures literally, viewed them as gods, pure and simple, in a way not so different from the *grama devata* worship of the villages. Indeed, one history of South India spoke of a "fusion of village deities and Vedic Brahminical deities" going back to around the beginning of the Christian era that had brought a comingling of different forms of worship.

But sophisticated Hindus, at least, understood that these stone "deities" merely represented forms or facets of a single godhead; in contemplating them, you were reawakened to the Oneness of all things. For those whose worship remained primitive, meanwhile, the garish stone figures could be seen as hooks by which to snare the spiritually unsophisticated and direct them toward something higher and finer.

The genius of Hinduism, then, was that it left room for everyone. It was a profoundly tolerant religion. It denied no other faiths. It set out no single path. It prescribed no one canon of worship and belief. It embraced everything and everyone. Whatever your personality there was a god or goddess, an incarnation, a figure, a deity, with which to identify, from which to draw comfort, to rouse you to a higher or deeper spirituality. There were gods for every purpose, to suit any frame of mind, any mood, any psyche, any stage or station of life. In taking on different forms, God became formless; in different names, nameless.

Among the thousands of deities, most South Indian families tended to invest special powers in a particular one—which became as much part of the family's heritage as stories passed down through the generations, or its treasured jewelry. This *kula devata* became the focal point of the family's supplications in time of trouble, much as some Roman Catholics invoke a particular patron saint. Things would go wrong, and you'd propitiate your family deity before you would any other. In South India, many a well-traveled Brahmin with wide knowledge of the world—perhaps a scholar, a professional, fluent in English, well-read in Sanskrit, who could intelligently discuss Indian nationalism, Tamil poetry, or mathematics—routinely and ardently prayed before the shrine of his family deity.

In Ramanujan's family, the family deity was the goddess Namagiri, consort of the lion-god Narasimha. Her shrine at Namakkal was about a hundred miles from Kumbakonam, about three-quarters of the way to Erode, near where Komalatammal's family came from. It was Namagiri whose name was always on his mother's lips, who was the object of those first devotions, whose assumed views on matters great and small were taken with the utmost seriousness.

It had been Namagiri to whom Ramanujan's mother and father, childless for some years after they married, had prayed for a child. Ramanujan's maternal grandmother, Rangammal, was a devotee of Namagiri and was said to enter a trance to speak to her. One time, a vision of Namagiri warned her of a bizarre murder plot involving teachers at the local school. Another time, many years earlier, before Ramanujan's birth, Namagiri revealed to her that the goddess would one day speak through her daughter's son. Ramanujan grew up hearing this story. And he, too, would utter Namagiri's name all his life, invoke her blessings, seek her counsel. It was goddess Namagiri, he would tell friends, to whom he owed his mathematical gifts. Namagiri would write the equations on his tongue. Namagiri would bestow mathematical insights in his dreams.

So he told his friends. Did he believe it?

His grandmother did, and so did his mother. Her son's birth, after long prayer to Namagiri, had only intensified her devotion, made her more fervent in her belief. That's how Komalatammal was: Why, she had practically willed herself a child. The whole force of her personality, her ferocious will, surged through all she did.

Ramanujan absorbed that from her; she never had to teach it to him, because it was imprinted in the example of her life. He learned from her to heed the voice within himself and to exert the will to act on it. His father was mired in the day-to-day, a slave to its routines, preoccupied with rupees and annas; he would want Ramanujan married off, bringing money into the family, well settled. But Komalatammal gave herself over to deeper forces, dwelt in a rich, inner world—and pulled Ramanujan into it with her.

So that when a powerful new influence on Ramanujan's young life came along, he had his mother's sanction to embrace it, to give his life over to it, to follow it with abandon.

CHAPTER TWO

Ranging with Delight

[1903 to 1908]

1. THE BOOK OF CARR

It first came into his hands a few months before he left Town High School, sometime in 1903. Probably, college students staying with Ramanujan's family showed him the book. In any case, its title bore no hint of the hold it would have on him: *A Synopsis of Elementary Results in Pure and Applied Mathematics.*

In essence, the book was a compilation of five thousand or so equations, written out one after the other—theorems, formulas, geometric diagrams, and other mathematical facts, marching down the page, tied together by topic, with big, bold-faced numbers beside each for cross-reference. Algebra, trigonometry, calculus, analytic geometry, differential equations—great chunks of mathematics as it was known in the late nineteenth century, ranged not over a whole shelf of textbooks, but compressed within two modest volumes (the second of which Ramanujan may not have seen until later).

"The book is not in any sense a great one," someone would later say of it, "but Ramanujan has made it famous."

The *Synopsis* was a product of the genius of George Shoobridge Carr. Except that Carr was no genius. He was a mathematician of distinctly middling rank who for years tutored privately in London; the book was a distillation of his coaching notes.

Mathematics students in England during the late nineteenth century were preoccupied to the point of obsession with a notoriously difficult examination, known as the Tripos, one's ranking on which largely determined one's career. The Tripos system encouraged what educators today might deride as "teaching to the test," and soon mathematicians would clamor for its reform. But back in the 1860s, in the period giving rise to the book, its hold went unchallenged. Not surprisingly, given the exam's importance, armies of private tutors had arisen to coach students for it. Carr was one of them.

Carr himself had a peculiar academic history. Born in 1837 in Teignmouth, near where the Pilgrims sailed for the New World, he attended school in Jersey, a Channel Island off the French coast, and later University College School in London. At least by 1866, and perhaps earlier, he started tutoring. Apparently he took it quite seriously and was forever updating his notes, refining his teaching methods, developing mnemonics to help his pupils cover the vast range of material they were supposed to master.

Then, at thirty-eight, more in the modern style than was common at the time, he decided to go back to school. Admitted to Gonville and Caius College, Cambridge University, he received his B.A. in 1880, and then—four years shy of fifty—his M.A.

He was no star student. In the Tripos, he was classed among the "Senior Optimes," not the higher-ranking "Wranglers," and only twelfth among them. He knew he was not the brightest light in the firmament of English mathematics. In the preface to the *Synopsis* he suggested that "abler hands than mine" might have done a better job with it, but that "abler hands might also, perhaps, be more usefully employed"—presumably in making the original mathematical discoveries to which his intellect or temperament failed to suit him.

But while Carr as a mathematician was no more than normally bright, he had the enthusiasm and love of subject to teach it to those abler than himself. In any case, it was just about the time he was granted his Cambridge B.A. that, on May 23, 1880, from his desk in Hadley, outside London, he put the finishing touches on the first volume—a second appeared in 1886—of the *Synopsis* which would link his name to Ramanujan's forever.

One strength of Carr's book was a movement, a *flow* to the formulas seemingly laid down one after the other in artless profusion, that gave the book a sly, seductive logic of its own.

Take, for example, the first statement on the first page:

$$a^2 - b^2 = (a-b)(a+b)$$

This is, first of all, an equation. It says—*any* equation says—that whatever is on the left-hand side of the equals sign is equivalent to what's on the right, as in $2+2=4$. Only in this case, it's not numbers, but symbols—the letters a and b—that figure in the equation. That they are symbols changes nothing. Some equations are true only when their variables take on certain values; the job, then, is to "solve" the equation, to determine those values—$x=3$, say, or $z=-8.2$—that make it valid. But this one, an "identity," is always true; *whatever* you make a and b, the statement still holds.

So, try it: Let $a=11$, say, and $b=6$. What happens?

Well, $a+b$ is just $11+6$, which is 17.

And $a-b$ is $11-6$, or 5.

Now, to set off quantities in parentheses, as they are in Carr's equation, means just to multiply them—$(a+b)$ and $(a-b)$—together. In this case, (17) (5) is just 17×5, or 85. That's the right-hand side of the equation.

Now for the left. a^2, of course, is just a times a, which is 11×11, or 121. b^2 is 36. $a^2 - b^2$, then, is just $121 - 36$, or 85. Which is just what the right-hand side of the equation comes to. Sure enough, the two sides match. The equation holds.

You could keep on doing this forever—verifying that the equation holds, with big numbers and little numbers, positive numbers and negative, fractions and decimals. You could do that, but who'd want to?

More sensible is to do what mathematicians do—prove the identity holds *generally,* for *any a* and *any b*. To do that, you dispense with particular numbers and manipulate instead the symbols themselves. You add and subtract the letters a and b, multiply and divide them, just as you would numbers.

In this case, the equation tells us to multiply $(a-b)$ times $(a+b)$. Doing that is about as simple as it looks. If you made $10 an hour, but then got a pay cut of $1 per hour, you could multiply the number of hours you worked by 10, then by 1, and subtract one product from the other. Or you could simply multiply the total hours worked by 9. Same thing. In this case, you can multiply the whole second term, $(a+b)$, by a, then by b, and then subtract one product from the other. Or, symbolically,

$$(a-b)(a+b) = a(a+b) - b(a+b)$$

What now? Well, $a(a+b)$ is just a^2+ab. And $b(a+b)$ is just $ba+b^2$. But ba (which means $b \times a$) is just the same as ab. So we get:

$$(a-b)(a+b) = (a^2+ab) - (ab+b^2)$$

In manipulating their equations, mathematicians often get caught in a clutter of numbers, letters, and symbols. And for the same reasons you do around the house, they periodically take time to tidy up—so they're not forever stepping over mounds of mathematical debris, and so any attractive qualities of their mathematical habitat are shown off to best advantage. "Grouping like terms" is one form housecleaning takes; you cluster mathematical entities in their appropriate categories. You place dirty clothes in the laundry bin, freshly laundered napkins in the linen drawer, cereal back in the pantry. You put things where they belong.

In this case, we bring everything out from behind the parentheses, add up the a^2 terms, and the b^2 terms, and the ab terms. And when we do, something interesting happens. The ab terms cancel each other; they "drop out." The $+ab$ and the $-ab$ add up to a grand total of zero, so that the quantity ab just disappears from the equation. Which leaves us with a^2-b^2, which is just what's on the left-hand side of the original equation—and just what we're supposed to prove.

What this simple exercise demonstrates is a "proof" of sorts, though a mathematician might shudder at the claim. But at least on casual inspection, it seems that for any a and any b, the two sides of Carr's equation are the same. We don't have to check $a = 735$ and $b = .0231$. We know it will work because we proved the general case.

So much for Carr's first equation. His second is this:

$$a^3-b^3 = (a-b)(a^2+ab+b^2)$$

Proving this differs little from proving the first. Working with the symbols, you multiply, add, and subtract, line up like terms—add apples to apples, and oranges to oranges, but never apples and oranges together—hope something cancels out, and soon are left with either side of the equals sign the same. Why, it's hardly worth the trouble to go through it. . . .

And right there, in the normal, natural—and appropriate—impulse to say it's ***hardly worth the trouble***, we gain a clue to Carr's pedagogical wisdom (and to how mathematicians, generally, think). The second equation, though different from the first, resembles it, seems an extension or natural progression from it: In following one with the other, Carr was *going somewhere*. There was a direction, a development, not within the mathematical statements he set down but implicit within the order in which he set them down. The first equation dealt with *a* and *b* "raised to the second power," in the form of $a^2 - b^2$; the second with *a* and *b* to the third power, as $a^3 - b^3$. What, one might now wonder, would be the equation for $a^4 - b^4$? Having worked out the first two, you'd suspect you could work it out easily, following the earlier examples. And you'd be right; the answer holds no surprises.

So Carr didn't set it down at all. That would have been tedious, and trivial. Instead, he generalized:

$$a^n - b^n = (a - b)\ (a^{n-1} + a^{n-2}b + \ldots b^{n-1})$$

This is the decisive step, for in taking it, the last ordinary number in the equation disappears. It's not the second power to which *a* and *b* are raised this time, or the third, or the eighth, but the *n*th.

Abruptly, we are in a new world. It's still simple algebra, but by daring to replace those safe 2s and 3s by the more mysterious *n*, the equation short-circuits routine mathematical manipulations. Now, you give me a number and I can just write out the equation, merely by substituting for the general *n*. The ellipsis appearing midway through the equation, the three little dots, just means you continue in the pattern the first two terms establish. O.K., so $n = 8$? Fine, plug it into Carr's general equation, and the equation writes itself. Where Carr's equation says *n*, you write 8. Where it says $n - 1$, you write 7, and so on.

As mathematicians might say, the equation with the *n*'s is more general than the previous two. Or, put another way, the first two equations were merely special cases of the third. Were Ramanujan not already familiar with it—and it's inconceivable that he wasn't—he could have confirmed it at a glance. Still, it suggests how he was guided through mathematical realms new to him; it wasn't just the statements Carr made that counted, but the path he nudged the student along in making them.

And the way in which he set about proving them. Or, rather, *not* proving them.

In fact, Carr didn't prove much in his book, certainly not as mathematicians normally do, and not even as we have here. Then, as now, the typical mathematics text methodically worked through a subject, setting out a theorem, then going through the steps of its proof. The student was expected to dutifully follow along behind the author, tracking his logic, perhaps filling in small gaps in his reasoning. "Oh, yes, that follows . . ." the student thinks. "Yes, I see . . ."

But mathematics is not best learned passively; you don't sop it up like a romance novel. You've got to go out to it, aggressive and alert, like a chess master pursuing checkmate. And mechanically following a proof laid out by another hardly encourages that, leaves scant opportunity to bring much of yourself to it. Whatever its other merits, the trigonometry text by S. L. Loney that Ramanujan had sailed through a few years before had clung to the mold; it was a text you followed rather than one which demanded you cut your own path.

Carr's was different.

There was no room for detailed proofs in the *Synopsis*. Many results were stated without so much as a word of explanation. Sometimes, a little note would be appended to the result. Theorem 245, for example, simply notes, "by (243), (244)." That is, one can arrive at the conclusion of no. 245 by extending the logic of 243 and 244. Theorem 2912 notes: "Proof—By changing x into πx in (2911)." In other words—mathematicians use this trick all the time—by an astute change of variable, the result assumes a clearer, more revealing form. In any case, Carr offered no elaborate demonstrations, no step-by-step proofs, just a gentle pointing of the way.

Scholars would one day probe Carr's book, searching for the elusive mathematical sophistication that might have inspired Ramanujan. Some would point to how it covered, or failed to cover, this or that mathematical topic. Some would point to its unusually helpful index, others to its broad compass.

But in fact it's hard to imagine a book *more* apt to influence a mathematically precocious sixteen-year-old, at least one like Ramanujan. For in baldly stating its results it almost dared you to jump in and prove them for yourself. To Ramanujan, each theorem was its own little research project. Or like a crossword puzzle, with its empty grid begging to be filled in. Or one of those irresistible little quizzes in popular magazines that invite you to rate your creativity or your sex appeal.

Nor was all this just an accident, or a by-product of the concision any compendium might demand; Carr had it in mind all along, and said so in the preface. "I have, in many cases," he explained,

> merely indicated the salient points of a demonstration, or merely referred to the theorems by which the proposition is proved. . . . The difference in the effect upon the mind between reading a mathematical demonstration,

and originating one wholly or partly, is very great. It may be compared to the difference between the pleasure experienced, and interest aroused, when in the one case a traveller is passively conducted through the roads of a novel and unexplored country, and in the other case he discovers the roads for himself with the assistance of a map.

But it wasn't even a map Carr supplied; rather, advice like, *Once out of town, turn left.*

A Western mathematician who knew Ramanujan's work well would later observe that the *Synopsis* had given him direction, but had "nothing to do with his *methods*, the most important of which were completely original." In fact, there *were* no methods, at least not detailed ones, in Carr's book. So Ramanujan, charging into the dense mathematical thicket of its five thousand theorems, had largely to fashion his own. That's what he now abandoned himself to doing. "Through the new world thus opened to him," two of his Indian biographers later wrote, "Ramanujan went ranging with delight."

2. THE CAMBRIDGE OF SOUTH INDIA

In 1904, soon after discovering Carr, Ramanujan graduated from high school and entered Kumbakonam's Government College with a scholarship awarded on the strength of his high school work. He was an F.A. student, for First Arts, a course of study that, by years in school, might today correspond to an associates degree but in India, then, counted for considerably more.

From the center of town, the college was about a twenty-minute walk—along the street that ran by Town High, down to the Cauvery's edge, then right, along the river to a point opposite the college. The bridge today spanning the river dates only to 1944; before that, a little boat ferried you across. Or else, you'd swim—a feat less daunting in March and April, when the river had dried to a trickle.

Government College was small, its faculty consisting of barely a dozen lecturers. And the best local students had begun to forsake it for larger schools in Madras. Still, for its time and place, it was pretty good—good enough, at any rate, to earn the moniker "the Cambridge of South India." Its link to the great English university rested in part on the campus's proximity to the Cauvery, which flowed beside it like the River Cam in Cambridge. But also playing a role was the repute of its graduates and the positions many of them held in South Indian life.

The year 1854 saw the college's founding on land given by the maharani of Tanjore; you could still see the steps, leading down from the dressing cabin, that royal princesses took down to the river to bathe. Beginning in 1871, existing buildings were repaired and enlarged, new ones built. In the 1880s its last secondary classes were dropped, and it became a full-blown college. Its grounds were enlarged and landscaped. A gymnasium was built. While Ramanujan was there, a hostel for seventy-two students was going up, complete with separate dining facilities for Brahmins.

The college occupied a site of considerable natural beauty. The river streamed by. Birds chirped. Groves of trees afforded shelter from the high, hot sun. Luxuriant vines crawled everywhere, forever threatening to overrun the college buildings. Even with the new construction since the maharani's time, the college did not dominate its site but rather clung there, at nature's sufferance. The place was lovely, idyllic, serene.

And the scene of Ramanujan's first academic debacle.

One can only guess at the effects of a book like Carr's *Synopsis* on a mediocre, or even normally bright student. But in Ramanujan, it had ignited a burst of fiercely single-minded intellectual activity. Until then, he'd kept mathematics in balance with the rest of his life, had been properly attentive to other claims on his energy and time. But now, ensnared by pure mathematics, he lost interest in everything else. He was all math. He couldn't get enough of it. "College regulations could secure his bodily presence at a lecture on history or physiology," E. H. Neville, an English mathematician who later befriended Ramanujan, would write, "but his mind was free, or, shall we say, was the slave of his genius."

As his professor intoned about Roman history, Ramanujan would sit manipulating mathematical formulas. "He was quite unmindful of what was going on around him," recalled one classmate, N. Hari Rao. "He had no inclination whatsoever for either following the class lessons or taking an interest in any subject other than mathematics." He showed Hari Rao how to construct "magic squares"—tic-tac-toe grids stuffed with numbers which, in every direction, add up to the same quantity. He worked problems in algebra, trigonometry, calculus. He played with prime numbers, the building blocks of the number system, and explored them for patterns. He got his hands on the few foreign-language math texts in the library and made his way through at least some of them; mathematical symbols, of course, are similar in all languages.

One math professor, P. V. Seshu Iyer, sometimes left him to do as he pleased in class, even encouraging him to tackle problems appearing in mathematics journals like the London *Mathematical Gazette*. One day Ramanujan showed him his work in an area of mathematics known as infinite series; "ingenious and original," Seshu Iyer judged it. But attention like that was rare, and Ramanujan's intellectual eccentricities were, on the whole, little indulged. More typical was the professor from whom Ramanujan borrowed a calculus book who, once he saw how it interfered with Ramanujan's other schoolwork, demanded its return. Even Seshu Iyer may not have been as solicitious as he later remembered; Ramanujan complained to one friend that he was "indifferent" to him.

Meanwhile, he ignored the physiology, the English, the Greek and Roman history he was supposed to be studying; he was no longer, if he'd ever been, "well-rounded." Back in 1897, his high standing on the Primary Examination had depended on excelling in many subjects, including English. Letters known to be written by him later, while showing no special grace, were competent enough, as were his mathematics notebooks when he used words, rather than symbols, to explain something. Yet now, at Government College, he failed English composition. "To the college authorities," E. H. Neville observed later, "he was just a student who was neglecting flagrantly all but one of the subjects he was supposed to be studying. The penalty was inevitable: his scholarship was taken away."

His mother, of course, was incensed and went to see the principal. How could he refuse her son a scholarship? He was unequaled in mathematics. They had never seen his like. The principal was polite, but firm. Rules were rules. Her son had failed the English composition paper, and miserably so. That was that.

Ramanujan's scholarship was no matter of mere prestige to him. Tuition was thirty-two rupees per term—as much as his father made in a month and a half. The scholarship insulated him from it; it enabled him to attend. He *needed* it.

Still, he managed to hang on for a few months, showing up for class enough to earn a certificate in July 1905 attesting to his attendance. The effort must have taxed him. He'd lost the scholarship, and everybody knew it. His parents were under a heavy financial burden; he knew that, too. He felt pressure to do well in his other subjects, yet he didn't *want* to lay mathematics aside for their sake. He was torn and miserable.

He endured the situation until he could endure it no longer. In early August 1905, Ramanujan, seventeen years old, ran away from home.

3. FLIGHT

As the hot breeze poured through the open windows of the railway car, Ramanujan watched the South Indian countryside slip by at twenty-five miles an hour. Villages of thatched roofs weathered to a dull barn-gray; intense pink flowers poking out from bushes and trees; palm trees, like exclamation points, punctuating the rice field flatness. From a distance, the men in the fields beside the tracks were little more than brown sticks, their dhotis and turbans white cotton puffs. The women were bright splashes of color, their orange and red saris set off against the startling green of the rice fields.

A snapshot might have recorded the scene as a charming bucolic tableau, but Ramanujan saw people everywhere engaged in purposeful activity. Men tending cattle. Women, stooped over in the fields, nursing the crops. Sometimes they worked alone, sometimes together in groups of a dozen or more, baskets perched atop their heads, fetching water from streams. Occasionally, a child with its mother would glance up from the surrounding fields and wave at the train bearing Ramanujan north to Vizagapatnam.

For eons, transportation in India, by bullock cart or the one-horse vehicle known as a *jutka*, had been painfully slow. Roads were terrible. Even by Ramanujan's time, only about an eighth of Tanjore District's seventeen hundred miles of road were "metalled," or paved with limestone or other rock. The difference was considerable. Cart drivers forced to travel on bumpy dirt roads thickly covered by dust or mud, rather than a metalled one, normally planned on carrying two-thirds the load, at two-thirds the speed. Twenty-five miles was a good day's journey.

The coming of the railroad had changed Indian life. It was the crowning engineering achievement of the British Raj, emerging in the mid-nineteenth century to knit the far-flung country together. In the South, the first lines had been laid in 1853, and in 1874 they began pushing south from Madras. In 1892, with the line to Vizagapatnam still unfinished, to get there from Kumbakonam could still take three weeks by train, bullock cart, and canal boat. By the following year, construction now complete, the trip took one day.

The railroads were the great leveler; everyone used them, irrespective of caste. "When you get to the third-class railway carriage you override even such a tough obstacle as caste," an English writer from this period noted. "Into it are bundled Brahmin and Pariah; they sit on the same seat; they rub shoulders who might not mingle shadows. 'You must drop your caste,' says the railway, 'if you want to travel at a farthing a mile'; and it is dropped—to be resumed again outside the station."

Ramanujan had grown up with the railways—as when, a child, he'd been shuttled among schools in Kumbakonam, Kanchipuram, and Madras. And now, in 1905, in the wake of losing his scholarship at Government College, the rails facilitated his flight.

Madras was 194 miles up the tracks of the South Indian Railways from Kumbakonam. And Vizagapatnam, following the main line along the coast, was 484 miles beyond that. A town of about forty thousand, it lay in an angle of the Bay of Bengal formed by a promontory known as the "Dolphin's Nose." Boasting the only natural harbor on India's east coast, it was a flourishing seaport; through it, yarn and piece goods entered India and manganese ore and raw sugar left. A new lighthouse had just been built near the anchorage. Now the engineers were planning to dredge the backwater and river and build new docks. Vizagapatnam was on the move.

And it was for this largely Telugu-speaking town halfway up the coast to Calcutta that Ramanujan, informing no one, set out. Fragmentary accounts from the period variously give as reasons the influence of a friend, the pursuit of a scholarship, the wish to find a patron, or—under pressure from his father—a job. But invariably, they also use language like "owing to disappointment," "ran away," and "too sensitive to ask his parents for help," and it's plain that whatever Ramanujan may have sought in Vizagapatnam, he was running *from* something, too.

There is evidence that the family, distraught over their son's disappearance, advertised in newspapers for him; that his father went house to house in Madras and Trichinopoly, looking for him. Otherwise, details of Ramanujan's impetuous flight are scanty. Except that soon, probably by September, his parents had him safely back in Kumbakonam.

It was the first of the Great Disappearances, the first of numerous such occasions on which Ramanujan would abruptly vanish, and about which little subsequently became known. But it was not the first time he'd taken abrupt and heedless action in the wake of what he deemed an intolerable blow to his self-esteem.

Back in 1897, aged nine, when he took his primary exam at Town Hall in Kumbakonam, he had scored a 42 out of 45 on the arithmetic portion, while a friend, K. Sarangapani Iyengar, got a 43. Hurt and angry, Ramanujan refused to speak to him. Sarangapani was mystified; what was the big deal? Trying to mollify him, he pointed out that in the other subjects Ramanujan had scored higher. Didn't matter, grumbled Ramanujan—in arithmetic he *always* scored highest. This time he had not, and everyone knew it. It was all too much to bear—whereupon he ran home crying to his mother.

Later, in high school, Ramanujan saw how trigonometric functions could be expressed in a form unrelated to the right triangles in which, superficially, they were rooted. It was a stunning discovery. But it turned out that the great Swiss mathematician Leonhard Euler had anticipated it by 150 years. When Ramanujan found out, he was so mortified that he secreted the papers on which he had recorded the results in the roof of his house.

Adolescent behavior quirks, irrelevant in the broad sweep of a genius's life? Perhaps. But together, and coupled with many other such instances later, they suggest an almost pathological sensitivity to the slightest breath of public humiliation. When, years later, Ramanujan stopped getting letters from a once-close friend, he wrote to the friend's brother that perhaps "he is too sorry for his failure in the Exam to write to me." Plainly, it was behavior to which he was keenly sensitive.

Shame is what psychologists call this sensitivity to public disgrace, something quite distinct from "guilt." Guilt, roughly speaking, comes from doing wrong, shame at being discovered, or at the prospect of being discovered, in some failure or vice; you're caught masturbating, say, or with your hand in the till. "An obligatory aspect of shame is the role discovery plays," writes Leon Wurmser, a University of Maryland psychiatrist, in *The Mask of Shame*. "It is usually a more or less sudden exposure, and exposure that abruptly brings to light the discrepancy between expectation and failure." The feeling is that of sudden, sharp, inescapable humiliation—of a yawning gap between who you say you are and who your failures reveal you to be, of an ugly stain upon your public face.

It is not necessary to actually be discovered, Wurmser points out; one can feel shame before oneself, at the mere thought of discovery. "We may wince at ourselves in the mirror and despise and degrade ourselves for the dishonor we feel within. . . . No one else has to see this stain—the shame remains."

The single most reliable marker of the shame syndrome is the impulse to flee. Writes Wurmser: "Hiding is intrinsic to and inseparable from the concept of shame." One experiences "the wish to hide, to flee, to 'cover one's face,' to 'sink into the ground.' " And that's just what Ramanujan did when faced with the ignominy of scoring only second in the arithmetic exam; in hiding evidence that his discovery was in fact rediscovery; and in running off to Vizagapatnam. One account has Ramanujan suffering a "mental aberration" during this period. Another calls it "a temporary unsoundness of mind." Whatever it was, acutely felt shame may have triggered it.

Years later, the memory of his school failure would make Ramanujan seek assurance that a scholarship he had been offered would not leave him with another examination to pass. The Government College fiasco humiliated him, apparently to the point of psychic trauma. His impulse, as it would be all his life, was to escape. And in fleeing to Vizagapatnam, he yielded to it.

Nor was it incongruous that one who, as mathematician, would prove so free from the intellectual blinders of the crowd should care so deeply how the crowd perceived him. Ramanujan was supremely self-assured about his mathematical gifts. Yet socially, he was a thoroughgoing conformist. If he cared not at all to follow mathematical paths others had trod, he cared deeply how others esteemed the path he had chosen.

Later, while in England and learning of a mathematics prize, he breathlessly inquired whether he might apply for it; formal, outward acknowledgment was no matter of indifference to him, and he never pretended otherwise. Similarly, when the British awarded him a high honor, his letter acknowledging word of it fairly bubbled over with excitement.

Was he respected as a mathematician? Was he deemed a dutiful son, a good Brahmin? Did he hold an important scholarship? Had he won a prize? The answers, as outward markers of acceptance or success, *counted*—and certainly never more so than now, as a teenager, at an age of exquisite sensitivity to the opinions of others.

Tales of Ramanujan's youth reveal a boy content to camp out on the *pial* of his house and work at mathematics, outwardly oblivious to the raucous play of his friends out on the street. Often, wrapped up in mathematics, he *was* oblivious. At other times, though, he must have wanted to be part of it. His thirst for public acknowledgment of his gifts, his pain when denied it, and his sensitivity to social slight, show how deeply, at another level, he really cared.

4. ANOTHER TRY

Pachaiyappa Mudaliar, born in 1784 of a destitute rural family, was a dubash, a master of two languages, who thereby served as a vital link in commerce with the British. By the time he was twenty-one he had amassed a fortune. At his death, aged forty-six, he left great heaps of it to charity. The college bearing his name, founded in 1889 and open only to Hindus, was by 1906 a respectable institution. Surely the building in which it was housed did nothing to sully its reputation—a great columned structure modeled on the Temple of Theseus in Athens, located on what was then known as China Bazaar Road in the busy Georgetown section of Madras.

It was to Pachaiyappa's—pronounced Pa-*shay*-a-pas—College that Ramanujan was bound when, one day early in 1906, he arrived at Egmore Station in Madras, so tired and disoriented that he fell asleep in the waiting room. A man woke him, took him back to his house, fed him, gave him directions, and sent him on his way to the college.

In India a college degree was no mere prerequisite for a good job; it virtually guaranteed you one, and a good start in your career. You earned a degree not by taking so many courses, or accumulating so many credits, but by passing an examination administered by the University of Madras; the "university" was not teachers and students, but merely an examining body. "To appear and succeed at the university examinations has been the ambition of every youth of promise," an English writer from the period noted. Some of Ramanujan's contemporaries at the college in Kumbakonam transferred to Presidency College in Madras, the crown jewel of the South Indian educational system, in hopes of better preparing for the all-important examination.

For most who sought a degree, though, it was all in vain. Of those taking the matriculation exam—equivalent to a high school diploma but more eagerly sought—half failed. A similar proportion fell out at each degree step along the way; failures of Pachaiyappa's students on the F.A. exam ran to 80 percent. In 1904, fewer than five thousand boys—and just forty-nine girls—were enrolled in the presidency's colleges and professional schools. And among all its forty-three million people, the number earning an F.A. degree each year came to barely a thousand.

Ramanujan, eighteen years old now, aimed to be one of them. A year after his failure in Kumbakonam, he was giving college another try in Madras.

For a time, he lived a few blocks away from Pachaiyappa's in a small lane off the fruit bazaar on Broadway in his grandmother's house. It was dingy and dark. And the air seemed to hang, static and close. But at least he was back in school.

Ramanujan's new math teacher, shown his notebooks, came away so impressed that he introduced him to the principal—who, on the spot, awarded him a partial scholarship. Though interrupted by a bad bout of dysentery that brought him back to Kumbakonam for three months, Ramanujan's early days at Pachaiyappa's College seemed filled with new promise.

N. Ramanujachariar, the math teacher, would take two sliding blackboards to work out a problem in algebra or trigonometry, reaching the solution in a dozen scrawled mathematical steps; Ramanujan would get up and show how to solve it in three or four. "Uh, what was that?" Ramanujachariar, who was a little deaf, would have to ask. So Ramanujan would obligingly run through it again. Sometimes the teacher would interrupt the lecture, turn to Ramanujan, and ask, "And what do you think, Ramanujan?" The prodigy from Kumbakonam tended to jump around the problem, working out key steps in his head but omitting them from his exposition—leaving his classmates thoroughly confused.

Sometimes he'd get together with the college's senior math professor, P. Singaravelu Mudaliar. Singaravelu—something of a catch for Pachaiyappa's, having formerly been an assistant professor of mathematics at the more prestigious Presidency College across town—was struck by Ramanujan's gifts. Together the two of them would tackle problems appearing in mathematical journals. If Ramanujan couldn't crack one of them, he'd give it to Singaravelu to work on overnight; invariably the professor couldn't solve it, either.

Everyone was struck by Ramanujan's gifts; but there was nothing new in that. Nor was there anything new in that nothing tangible came of it. For his experience in Kumbakonam now repeated itself at Pachaiyappa's. At Government College, it was English that had been his undoing. Now, among other subjects remote from mathematics he had to master, there was physiology. And this he found not merely boring, but repellent.

The text was a small book, *Physiology for Beginners,* written by two Cambridge dons, Michael Foster and Lewis E. Shore, published in 1894, and consisting mostly of the kind of flat descriptive accounts that passed for science in the late nineteenth century: "At the upper left-hand part of the stomach is the opening into it of the esophagus, a tube which passes from the mouth down the neck, through the thorax, and piercing the diaphragm, enters the stomach." It was full of elaborate drawings showing a rabbit with its skin peeled back, its internal organs revealed in graphic detail; a sheep's heart filling most of one page, a cutaway of a human mouth and tongue on another.

This was as far from the abstract heights of mathematics as you could get; if mathematics was art deco, with its cool geometric elegance, physiology was a kind of art nouveau, fluid and sumptuous. It was a world for which Ramanujan, as a strict vegetarian, could scarcely have had much taste: "Procure a rabbit which has been recently killed, but not skinned," chapter 3 of the text began. "Fasten the rabbit on its back by its four limbs to a board, and then, with a small sharp and pointed knife and a pair of scissors . . ."

Ramanujan reacted to all this with a skittish—and uncharacteristic—sarcasm. The professor would dissect a big, anesthetized frog, earnestly pointing out physiological similarities to humans, only to have Ramanujan pipe up with, And where is the serpent in this frog?—apparently a reference to the *nade,* or serpent power, that Hindu tradition ascribes to human nature. Another time, on an exam covering the digestive system, Ramanujan simply wrote a few lines in the answer book and handed it back unsigned: "Sir, this is my undigested product of the Digestion chapter." The professor had no trouble figuring whose it was.

Ramanujan, it need hardly be said, flunked physiology. Except for math he did poorly in all his subjects, but in physiology he reached particularly impressive lows, often scoring less than 10 percent on exams. He'd take the three-hour math exam and finish it in thirty minutes. But that got him exactly nowhere. In December 1906, he appeared again for the F.A. examination and failed. The following year, he took it again. And failed again.

Government College, Kumbakonam, 1904 and 1905 . . . Pachaiyappa's College, Madras, 1906 and 1907 . . . In the first decade of the twentieth century, there was no room for Srinivasa Ramanujan in the higher education system of South India. He was gifted, and everyone knew it. But that hardly sufficed to keep him in school or get him a degree.

The System wouldn't budge.

Describing the obsession with college degrees among ambitious young Indians around this time, an English writer, Herbert Compton, noted how "the loaves and fishes fall far short of the multitude, and the result is the creation of armies of hungry 'hopefuls'—the name is a literal translation of the vernacular generic term *omedwar* used in describing them—who pass their lives in absolute idleness, waiting on the skirts of chance, or gravitate to courses entirely opposed to those which education intended." Ramanujan, it might have seemed in 1908, was just such an *omedwar*. Out of school, without a job, he hung around the house in Kumbakonam.

Times were hard. One day back at Pachaiyappa's, the wind had blown off Ramanujan's cap as he boarded the electric train for school, and Ramanujan's Sanskrit teacher, who insisted that boys wear their traditional tufts covered, asked him to step back out to the market and buy one. Ramanujan apologized that he lacked even the few annas it cost. (His classmates, who'd observed his often-threadbare dress, chipped in to buy it for him.)

Ramanujan's father never made more than about twenty rupees a month; a rupee bought about twenty-five pounds of rice. Agricultural workers in surrounding villages earned four or five annas, or about a quarter rupee, per day; so many families were far worse off than Ramanujan's. But by the standards of the Brahmin professional community in which Ramanujan moved, it was close to penury.

The family took in boarders; that brought in another ten rupees per month. And Komalatammal sang at the temple, bringing in a few more. Still, Ramanujan occasionally went hungry. Sometimes, an old woman in the neighborhood would invite him in for a midday meal. Another family, that of Ramanujan's friend S. M. Subramanian, would also take him in, feeding him *dosai,* the lentil pancakes that are a staple of South Indian cooking. One time in 1908, Ramanujan's mother stopped by the Subramanian house lamenting that she had no rice. The boy's mother fed her and sent her younger son, Anantharaman, to find Ramanujan. Anantharaman led him to the house of his aunt, who filled him up on rice and butter.

To bring in money, Ramanujan approached friends of the family;

perhaps they had accounts to post, or books to reconcile? Or a son to tutor? One student, for seven rupees a month, was Viswanatha Sastri, son of a Government College philosophy professor. Early each morning, Ramanujan would walk to the boy's house on Solaiappa Mudali Street, at the other end of town, to coach him in algebra, geometry, and trigonometry. The only trouble was, he couldn't stick to the course material. He'd teach the standard method today and then, if Viswanatha forgot it, would improvise a wholly new one tomorrow. Soon he'd be lost in areas the boy's regular teacher never touched.

Sometimes he would fly off onto philosophical tangents. They'd be discussing the height of a wall, perhaps for a trigonometry problem, and Ramanujan would insist that its height was, of course, only relative: who could say how high it seemed to an ant or a buffalo? One time he asked how the world would look when first created, before there was anyone to view it. He took delight, too, in posing sly little problems: If you take a belt, he asked Viswanatha and his father, and cinch it tight around the earth's twenty-five-thousand-mile-long equator, then let it out just 2π feet—about two yards—how far off the earth's surface would it stand? Some tiny fraction of an inch? Nope, one foot.

Viswanatha Sastri found Ramanujan inspiring; other students, however, did not. One classmate from high school, N. Govindaraja Iyengar, asked Ramanujan to help him with differential calculus for his B.A. exam. The arrangement lasted all of two weeks. You can think of calculus as a set of powerful mathematical tools; that's how most students learn it and what most exams require. Or else you can appreciate it for the subtle questions it poses about the nature of the infinitesimally small and the infinitely large. Ramanujan, either unmindful of his students' practical needs or unwilling to cater to them, stressed the latter. "He would talk only of infinity and infinitesimals," wrote Govindaraja, who was no slouch intellectually and wound up as chairman of India's public service commission. "I felt that his tuition [teaching] might not be of real use to me in the examination, and so I gave it up."

Ramanujan had lost all his scholarships. He had failed in school. Even as a tutor of the subject he loved most, he'd been found wanting.

He had nothing.

And yet, viewed a little differently, he had everything. For now there was nothing to distract him from his notebooks—notebooks, crammed with theorems, that each day, each week, bulged wider.

5. THE NOTEBOOKS

"In proving one formula, he discovered many others, and he began to compile a note-book" to record his results. That's how Ramanujan's friend Neville put it many years later, and it remains as concise a distillation as any of how his notebooks came to be. Certainly, it was in working through Carr's *Synopsis,* as he tottered through college during the years from 1904 to 1907, that he began keeping them in earliest form.

After Ramanujan's death, his brother prepared a succession of handwritten accounts of the raw facts, data, and dates of his life. And preserved in their original form as they are, they remind us of a world before computers and word processors made revision easy and routine: we see rude scrawls growing neater, more digested and refined, as they are copied and recopied through successive versions.

Such was the likely genesis of Ramanujan's notebooks.

The first of the published *Notebooks* that come down to us today, which Ramanujan may have prepared around the time he left Pachaiyappa's College in 1907, was written in what someone later called "a peculiar green ink," its more than two hundred large pages stuffed with formulas on hypergeometric series, continued fractions, singular moduli . . .

But this "first" notebook, which was later expanded and revised into a second, is much more than mere odd notes. Broken into discrete chapters devoted to particular topics, its theorems numbered consecutively, it suggests Ramanujan looking back on what he has done and prettying it up for formal presentation, perhaps to help him find a job. It is, in other words, edited. It contains few outright errors; mostly, Ramanujan caught them earlier. And most of its contents, arrayed across fifteen or twenty lines per page, are entirely legible; one needn't squint to make out what they say. No, this is no impromptu record, no pile of sketches or snapshots; rather, it is like a museum retrospective, the viewer being guided through well-marked galleries lined with the artist's work.

Or so they were intended. At first, Ramanujan proceeded methodically, in neatly organized chapters, writing only on the right-hand side of the page. But ultimately, it seems, his resolve broke down. He began to use the reverse sides of some pages for scratch work, or for results he'd not yet categorized. Mathematical jottings piled up, now in a more impetuous hand, with some of it struck out, and sometimes with script marching up and down the page rather than across it. One can imagine Ramanujan vowing that, yes, this time he is going to keep his notebook pristine . . . when, working on an idea and finding neither scratch paper nor slate at hand, he abruptly reaches for the notebook with its beckoning blank sheets—the result coming down to us today as flurries of thought transmuted into paper and ink.

In those flurries, we can imagine the very earliest notebooks, those predating the published ones, coming into being. Ramanujan had set out to prove the theorems in Carr's book but soon left his remote mentor behind. Experimenting, he saw new theorems, went where Carr had never—or, in many cases, no one had ever—gone before. At some point, as his mind daily spun off new theorems, he thought to record them. Only over the course of years, and subsequent editions, did those early, haphazard scribblings evolve into the published *Notebooks* that today sustain a veritable cottage industry of mathematicians devoted to their study.

"Two monkeys having robbed an orchard of 3 times as many plantains as guavas, are about to begin their feast when they espy the injured owner of the fruits stealthily approaching with a stick. They calculate that it will take him 2¼ minutes to reach them. One monkey who can eat 10 guavas per minute finishes them in ⅔ of the time, and then helps the other to eat the plantains. They just finish in time. If the first monkey eats plantains twice as fast as guavas, how fast can the second monkey eat plantains?"

This charming little problem had appeared some years before Ramanujan's time in an Indian mathematical textbook. Exotic as it might seem at first, one has but to change the monkeys to foxes, and the guavas to grapes, to recognize one of those exercises, beloved of some educators, supposed to inject life and color into mathematics' presumably airless tracts. Needless to say, this sort of trifle, however tricky to solve, bears no kinship to the brand of mathematics that filled Ramanujan's notebooks.

Ramanujan needed no vision of monkeys chomping on guavas to spur his interest. For him, it wasn't what his equation stood for that mattered, but the equation itself, as pattern and form. And his pleasure lay not in finding in it a numerical answer, but from turning it upside down and inside out, seeing in it new possibilities, playing with it as the poet does words and images, the artist color and line, the philosopher ideas.

Ramanujan's world was one in which numbers had properties built into them. Chemistry students learn the properties of the various elements, the positions in the periodic table they occupy, the classes to which they belong, and just how their chemical properties arise from their atomic structure. Numbers, too, have properties which place them in distinct classes and categories.

For starters, there are even numbers, like 2, 4, and 6; and odd numbers, like 1, 3, and 5.

There are the integers—whole numbers, like 2, 3, and 17; and nonintegers, like 17¼ and 3.778.

Numbers like 4, 9, 16, and 25 are the product of multiplying the integers 2, 3, 4, and 5 by themselves; they are "squares," whereas 3, 10, and 24, for example, are not.

A 6 differs fundamentally from a 5, in that you can get it by multiplying two other numbers, 2 and 3; whereas a 5 is the product only of itself and 1. Mathematicians call 5 and numbers like it (2, 3, 7, and 11, but not 9) "prime." Meanwhile, 6 and other numbers built up from primes are termed "composite."

Then, there are "irrational" numbers, which can't be expressed as integers or the ratio of integers, like $\sqrt{2}$, which is approximately 1.414 . . . , but which, however many decimal places you take to express it, remains approximate. Numbers like 3, ½, and $9^{11}/_{16}$, on the other hand, are "rational."

And what about numbers, like the square root of -1, which seem impossible or absurd? A negative number times a negative number, after all, by mathematical convention is positive; so how can *any* number multiplied by itself give you a negative number? No ordinary number, of course, can; those so defined are called "imaginary," and assigned the label *i;* $\sqrt{-16} = 4\sqrt{-1} = 4i$. Such numbers, it turns out, can be manipulated like any other and find wide use in such fields as aerodynamics and electronics.

That happens often in mathematics; a notion at first glance arbitrary, or trivial, or paradoxical turns out to be mathematically profound, or even of practical value. After an innocent childhood of ordinary numbers like 1, 2, and 7, one's initial exposure to negative numbers, like -1 or -11, can be unsettling. Here, it doesn't require much arm-twisting to accept the idea: If t represents a temperature rise, but the temperature *drops* 6 degrees, you certainly couldn't assign the same $t = 6$ that you would for an equivalent temperature rise; some other number, -6, seems demanded. Somewhat analogously, imaginary numbers—as well as many other seemingly arbitrary or downright bizarre mathematical concepts—turn out to make solid sense.

Ramanujan's notebooks ranged over vast terrain. But this terrain was virtually all "pure" mathematics. Whatever use to which it might one day be put, Ramanujan gave no thought to its practical applications. He might have laughed out loud over the monkey and the guava problem, but he thought not at all, it is safe to say, about raising the yield of South Indian rice. Or improving the water system. Or even making an impact on theoretical physics; that, too, was "applied."

Rather, he did it just to do it. Ramanujan was an artist. And numbers—and the mathematical language expressing their relationships—were his medium.

Ramanujan's notebooks formed a distinctly idiosyncratic record. In them even widely standardized terms sometimes acquired new meaning. Thus, an "example"—normally, as in everyday usage, an illustration of a general principle—was for Ramanujan often a wholly new theorem. A "corollary"—a theorem flowing naturally from another theorem and so requiring no separate proof—was for him sometimes a generalization, which *did* require its own proof. As for his mathematical notation, it sometimes bore scant resemblance to anyone else's.

In mathematics, the assignment of x's and y's need conform to no particular rule; while an equation may reveal profound mathematical truths, just how it is couched—the letters and symbols assigned to its various entities, for example—is quite arbitrary. Still, in a mature field, one or very few notational systems normally take hold. A mathematician laying open a new field picks the Greek letter π, say, to stand for a certain variable; soon, through historical accident or force of habit, it's become enshrined in the mathematical literature.

To pick an example familiar from high school algebra, the two roots of a quadratic equation (which describes the geometric figure known as a parabola) are given by

$$x = \frac{-b \pm \sqrt{b^2 - 4ac}}{2a}$$

where a, b, and c are constants, x a variable. So entrenched is this form of the equation that it's hard to imagine anything else. And yet, there's no reason why the constants couldn't be p, q, and r. Or m_1, m_2, and m_3. And the quantity within the square root sign could be seen as the difference of two squares and broken up into two terms. And the square root itself could be expressed as a fractional power. And each of the two roots could get its own equation. The result would be:

$$x_1 = -\tfrac{1}{2}\frac{m_2}{m_1} + \frac{[(m_2+2\sqrt{m_1m_3})\ (m_2-2\sqrt{m_1m_3})]^{1/2}}{2m_1}$$

$$x_2 = -\tfrac{1}{2}\frac{m_2}{m_1} - \frac{[(m_2+2\sqrt{m_1m_3})\ (m_2-2\sqrt{m_1m_3})]^{1/2}}{2m_1}$$

The mathematical gymnastics don't matter here, only that this is identical to the more canonical version—and yet, on its face, unrecognizable. Someone coming up with the result on his own, and expressing it in this alien notation because he did not know the established one, would face extra roadblocks to being understood and might be written off as unorthodox or strange.

Which is just how Ramanujan's notebooks would tend to be regarded by mathematicians, both of his own day and of our own. In the area of elliptic functions, where everybody used k for the modulus, an important constant, Ramanujan used the Greek letter $\sqrt{\alpha}$, or $\sqrt{x}$. Sometimes n was, in his notebooks, a continuous variable, which for professional mathematicians it never was. As for the quantity $\pi(x)$, by which everyone else meant the number of prime numbers among the first x integers, it never appeared at all.

There was nothing "wrong" in what Ramanujan did; it was just *weird.* Ramanujan was not in contact with other mathematicians. He hadn't read last month's *Proceedings of the London Mathematical Society.* He was not a member of the mathematical community. So that today, scholars citing his work must invariably say, "In Ramanujan's notation," or "Expressing Ramanujan's idea in standard notation," or use similar such language.

He was like a species that had branched off from the main evolutionary line and, like an Australian echidna or Galápagos tortoise, had come to occupy a biological niche all his own.

If offbeat to other mathematicians, the parade of symbols in Ramanujan's notebooks amounted to a foreign language to most lay people. And yet, as arcane a language as it was, the concepts it expressed often turned out to be surprisingly straightforward.

Take, for example, the $f(x)$'s and other examples of "functional notation" that litter Ramanujan's notebooks. Here, $f(x)$, read "ef of ex," doesn't mean f times x, but rather some unspecified function of x; something, in other words, depends on x. Without defining the function we don't know *how* it depends. Later, we may specify, for example, that $f(x)$

$= 3x + 1$. Then we *do* know how it depends on x; the algebraic formula tells us, describing its mathematical behavior: In this case, when $x = 1$, $f(x) = 4$; when $x = 2, f(x) = 7$; and so on. But often, the mathematician doesn't *want* to get down to specifics. Functional notation lets him work in the more abstract realms he prefers, free from slavery to particular cases.

In functional notation, $\phi\ (a,b)$, read "phi of ay and bee," just means some unspecified function, ϕ, that depends on the variables a and b. And $f(3)$ just means $f(x)$ evaluated when $x = 3$. And $g(-x)$ just means $g(x)$ with -1 plugged into the equation whenever $x = 1$, -2 when $x = 2$, and so on. With such broad brush strokes, sometimes never stooping to particular functions at all, the mathematician fashions his world.

Or sometimes he does make $f(x)$ a specific function, then goes on to discover its odd or revealing properties. On page 75 of the first notebook, for example, Ramanujan writes

$$\phi(x) + \phi(-x) = \tfrac{1}{2}\phi(-x^2)$$

for a particular function defined previously. Evaluate $\phi(x)$ at, say, $x = \frac{1}{2}$, then at $x = -\frac{1}{2}$. Add up the two results. And that will equal half of what you get if you evaluate the function at $x = -\frac{1}{4}$. But Ramanujan's equation says it more generally, reveals the function's mathematical idiosyncrasies. And says it without so many words.

Which is one reason why Ramanujan, and all mathematicians, use their seemingly alien language in the first place—as a stand-in for long-winded verbiage. When on page 86 of the first notebook, and in many other places, Ramanujan writes Σ, the Greek letter sigma, he means, simply, "the sum of . . ." A notational fragment like

$$\sum_{k=1}^{\infty} \frac{x^k}{k}$$

may be read as "the sum of all the terms of the form x to the kth power divided by k, when k goes from 1 to infinity." That means, "whenever you see a k, replace it by 1, and note it; then by a 2, and add it to the first. . . . Continue in this way forever." And *that* is equivalent to

$$x + \frac{x^2}{2} + \frac{x^3}{3} + \frac{x^4}{4} + \ldots$$

Mathematics is full of similarly simple ideas lurking behind alien terminology. Want to specify a series of terms that alternate between positive and negative? That's easy: Just include the fragment $(-1)^k$. As k marches up through the integers one by one, the sign alternates automatically between plus and minus, because minus-times-minus is plus and minus-times-plus is minus. Or maybe you wish to specify only odd numbers, like 1, 3, 5, 7, and so on? The expression $2n+1$ churns them out. For n equals 0, $2n+1$ equals 1. For $n=1$, $2n+1=3$. For $n=2$, $2n+1=5$ . . .

Short, sweet, concise.

If you don't know English, you can't write a job application, and you can't write *King Lear*. But just knowing English isn't enough to write Shakespeare's play. The same applies to Ramanujan's notebooks: its pages of mathematical scrawl were, to professional mathematicians, what was *least* difficult about them. As with the English of *Lear*, it was what they *said* that took all the work.

And work it was—in expressing mathematical entities, performing operations on them, trying special cases, applying existing theorems to new realms. But some of the work, too, was numerical computation. "Every rational integer was his personal friend," someone once said of Ramanujan; as with friends, he liked numbers, enjoyed being in their company.

Even in the published notebooks, you can see Ramanujan giving concrete numerical form to what others might have left abstract—plugging in numbers, getting a feel for how functions "behaved." Some pages, with their dearth of Σ's and $f(x)$'s, and their profusion of 61s and 3533s, look less like mathematical treatise, more like the homework assignment of a fourth-grader. Numerical elbow grease it was. And he put in plenty of it. One Ramanujan scholar, B. M. Wilson, later told how Ramanujan's research into number theory was often "preceded by a table of numerical results, carried usually to a length from which most of us would shrink."

From which most of us would shrink. There's admiration there, but maybe a wisp of derision, too—as if in wonder that Ramanujan, of all people, could stoop so willingly to the realm of the merely arithmetical. And yet, Ramanujan was doing what great artists always do—diving into his material. He was building an intimacy with numbers, for the same reason that the painter lingers over the mixing of his paints, or the musician endlessly practices his scales.

And his insight profited. He was like the biological researcher who sees things others miss because he's there in the lab every night to see them.

His friends might later choose to recall how he made short work of school problems, could see instantly into those they found most difficult. But the problems Ramanujan took up were as tough slogging to him as school problems were to them. His successes did not come entirely through flashes of inspiration. It was hard work. It was full of false starts. It took time.

And that was the irony: in the wake of his failure at school, time was one thing he had plenty of.

6. A THOUGHT OF GOD

In 1807, a hundred years before Ramanujan was to fail his F.A. exam for the last time and experience India's educational system in all its oppressive rigidity, William Thackeray, an Englishman with experience in India as translator, judge, and civil servant, concluded his *Report on Canara, Malabar, and Ceded Districts*. In it, he wrote:

> It is very proper that in England, a good share of the produce of the earth should be appropriated to support certain families in affluence, to produce senators, sages and heroes for the service and defense of the state; or in other words, that a great part of the rent should go to opulent nobility and gentry, who are to serve their country in Parliament, in the army, in the navy, in the departments of science and liberal professions. The leisure, independence and high ideals which the enjoyment of this rent affords has enabled them to raise Britain to pinnacles of glory. Long may they enjoy it. But in India that haughty spirit, independence and deep thought which the possession of great wealth sometimes gives ought to be suppressed. They are directly averse to our power and interest. The nature of things, the past experience of all governments, renders it unnecessary to enlarge on this subject. We do not want generals, statesmen and legislators; we want industrious husbandmen. If we wanted restless and ambitious spirits there are enough of them in Malabar to supply the whole peninsula.

If Thackeray's sentiments mirrored British educational policy in India, no better evidence for it could be found than Ramanujan, for whom college might have seemed aimed at suppressing "haughty spirit, independence, and deep thought." Indeed, Indian higher education's failure to nurture one of such undoubted, but idiosyncratic, gifts could serve as textbook example of how bureaucratic systems, policies, and rules really *do* matter. People, as individuals, appreciated and respected Ramanujan;

but the System failed to find a place for him. It was designed, after all, to churn out bright, well-rounded young men who could help their British masters run the country, not the "restless and ambitious spirits" Thackeray warned against.

Viewed one way, then, for at least the five years between 1904 and 1909, Ramanujan floundered—mostly out of school, without a degree, without a job, without contact with other mathematicians.

And yet, was the cup half-empty—or half-full?

The great nineteenth-century mathematician Jacobi believed, as E. T. Bell put it in *Men of Mathematics,* that young mathematicians ought to be pitched "into the icy water to learn to swim or drown by themselves. Many students put off attempting anything on their own account till they have mastered everything relating to their problem that has been done by others. The result is that but few ever acquire the knack of independent work."

Ramanujan tossed alone in the icy waters for years. The hardship and intellectual isolation would do him good? They would spur his independent thinking and hone his talents? No one in India, surely, thought anything of the kind. And yet, that was the effect. His academic failure forced him to develop unconventionally, free of the social straitjacket that might have constrained his progress to well-worn paths.

For five solid years, Ramanujan was left alone to pursue mathematics. He received no guidance, no stimulation, no money beyond the few rupees he made from tutoring. But for all the economic deadweight he represented, his family apparently discouraged him little—not enough, in any case, to stop him. India, it might be said, left room for the solitary genius in him as it would for the sage, the mystic, the sanyasi. His friends, his mother, and even his father tolerated him, made no unduly urgent demands that he find work and make something of himself. Indeed, in looking back to Ramanujan's early years, Neville would refer to "the carefree days before 1909." And, in a sense, they were. In some ways, they were the most productive of his life. Ramanujan had found a home in mathematics, one so thoroughly comfortable he scarcely ever wished to leave it. It satisfied him intellectually, aesthetically, emotionally.

And, the evidence suggests, spiritually, as well. Countless stories would later attest to how, in Ramanujan, the mathematical and the metaphysical lay side by side, inextricably intertwined. Once, while a student at Pachaiyappa's College, he is said to have warned the parents of a sick child to move him away; "the death of a person," he told them, "can occur only in a certain space-time junction point." Another time, in a dream, he saw a hand write across a screen made red by flowing blood, tracing out elliptic integrals.

One idea Ramanujan bruited about dealt with the quantity $2^n - 1$. That, a friend remembered him explaining, stood for "the primordial God and several divinities. When n is zero the expression denotes zero, there is nothing; when n is 1 the expression denotes unity, the Infinite God. When n is 2, the expression denotes Trinity; when n is 3, the expression denotes 7, the Saptha Rishis, and so on."

Ramanujan was unfailingly congenial to metaphysical speculation. In Kumbakonam, there was a gymnastics teacher, Satyapriya Rao, whose fevered outpourings even tolerant South Indians dismissed. He would stand there, by the Cauvery, staring into the sun, raving; sometimes he'd have to be chained up when he got too hysterical. Most people ignored him. But not Ramanujan, who would sometimes collect food for him; some thought *he* must be mad to indulge him so. Yes, Ramanujan explained, he knew the man had visions, saw tiny creatures. But in an earlier birth, he was sure, Satyapriya had earned great merit. What others wrote off as the ravings of a madman was actually a highly evolved vision of the cosmos.

Later, in England, Ramanujan would build a theory of reality around Zero and Infinity, though his friends never quite figured out what he was getting at. Zero, it seemed, represented Absolute Reality. Infinity, or ∞, was the myriad manifestations of that Reality. Their mathematical product, $\infty \times 0$, was not one number, but all numbers, each of which corresponded to individual acts of creation. To philosophers, perhaps—and to mathematicians, certainly—the idea might have seemed silly. But Ramanujan found meaning in it. One friend, P. C. Mahalanobis—the man who discovered him shivering in his Cambridge room—later wrote how Ramanujan "spoke with such enthusiasm about the philosophical questions that sometimes I felt he would have been better pleased to have succeeded in establishing his philosophical theories than in supplying rigorous proofs of his mathematical conjectures."

In the West, there was an old debate as to whether mathematical reality was made by mathematicians or, existing independently, was merely discovered by them. Ramanujan was squarely in the latter camp; for him, numbers and their mathematical relationships fairly threw off clues to how the universe fit together. Each new theorem was one more piece of the Infinite unfathomed. So he wasn't being silly, or sly, or cute when later he told a friend, "An equation for me has no meaning unless it expresses a thought of God."

7. ENOUGH IS ENOUGH

At the age of twenty, Ramanujan was, as he'd been most of his life, fat. He was short and squat, with a full nose set onto a fleshy, lightly pock-marked face, bare of mustache or beard. His shaved forehead, with its prominent red and white caste mark, and the rest of his full black hair gathered behind his head into a tuft, made him seem even rounder, fleshier, and fuller than he was.

But there was no thick, lumbering sluggishness to Ramanujan's bulk; if anything it was more like that of a sumo wrestler, or a Buddha, with a lightness to it, even a delicacy. He walked with head erect, a sprightliness in his gait, body pitched forward onto his toes. He had long arms, hands of a surprising, velvety smoothness, and slender, tapering fingers forever in motion as he talked.

When he grew animated, the words tumbled out. Even eating, which he did with gusto, rarely staunched the flow; he'd go on with an idea or a joke even with his mouth full. And always, his dark eyes glowed; the rest of him could sometimes seem to fall away, leaving only the light in his eyes.

Occasionally he'd drop by the college that had flunked him, to borrow a book, or see a professor, or hear a lecture. Or he'd wander over to the temple. But mostly, Ramanujan would sit working on the *pial* of his house on Sarangapani Sannidhi Street, legs pulled into his body, a large slate spread across his lap, madly scribbling, seemingly oblivious to the squeak of the hard slate pencil upon it. For all the noisy activity of the street, the procession of cattle, of sari-garbed women, of half-naked men pulling carts, he inhabited an island of serenity. Human activity passed close by, yet left him alone, and free, unperturbed by exams he had no wish to take, or subjects he had no wish to study.

The *Hindu,* South India's premier English language newspaper, had observed in an 1889 editorial that "the Indian character has seldom been wanting in examples of what may be called passive virtues. Patience, personal attachment, gentleness and such like have always been prominent. But for ages together, India has not had amongst her sons one like Gordon, Garibaldi, or Washington. . . . In all departments of life," it went on, "the Hindus require a vigorous individuality, a determination to succeed and to sacrifice everything in the attempt." And despite his seeming indifference to worldly success, Ramanujan, inwardly, was a model of all the *Hindu* editorial writer could have wanted.

A determination to succeed and to sacrifice everything in the attempt. That could be a prescription for an unhappy life; certainly for a life out of balance, sneering at timidity and restraint. Sometimes, as Ramanujan sat or squatted on the *pial,* he'd look up to watch the children playing in the street with what one neighbor remembered as "a blank and vacant look." But inside, he was on fire.

When he thought hard, his face scrunched up, his eyes narrowed into a squint. When he figured something out, he sometimes seemed to talk to himself, smile, shake his head with pleasure. When he made a mistake, too impatient to lay down his slate pencil, he twisted his forearm toward his body in a single fluid motion and used his elbow, now aimed at the slate, as an eraser.

Ramanujan's was no cool, steady Intelligence, solemnly applied to the problem at hand; he was all energy, animation, force.

He was also a young man who hung around the house, who had flunked out of two colleges, who had no job, who indulged in mystical disquisitions that few understood, and in mathematics that no one did. What value was his work to anyone? Maybe he was a genius, maybe a crank. But in any case, why waste one's time and energy in activity so divorced from the common purposes of life? Didn't his father, working as a lowly clerk in a silk shop, do the world and himself more good than he?

For a long time his parents put up with him. But in the end they too reached their limits, grew irritated and impatient. Enough is enough, his mother decided. And sometime probably late in 1908 she moved decisively to invoke what the Indian psychologist Ashis Nandy has called "that time-tested Indian psychotherapy"—an arranged marriage.

CHAPTER THREE

The Search for Patrons

[1908 to 1913]

1. JANAKI

One day late in 1908, Ramanujan's mother was visiting friends in the village of Rajendram, about sixty miles west of Kumbakonam. There she spied a bright-eyed wisp of a girl, Janaki, daughter of a distant relative. She asked for the girl's horoscope—the first step in virtually every arranged marriage in India—drew her son's horoscope on the wall of the house, compared it to that of the girl, and concluded that yes, this would make a good match. Negotiations ensued for the marriage of Ramanujan and Janaki, then about nine years old.

It was in many ways an apt match, between two persons of equally meager social standing. Janaki was an unassuming, only ordinarily pretty girl from a village so tiny it appears on none but the most detailed maps. The family had once been better off; her father dealt in jewelry-making supplies and had once owned a little property. But now, fallen on harder times, they could offer only a modest dowry, perhaps a few polished copper vessels. They could not afford to be choosy about a husband, especially since Janaki was but one of five daughters (along with one son). Most of all, they sought a family for their daughter apt to treat her kindly during those early years when, still without children, she toiled under the imperious eye and unquestioned authority of her mother-in-law.

Ramanujan, meanwhile, was no great catch. Outwardly, he was a total failure, lacking degree, job, or prospects. Janaki knew nothing of the man who was to become her husband; she would not so much as glimpse his face until their wedding. He was an ordinary young man from an ordinary family. Maybe, she thought later, her parents had heard Komalatammal tout her son as a mathematical genius; if so, she'd known nothing of it.

So far as Komalatammal was concerned it was all set, which in Ramanujan's family meant it was. But when husband Srinivasa learned of the arrangements, he fumed. The boy can do better than that, he protested. Many families in Kumbakonam would be proud to count him as son-in-law; in fact, two years before, when Ramanujan was off at Pachaiyappa's College, a family in Kanchipuram had come forward with an offer, and only a death in the bride's family had gotten in the way. What really riled Srinivasa, though, was that he'd had no say in the plans. Nothing mollified him. So the following July, when it came time to travel to Rajendram for the wedding, he stayed home.

In these events, Srinivasa's exclusion was unusual. Arranged marriages, without a say for the bride or groom, were virtually universal, the institution of child-brides almost as much so; most girls married before puberty, though they didn't actually live with their husbands, consummating the marriage, until later. The practice was repugnant to most Europeans, but the British, ever sensitive to local custom, did nothing to change it. In 1894, the state of Mysore had passed a law barring the marriage of girls younger than eight; a similar provision in Madras had failed.

Extending over four or five days, an Indian wedding was a glory of color and tinsel, music and ceremony. The whole economy was influenced by the scale and expense of these grand affairs, on which six months' income might be blown with scarcely a thought. Even the poorest families unblinkingly assumed every burden—saved every spare rupee, indebted themselves to local usurers—to provide their daughters' dowries, to buy new saris, and to pay for the meals and music of the wedding itself.

Ramanujan's was a double wedding, Janaki's sister Vijayalakshmi being set to marry the same day. (By December, she would be dead, prey to a severe fever.) The other bridegroom showed up on schedule. But long into the day before the wedding, and then into the night, Ramanujan and his family failed to appear. Janaki's father, Rangaswamy, had never been entirely won over by the prospect of the match. Now, Janaki heard him say, if Ramanujan didn't show up soon, they'd marry her off, then and there, to someone else, maybe his nephew. . . .

The train from Kumbakonam rolled into Kulittalai, the station nearest Rajendram, hours late. And so it was long past midnight before Ramanujan and his mother, on a bullock cart from the station, reached the village. Rangaswamy, his nerves stretched to the breaking point, railed, spoke of calling off the wedding. But Komalatammal, marshaling her considerable persuasive powers, wondered out loud whether a father of five daughters ought to hesitate when opportunity knocked. . . .

As usual, Komalatammal had her way. The bridegroom's reception took place at one o'clock in the morning. Then came the *kasi yatra,* in which the bridegroom makes a show of renouncing domestic pleasures, even starts off for Benares, the sacred city of the North, to become a sanyasi; he gets maybe a hundred yards before being headed off by the bride's family, who wash his feet in supplication and beg him to return. Finally, on July 14, 1909 Janaki took the *saptapadi,* or seven steps, that made the marriage irrevocable.

Inauspicious incidents, however, marred the wedding. While Ramanujan and Janaki, in the finest silk sari her family could afford, sat together on the traditional swing being serenaded by singers, the screams of a retarded girl from town shattered the moment's harmony. At another point, a garland Janaki sought to place around Ramanujan's neck fell to the ground. Finally, as drummers and musicians entertained them, a fire broke out in a corner of the choultry where the wedding was being held. Though quickly extinguished, it was deemed an ill omen.

Through it all, the doughty Komalatammal remained cheerful, her unflappability winning her sympathy and not a little wonder.

At first, Ramanujan's marriage changed nothing, at least outwardly. Janaki wouldn't actually join him for three years, until after she'd reached puberty. Rather, after a brief spell with his family in Kumbakonam, she would return to her own in Rajendram, there to work with her mother around the kitchen, learn cooking and domestic chores, and be further schooled in the arts of obedience and respect for her parents-in-law and husband.

But though outward circumstances had changed little, Ramanujan had entered a new stage of life. Hindu thinking sees life passing through four stages. As brahmacharya, you are a student, learning the spiritual and intellectual ropes. As grihasta, occupying the longest span, you are a householder, with responsibilities to home and family. As vana prastha, or "inhabitant of the forest," you begin to throw off the bustle of family life and seek solitude, introspective calm. Finally, as sanyasi, you relinquish everything—family, possessions, attachments—in pursuit of spiritual fulfillment. At his wedding, in heading off for Benares, Ramanujan had ritually opted for this last stage. But in fact, he was now a grihasta. He had responsibilities now. He had a wife. His father was pushing fifty. No longer was he a free spirit, left "ranging with delight" through mathematics, happily on his own. It was time that he assume the mantle of adulthood.

But now a medical problem intervened. Some accounts later found it more delicate to refer vaguely to "kidney trouble," but in fact Ramanujan had developed a hydrocele, an abnormal swelling of the scrotal sac.

"Hydrocele" is a physical finding, not some particular illness. A subtle, and otherwise harmless, imbalance in the rate of absorption of scrotal fluid can cause it. So can filariasis, endemic in South India, an infection of the lymph system by mosquito-borne parasites. So can other infections, among them tubercular. Usually, there are no symptoms, not even sexual; men sometimes carry a small hydrocele around with them for years. Only when one reaches the size of, say, a tennis ball, does sheer mechanical inconvenience make it a problem and demand surgery. The operation is simple; an incision is made in the scrotal sac to release the blocked fluid. Because the area is so rich in blood vessels, healing is normally rapid, and infection rare, even under poor sanitary conditions.

There was one problem; the family had no money for the operation. Komalatammal asked friends for help, but none was forthcoming. Finally, in January 1910, a certain Dr. Kuppuswami volunteered to do the surgery for free. As the chloroform was being administered, a friend later recalled in wonder, Ramanujan noted the order in which his five senses were blocked.

For a time, Ramanujan was left prostrate. One day, hurrying onto his legs again too soon, he walked with his friend Anantharaman to a village a few miles out of town; the wound began to bleed. But soon, recovered, fueled by the new resolve a long rest can bring, Ramanujan began to go out to the wider world beyond the *pial.*

Since discovering Carr, he had turned his back on anything—school, family, friends—that took him away from mathematics. During that period, he'd probably needed to be left alone, undistracted, free to follow his mathematical muse. But now, after six years, maybe it was time to stitch himself into the broader social fabric again. It is tempting, of course, to see the hand of his mother in all this—that, consciously or not, she'd realized that if her son was to achieve anything, he had to reach out to the world, and that his marriage would force him to do that. In any event, that's what happened. Ramanujan was a grihasta now, and even if inwardly kicking and screaming, he gave up the social wilderness that had long been his home.

2. DOOR-TO-DOOR

Ramanujan sought now not a scholarship, nor even the chance to be a mathematician, but just a job, a chance at a future, a new life. For the next two years, the sheer desperation of his lot sent him across South India, first from Kumbakonam as his base and then, increasingly, Madras.

Once again, he took to the rails, though he would often have to depend for his ticket on friends and well-wishers. To the English, even first-class seats were a bargain. But for Ramanujan, round-trip to Madras at a quarter-anna or so per mile for the crowded third-class carriage was worth more than a week's pay to his father, the equivalent of more than a hundred pounds of rice.

Early during this period, at least, he had no real home, but camped out with friends. At one point, he showed up at the house of a friend begging for a place to stay and was directed to quarters he might share with an old monk. For a while in 1910, he stayed with Viswanatha Sastri, whom he had tutored in Kumbakonam and who was now a student at Presidency College in Madras. Viswanatha lived at the Victoria Student Hostel near the college, a large red and black brick structure whose turrets and three stories of brick-columned arches looked as if they had been transplanted intact from England. Ramanujan joined him there, heading out each morning in search of students to tutor.

But apparently his reputation as a tutor of unworldly bent preceded him, for he drew few students. At night, Viswanatha Sastri recalled later,

> he used to bemoan his wretched condition in life. When I encouraged him by saying that being endowed with a valuable gift he need not be sorry but only had to wait for recognition, he would reply that many a great man like Galileo died in inquisition and his lot would be to die in poverty. But I continued to encourage him that God, who is great, would surely help him and he ought not to give way to sorrow.

It was an emotionally fragile period; even so thin a ray of pleasure as the thin, peppery soup, or *rasam,* the hostel served, would loom large in Ramanujan's memory across the years.

Later in 1910 and on into the following year, Ramanujan lived on Venkatanarayan Lane, in a neighborhood called Park Town near the red buildings of Central Railway Station. This time he lived on the sufferance of two old Kumbakonam friends, K. Narasimha Iyengar and his brother, K. Sarangapani Iyengar (whom he'd apparently forgiven for scoring higher than he on the arithmetic exam ten years before). Back in Kumbakonam, the brothers had sometimes footed the bill for his clothes and rail fares. Now, they were helping him again.

Narasimha was a student at Madras Christian College, a school run by Scottish missionaries, and Ramanujan tutored him in math. As the F.A. examination approached, Narasimha, who was no mathematician, became nervous and depressed, even weighed skipping it altogether. On the day of the exam, Ramanujan walked the four miles from Park Town to Presidency College, where it was being held. There, he located his friend, convinced him to take the exam, gave him a little pep talk, and supplied a few last-minute tips. Whatever he said worked, if only barely: Narasimha squeaked by with the lowest passing score.

One day probably soon after this Ramanujan appeared, delivered by horse cart, at the doorstep of his friend from Pachaiyappa's days, R. Radhakrishna Iyer. He was sick again, perhaps suffering the effects of his operation earlier that year. Radhakrishna took him in, saw that he was properly fed, and called in a doctor. Ramanujan, advised Dr. Narayanaswami, needed constant nursing. So Radhakrishna took Ramanujan to Beach Station, near the harbor, and put him on the train back to his family in Kumbakonam. But before he left, in a moment that Radhakrishna would remember always, Ramanujan turned to him and said, "If I die, please hand these over to Professor Singaravelu Mudaliar [from Pachaiyappa's] or to the British professor, Edward B.

Ross, of the Madras Christian College," to whom he'd recently been introduced.

And with that, Ramanujan handed him two large notebooks stuffed with mathematics.

Ramanujan's notebooks were no longer for him merely a private record of his mathematical thought. As the preceding incident suggests, they were his legacy. And they were a selling document, his ticket to a job—"evidence," as his English friend Neville would later put it, "that he was not the incorrigible idler his failures seemed to imply." Propelled by necessity, he had begun calling on influential men who, he thought, could give him a job. And slung under his arm as he called were—just as photographers have their portfolios, or salesmen their display cases—his notebooks. Ramanujan had become, in the year and a half since his marriage, a door-to-door salesman. His product was himself.

In India more than elsewhere, it wasn't elaborate correspondence and formal application to anonymous bureaucrats that got you a hearing or landed you a job, but personal connections to someone at the top. Armed with an introduction from a friend of the family, or the family of a friend, you'd plunk yourself down at his doorstep. The physical blurring of the line between inside and outside in South Indian homes was matched by the permeability of South Indian social life; private and public realms were not so rigidly walled off as in the West. Often, you'd be admitted into the Great Man's presence. No day was so crowded, it seemed, no time so squeezed, that it couldn't accommodate one more job-seeker.

Ramanujan's refrain was always the same—that his parents had made him marry, that now he needed a job, that he had no degree but that he'd been conducting mathematical researches on his own. And here . . . well, why didn't the good sir just examine his notebooks.

His notebooks were his sole credential in a society where, even more than in the West, credentials mattered; where academic degrees usually appeared on letterheads and were mentioned as part of any introduction; where, when they were not, you'd take care to slip them into the conversation. "Like regiments we have to carry our drums, and tambourinage is as essential a thing to the march of our careers as it is to the march of soldiers in the West," Indian novelist and critic Nirad C. Chaudhuri has written of his countrymen's bent for self-promotion. "In our society, a man is always what his designation makes him." Ramanujan's only designations were *unemployed,* and *flunk-out.* Without his B.A., one prominent mathematics professor told him straight out, he would simply never amount to anything.

Ramanujan, then, had the toughest sort of selling job. But he brought to it qualities that, as he hawked his wares across South India, brought him a warm reception. People *liked* him.

"He was so friendly and gregarious," one who knew him later in Madras would say of him. He was "always so full of fun, ever punning on Tamil and English words, telling jokes, sometimes long stories, and going into fits of laughter when relating them. His tuft would come undone and he would try to knot it back as he continued to tell the story." Sometimes he'd start laughing before reaching the punch line, garble its telling, and have to repeat it. "He was so full of life and his eyes were mischievous and sparkling. . . . He could talk on any subject. It was hard not to like him."

Not that Ramanujan was the hail-fellow-well-met type. More often he seemed shy, his geniality emerging more among a few friends than in a crowd. Nor was he particularly attuned to interpersonal nuance. More than one otherwise fond reminiscence is like that of N. Hari Rao, a college classmate from Kumbakonam, who visited Ramanujan in Madras during this period. "He would open his notebooks and explain to me intricate theorems and formulae without in the least suspecting that they were beyond my understanding or knowledge." He just didn't *pick up.* Once Ramanujan was lost in mathematics, the other person was as good as gone.

And yet, ironically, this same want of social sensitivity conferred on him a species of charm. For its flip side was an innocence, a sincerity, upon which all who knew him invariably remarked.

"Ramanujan was such a simple soul that one could never be unfriendly toward him," recalled N. Raghunathan, a high school classmate who himself went on to become a mathematics professor. His humor ran toward the obvious. His puns were crude. His idea of entertainment was puppet shows, or *bommalattam;* or else simple street dramas, *terrukutu,* that ran all night during village festivals and to which Ramanujan would go with friends, cracking jokes and telling stories along the way. Ramanujan wore his spirits on his sleeve. There was something so direct, so unassuming, so transparent about him that it melted distrust, made you *want* to like him, made you *want* to help him.

In Kumbakonam a few years before, an old woman from the neighborhood had taken Ramanujan under her wing, often inviting him in for midday snacks. "She knew nothing of mathematics," one of Ramanujan's Indian biographers would note. But it was "the gleam in the eyes of Ramanujan and his total absorption in something—it is these that had endeared Ramanujan to her." And that unstudied absorption drew others to him, too.

People didn't take to Ramanujan because he was sensitive to them, or because he was especially considerate. They might not understand his mathematics. They might even flirt with the idea that he was a crank. And they might, in the end, be unable to help him. But, somehow, they couldn't help but like him.

Sometime late in 1910, Ramanujan boarded a northbound train from Kumbakonam and, about halfway to Madras, got off at Villupuram, just west of Pondicherry, the coastal city then still in French hands. At Villupuram, he changed trains for the twenty-mile trip west to Tirukoilur, a town of about nine thousand that was headquarters of its district. In Tirukoilur, V. Ramaswami Iyer held the midlevel government post of deputy collector. (Iyer, also spelled Aiyar, was the caste name of Brahmins who worshipped Siva, and was ubiquitous in South India.)

What made Ramaswami especially worth traveling to see was that he was a mathematician; in particular, he had recently founded the Indian Mathematical Society. Everyone called him "Professor," though he held no academic post. Back while a student at Presidency College, it seems, he had contributed mathematical articles to the *Educational Times* in England. Its editors, assuming he was a college professor, addressed him as such, and the name stuck.

Now, as ever, Ramanujan came armed with his notebook. The Professor looked at it. He was a geometer, and the mathematics he saw before him was mostly unfamiliar. Still, at least in the glow of memory, "I was struck by the extraordinary mathematical results contained in it." Did that mean he would give Ramanujan a job in the *taluk* office? Hardly. "I had no mind to smother his genius by an appointment in the lowest rungs of the revenue department," he wrote later. So he sent him on his way, with notes of introduction, to mathematical friends in Madras.

One of them, a charter member of the Mathematical Society, was P. V. Seshu Iyer, a pinch-faced man with glasses who'd been one of Ramanujan's professors at Government College. Since about 1906, they'd not seen one another. Now, four years later, Seshu Iyer had moved up to

Presidency College in Madras. Ramanujan met him there, notebooks in hand, but also this time with Ramaswami Iyer's recommendation. He left with leads and yet other notes of introduction.

He went to see S. Balakrishna Iyer, then himself just starting his career as a mathematics lecturer at Teachers' College in the Madras suburb of Saidapet. Would he, Ramanujan asked, recommend him to his English boss, a certain Dodwell, for a job as a clerk? It didn't matter how poorly it paid; anything would do. Balakrishna served him coffee, looked at his notebooks, which he didn't understand, and later went to see Dodwell three or four times on Ramanujan's behalf. Nothing came of it. "I was not big enough," apologized Balakrishna Iyer later—not important enough to exert any influence.

In December, Ramanujan went to see R. Ramachandra Rao, who was indeed "big enough." Educated at Madras's Presidency College, he had joined the provincial civil service in 1890, at the age of nineteen, and in time rose to become registrar of the city's Cooperative Credit Societies. Now he was district collector of Nellore, a town of about thirty-five thousand, a hundred miles up the East Coast Railway from Madras. Earlier in the year, he had been named "Dewan Bahadur," which was something like a British knight. All this, and he was a mathematician, too, serving as secretary of the Indian Mathematical Society, the group Ramaswami Iyer had founded four years earlier, and even sometimes contributing solutions to problems posed in its *Journal*. Intelligent, wealthy, and well connected, R. Ramachandra Rao was just the kind of paternal figure, at the head of a retinue of family and friends, through whose offices one got things done in India.

Just how Ramanujan got an audience with him is unclear, though accounts agree that Ramachandra Rao's nephew, R. Krishna Rao, was the final intermediary. Ramanujan's friend, Radhakrishna Iyer, to whom he'd earlier given his notebooks for safekeeping, said he wrote his father-in-law, an engineer in Nellore, to arrange the meeting. Seshu Iyer said later that he paved the way. Ramanujan's English friend, Neville, later speculated that Seshu Iyer did indeed supply Ramanujan with a letter of introduction to Ramachandra Rao—but that Ramanujan was "too timid" to use it. He may, then, have needed some extra push to go and meet this powerful man. If so, he got it from C. V. Rajagopalachari.

Rajagopalachari was just a few months older than Ramanujan, had grown up in the same town, frequented the same temple, attended Town High with him. One afternoon back in 1902, during recess, an older student, said to be the smartest in his class, handed him a math problem. Ramanujan was so smart? Well, then, let him solve this:

$$\sqrt{x} + y = 7$$

$$\sqrt{y} + x = 11$$

At first glance falling under the familiar heading of "two simultaneous equations in two unknowns," the problem actually confronted Ramanujan with a difficult fourth-degree equation and meant recalling a theorem applicable to a particular class of them. To any ordinarily smart fourteen-year-old, it would be exceedingly difficult. "To my astonishment," Rajagopalachari remembered later, "Ramanujan worked it out in half a minute and arrived at the answer by two steps."

In fact, he probably didn't "work it out" at all, but simply looked at it, guessed the answer might be one where each was a square, tried a couple of possibilities in his head, and saw the solution, $x=9$ and $y=4$, jump out at him; in other words, it was a piece of fancy footwork, nothing mathematically profound. Still, it impressed Rajagopalachari, and he and Ramanujan became friends.

Over the years, Rajagopalachari had followed a straight career trajectory toward becoming a lawyer, while Ramanujan floundered. The two lost contact. But now, in 1910, almost a decade later, they met again by chance in Madras. Despondent, Ramanujan told Rajagopalachari about his school failure. He had no future, he said. No one appreciated him. He'd written a famous mathematician in Bombay, Professor Saldhana, with samples of his work. He'd written the Indian Mathematical Society. Nothing had come of any of it. So, thanks to a friend who was supplying the ticket, he was taking a train back to Kumbakonam that very night.

Don't go, said Rajagopalachari. Ramanujan may have mentioned he had a letter of introduction to Ramachandra Rao, but had not yet acted on it. In any case, Rajagopalachari said that *he* would take him to meet Ramachandra Rao. When Ramanujan protested that he had no money to remain in Madras, his friend said he'd foot the expenses.

The meeting occurred. Ramachandra Rao wrote about it later, in these words:

> Several years ago, a nephew of mine, perfectly innocent of mathematics, spoke to me, "Uncle, I have a visitor who talks of mathematics; I do not understand him; can you see if there is anything to his talk?" And in the

> plenitude of my mathematical wisdom, I condescended to permit Ramanujan to walk into my presence. A short uncouth figure, stout, unshaved, not overclean, with one conspicuous feature—shining eyes—walked in, with a frayed notebook under his arm.

Three times, according to Rajagopalachari, Ramanujan met with the great man. The first time, Ramachandra Rao asked to keep Ramanujan's papers a few days. The second time, having perused them, he said he'd never seen anything like Ramanujan's theorems, but since he could make nothing of them, he hoped they would not trouble him again. They did, of course, so now, on this third occasion, Ramachandra Rao put things more plainly. Perhaps Ramanujan was sincere, he allowed; but if no moral fraud, he was more than likely an intellectual one. In other words, he doubted that Ramanujan knew what he was talking about.

As the two friends left, Ramanujan mentioned that with him he had his correspondence with Professor Saldhana, the eminent Bombay mathematician. Saldhana, too, had concluded that he couldn't help him. But many of Ramanujan's formulas, he'd written in the margins of the sheet of paper Ramanujan had sent him, seemed intriguing indeed. It was just that he could hardly throw the weight of his reputation behind someone working in areas so unfamiliar to him.

This was hardly a ringing endorsement; indeed it differed only slightly from what Ramanujan would hear all through his early years—that his work was not well enough understood to classify as either the fulminations of a crank or the outpourings of a genius. Ramachandra Rao himself, in so many words, had said that; dubious, he'd erred on the side of caution, and decided not to take up Ramanujan's case. But Saldhana, erring even further on the side of caution, had at least made clear that, whatever else he was, Ramanujan was no crank.

That was enough for the tenacious Rajagopalachari, who saw in Saldhana's comments a way to allay Ramachandra Rao's doubts. Back they went—on so fine a knife edge did Ramanujan's fate hinge—a fourth time. At first, Ramachandra Rao was angry. Here *again?* Just a few minutes later? But then he was shown the Saldhana correspondence, as well as some of Ramanujan's easier, more accessible results. "These," he wrote later, "transcended existing books and I had no doubt that he was a remarkable man. Then, step by step he led me to elliptic integrals, and hypergeometric series. At last, his theory of divergent series, not yet announced to the world, converted me. I asked him what he wanted."

What he wanted, Ramanujan replied, was a pittance on which to live and work. Or, as Ramachandra Rao later put it, "He wanted leisure, in other words, simple food to be provided to him without exertion on his part, and that he should be allowed to dream on."

3. "LEISURE" IN MADRAS

He wanted leisure . . .

The word *leisure* has undergone a shift since the time Ramachandra Rao used it in this context. Today, in phrases like leisure activity or leisure suit, it implies recreation or play. But the word actually goes back to the Middle English *leisour,* meaning freedom or opportunity. And as the *Oxford English Dictionary* makes clear, it's freedom not *from* but "*to* do something specified or implied" [emphasis added]. Thus, E. T. Bell writes of a famous seventeenth-century French mathematician, Pierre de Fermat, that he found in the King's service "plenty of leisure"—leisure, that is, for mathematics.

So it was with Ramanujan. It was not self-indulgence that fueled his quest for leisure; rather, he sought freedom to employ his gifts. In his *Report on Canara, Malabar and Ceded Districts,* Thackeray spoke of the "leisure, independence and high ideals" that had propelled Britain to its cultural heights. The European "gentleman of leisure," free from the need to earn a livelihood, presumably channeled his time and energy into higher moral and intellectual realms. Ramanujan did not belong to such an aristocracy of birth, but he claimed membership in an aristocracy of the intellect. In seeking "leisure," he sought nothing more than what thousands born to elite status around the world took as their due.

And remarkably—in a testament to his stubbornness as much as his brains—he found it.

That he was a Brahmin probably helped. Ramanujan was poor, from a family that sometimes lacked enough to eat. But in India, economic class counted for less than caste. Being a Brahmin gave him access to circles otherwise closed to him. In fact, virtually all those whom Ramanujan met during these years were Brahmins. Ramaswami Iyer was a Brahmin. So was Seshu Iyer. So was Ramachandra Rao. Had Ramanujan been of another caste, he might likewise have sought, and received, help from wealthy and influential castemen. But in no other caste did prestige, connections, *and* a taste for the life of the mind merge so naturally as they did among Brahmins.

As a Brahmin, Ramanujan may also have felt freer to seek the sort of constructive idleness he thought he needed—and perhaps even, in some measure, conceived as his due. Traditionally, Brahmins were recipients of alms and temple sacrifices; earning a livelihood was for them never quite the high and urgent calling it was for others. Uncharitably, it might be said that Ramanujan exhibited a prima donna–like self-importance that left him unwilling to study what he had no wish to study, or to work for any reason but to support his mathematics. Less harshly—and, on balance, with greater justice—he was a secular sanyasi.

Ramachandra Rao sent Ramanujan back to Seshu Iyer, saying it would be cruel to let him rot in a backwater like Nellore. No, he would not give him a job in the local *taluk* office but rather would seek for him some scholarship to which, despite his penchant for failing examinations, he might be eligible. Meanwhile, let him stay in Madras; he, Ramachandra Rao, would pay his way.

Monthly, from then on, Ramanujan began receiving a money order for twenty-five rupees. It wasn't much. But it was enough to free him from economic cares. Life opened up for him. Now, more decisively than before, he left the Kumbakonam of his youth behind and, from early 1911 and for the next three years, stepped into the wider world of South India's capital, Madras.

It was the fifth-largest city in the British Empire and, after Calcutta and Bombay, third-largest on the subcontinent. Some traced its name to the legend of a fisherman named Madarasen; others to a corruption of Mandarajya, meaning realm of the stupid, or even Madre de Dios, Portuguese for mother of God. The city itself, however, was an invention of British colonial policy. The British East India Company bought land at the mouth of the Cooum River, and Fort St. George, which they constructed there in 1642, became the administrative hub of the British presence in South India.

Madras was not a compact city. The 550,000 people who inhabited it in 1910 were spread up and down along the Bay of Bengal for miles, dispersed in quite distinct population centers—Georgetown, Triplicane, Mylapore, Chepauk, and others. Many of these places went back hundreds or thousands of years. Three and a half miles south of the ragged center of town, for example, was Mylapore, site of the revered Kapalaswara Temple. There, St. Thomas the Apostle, patron saint of India, had settled in the first century A.D. But the area was known to the ancient Greeks and Romans, as a port, long before that.

The modern city of Madras slung low over the land, only the occasional gopuram of a thousand-year-old temple punctuating the flatness; no part of the city rose more than fifty feet above sea level. Spread all across it, especially at the sites of old villages, were clusters of "hutments," one-room dwellings of mud and thatch, tens of thousands of them. But even the more substantial structures with red tile roofs almost never rose higher than the second floor. Over the years, the city had expanded horizontally, not vertically; you'd add an extension to the front or back of the house rather than build another story. Madras, then, was more like a leisurely, sprawling Phoenix or San Diego than a restless, densely packed New York.

There were still large rural tracts within the city, with palm trees and paddy fields, buffalo and washermen in rivers and lagoons, fishermen's thatched huts and catamarans idled on the beach. Save for a few more crowded districts, the crush of people squeezed onto every square inch that the Westerner today associates with Indian cities was still in the future. The city retained an easygoing village slowness.

It was possible to gaze down from the top of the lighthouse overlooking the harbor and note, as one English visitor did around the turn of the century, that

> Madras is more lost in green than the greenest city further north. Under your feet the red huddled roofs of the Black Town [the adjacent native quarter] are only a speck. On one side is the bosom of the turquoise sea, the white line of surf, the leagues of broad, empty, yellow beach; on the other, the forest of European Madras, dense, round-polled green rolling away southwards and inland till you can hardly see where it passes into the paler green of the fields.

That was a European perspective, of course. But among Indians, too, Madras was regarded as slower and more congenial, greener and more spacious than a Calcutta or Bombay. It was hard being poor anywhere in India. But it was a little easier in Madras. There was never the cold to bear. And being so removed from the North, so much a regional capital, so much *South* Indian, the city felt comfortable and familiar to the thousands who, like Ramanujan, had moved to it from towns and villages across the South.

* * *

In May 1911, Ramanujan left the place he shared on Venkatanarayan Lane and moved to a little alley boarding house, on Swami Pillai Street, bearing the inflated name "Summer House." There he lived for the rest of the year and much of 1912 with close to a dozen others, mostly students, who frequented a Brahmin-run restaurant on Pycroft's Road, the main street of a neighborhood known as Triplicane.

A few minutes' walk down Pycroft's, right beside Presidency College, lay the beach. Even then it was a Madras landmark, a place anyone who'd visited the city for even a few days always remembered. It was not just a beach, but a freak of nature, a sweep of sand piled up by the roaring surf over the eons, that then had been refined, manicured, and developed by an otherwise obscure Madras governor, one Mountstuart George Grant-Duff, back in the 1880s. At the end of the long sloping sand, the breakers rumbled. Yet so deep was the beach that, having once stepped onto it, it was as if you still had a great desert to cross to reach them.

It was here that Ramanujan would come, letting his mathematical ruminations percolate as he strolled along by the sea. Or else, come the cooler hours of the evening, he would come with his friends, plunking himself down on the light brown sand, flecked with tiny fragments of seashell, and spin occult stories until long after dark.

There was an openness out here, away from the hot, dusty streets of Triplicane, a delightful coolness. Looking inland, Ramanujan could see the domed clock tower of Presidency College, made gold by the setting sun. Looking out to sea, he could spy merchant vessels—distant gray shapes, and others, closer to shore, all cargo booms and bright paint—plying their way up the coast from Colombo, in Ceylon, and from around the southern tip of India, bound for Madras.

Lightened by the load Ramachandra Rao's generosity had lifted from his shoulders, Ramanujan was happy, or something close to it. Now, after that anxious, groping two years following his marriage, he was surrounded by friends, doing what he liked to do, carefree and cheerful. C. R. Krishnaswami Iyer, who'd known him at Pachaiyappa's and now shared a room with him in Summer House, remembered how once Ramanujan stayed up exclaiming on astronomical wonders till late into the night. Finally, Krishnaswami's cousin, his sleep shattered by Ramanujan's monologue, poured a pot of water over him; *that* would cool his fevered brain, he said. But Ramanujan took it all in stride. Ah, yes, a refreshing *Gangasnanam*—a purging bath in the River Ganges; could he *have another?*

1911 was a good and hopeful year. It was the year the capital of India was shifted, with great pomp and ceremony, from Calcutta to Delhi. The year a new sewer system, complete with underground conduits, sand filters, and pumps, was being installed in Madras. The year its oil-lit streets began to give way to electricity. And it was the year Srinivasa Ramanujan's first paper appeared, in the *Journal of the Indian Mathematical Society*—representing his initial step onto the stage of Indian mathematics, and the world's.

4. JACOB BERNOULLI AND HIS NUMBERS

Five years before, in late 1906, several dozen professors at colleges in Madras, Mysore, Coimbatore, and elsewhere in South India received a letter from V. Ramaswami Iyer, in which he proposed the formation of a mathematical society. Behind the idea lay simple want. Just as Ramanujan had so depended on whatever few mathematical books had come his way, so did Indian mathematicians generally suffer a lack of books and journals from Europe and America. The society, in Ramaswami's conception, would subscribe to journals and buy books, then circulate them to members. Twenty-five rupees per year from even half a dozen members would be enough to get the society off the ground.

He wound up with 20 founding members, all hungry for mathematical fellowship, and what was known first as the Analytical Club, then the Indian Mathematical Society, was born. Soon it was publishing a journal of its own. Just a dozen years later, at its second conference in Bombay, it would claim 197 members and be circulating 35 European and American journals.

These events awaited modern times. But a thousand years before the British came, Indians were doing mathematics. Before the seventh century, while the West was still mired in awkward Roman numerals, India had introduced the numerals we use today. The zero, a symbol expressing nothingness, represented a particular triumph; it may go back to as early as the second century B.C. but definitely appeared in a book in the third century and on the wall of a temple near Gwalior, in central India, in the ninth (where it helped specify a flower garden as 270 units long).

Many of India's contributions to mathematics were spurred by the need to know, based on astronomical factors, the correct times for Vedic ceremonies. Algebra, geometry, and trigonometry were all thereby enriched. Figures like Aryabhata, born in A.D. 476, who established one of the earliest and best values for π, and Brahmagupta, 150 years later, left theorems even now associated with their names.

It was a rich tradition, but one quite different from that of Greece, the cradle of Western mathematics. Whereas the Greeks, especially Euclid, emphasized formal proof, as in the step-by-step process high school students first encounter in geometry, Indian mathematics stressed the results themselves, however obtained. And without that winnowing out of mathematical dross that formal proof achieved, Indian mathematics was wildly uneven; some of it was just plain wrong. One Muslim writer noted in a book about India that Hindu mathematics was "a mixture of pearl shells and sour dates . . . of costly crystal and common pebbles."

By the twentieth century, the pearl shells and crystal had long lain buried in the dust of time. For centuries, India had stood its mathematical ground against the rest of the world. But now, that was ancient history; of late it had added little to the world's mathematical treasure. Only a line of brilliant mathematicians in Kerala, on the subcontinent's southwest tip, broke the gloom that otherwise extended back to the great Bhaskara of the twelfth century. The birth of the Mathematical Society could not ensure a rebirth. But its founders—hungry to connect with the West, proud of their country's heritage yet soberly aware that reverence for the past was no substitute for present achievement—surely hoped it did.

It was into this nascent new world that Ramanujan "came out," as it were, as a mathematician in 1911. He had met Ramaswami Iyer, the society's founder, the previous year when, in search of a job, he had traveled to Tirukoilur. Now Ramanujan's work was appearing in volume 3 of Ramaswami Iyer's new *Journal*—which, like most mathematics publications, opened its pages to provocative or entertaining problems from its readers.

One of two problems Ramanujan posed, as question 289, simply asked the reader to evaluate

$$\sqrt{1+2\sqrt{1+3\sqrt{1+\ldots}}}$$

Seemingly straightforward arithmetic, with not so much as an x or y to complicate it? Well, three issues of the *Journal* came and went—six months—with no solution offered; in the end, Ramanujan supplied it himself. The problem was those three little dots, indicating that the nesting of square roots, and the sequence of numbers begun, was to continue ad infinitum.

Ramanujan had generated the problem years before in the form of an example illustrating a more general theorem. The fourth equation down in chapter 12, on page 105 of his first notebook, read:

$$x+n+a = \sqrt{ax + (n+a)^2 + x\sqrt{a(x+n) + (n+a)^2 + (x+n)\sqrt{\text{etc.}}}}$$

Break any number into three components, *x*, *n*, and *a*, the equation said, and you could represent it in the form of those endlessly nesting square roots. For example, 3 could be imagined as $(x+n+a)$ in which $x = 2$, $n = 1$, and $a = 0$. Plug those values into the equation and you wind up with just what Ramanujan had, in question 289, asked to have evaluated. The answer, in other words, was just plain 3. Of course, how you'd figure it out without Ramanujan's equation was scarcely obvious.

In its small way, Ramanujan's deceptively difficult problem rose up from mathematical terrain that had fascinated him for as long as he had worked in mathematics. A superficially similar problem *might* have asked the reader to take the nested square roots only so far, out to, say, 10, or 100, or 1000. That way, you could plug in the numbers, run through the computation, and be finished. But this wasn't the problem Ramanujan posed. He asked, what happened if you were never finished? What if the number of nested square roots was infinite?

This was, in a sense, a contradiction in terms: how could any number "equal" infinity? Infinity was no place you could reach, no quantity you could plug into an equation; there was no "last number." So to understand how a mathematical expression behaved "at" infinity was to explore an elusive and mysterious terrain out beyond all seeing.

And no one explored this terrain more ardently, or knew it more intimately, than Ramanujan.

Many kinds of mathematical processes can be ordered to proceed ad infinitum. Ramanujan's problem in the Mathematical Society *Journal* was built around "nested radicals"—square roots of square roots of square roots of . . . , an area little studied then or now. But Ramanujan also studied continued fractions—fractions of fractions of fractions of . . . Most of all, he explored infinite series, which appeared on virtually every page of his notebooks and would fairly litter his first substantial paper in the *Journal* later in 1911. "Infinite series," one mathematician has written, "were Ramanujan's first love."

The telltale three dots heralding their presence showed up early in his notebooks, though more often he simply wrote "&c," which meant the same thing: numbers or algebraic terms, following some particular pattern, were to be added to one another forever. Thus,

$$1+2+3+\ldots$$

clearly suggests that the next term is to be 4, then 5, and so on.

Of course, this is not very interesting because, here, an infinite number of terms just adds up to infinity. What makes infinite series so intriguing, so valuable, and the object of so much study, is when they *don't* build up without bound, when they add up to something finite. As mathematicians put it, the series "converges" to a particular value. For example,

$$1+\tfrac{1}{2}+\tfrac{1}{4}+\tfrac{1}{8}+\ldots$$

Here, the next term is $\tfrac{1}{16}$, the next $\tfrac{1}{32}$, and so on. And the curious thing, known even to the Greeks, is that even though you add terms forever, each term diminishes so rapidly from the preceding one that the sum, even after an infinite number of terms, is a quite manageable 2—the value to which the series is said to converge. The more terms you add, the closer you get to 2.

But just because each successive term in a series is smaller than the one before doesn't mean it converges. For example,

$$1+\tfrac{1}{2}+\tfrac{1}{3}+\tfrac{1}{4}+\ldots$$

is superficially similar to the earlier convergent series, but doesn't converge; go as far out into the series as you like, but so soon as you think it's adding up to something, more terms always take you beyond it. For example, is the sum of this series perhaps 2, as it was for the previous one? No—four terms are enough to exceed it. Maybe 3? Here it takes a little longer, but already by the eleventh term, you've passed it. Perhaps 10? Twelve thousand, three hundred ninety terms add up to more than 10. It turns out—and can be proven—that whatever number you pick, the same holds: The sum of the series is infinite; it doesn't converge.

The series of interest to mathematicians, then, are those that do converge, or that converge under certain circumstances. And it's *what* they converge *to* that is often the wonderful thing about them.

Take, for example, the trigonometric functions that many remember from high school, which teachers practically always first introduce as the ratios of the legs of a right triangle. These functions—named sine, cosine, tangent, and so on—are all computed by taking the length of one or another side of a right triangle and dividing it by the length of another. The backs of trigonometry books are stuffed with long lists of angles with the corresponding values of their trig functions. Give me an angle, the tables say, and I'll give you the value, for example, of the sine of that angle. The angle is 30 degrees? Its sine is .5000. And so on. Such tables in hand, navigators cross oceans, engineers design machines.

And yet these same trigonometric functions, historically rooted in right triangles and ratios, can be evaluated in a way seemingly unrelated—as the sums of infinite series. If, say, the angle θ (theta, the Greek letter) is expressed not in degrees but in another measure of angularity that mathematicians find more convenient (radians), then:

$$\sin \theta = \theta - \frac{\theta^3}{3!} + \frac{\theta^5}{5!} - \frac{\theta^7}{7!} + \ldots$$

(Here, 5!, read "factorial five," just means $5 \times 4 \times 3 \times 2 \times 1 = 120$.)

Want the sine of 30 degrees? Just plug into the equation its radian equivalent ($\pi/6$, or about .5236), and add up as many terms as you want to get a value as accurate as you want. Here, even three terms are enough to get you to .500002—quite close to the correct .500000; the series converges rapidly.

Thus, this alternating infinite series—adding a bit here, subtracting something a little smaller there, and so on through an infinite number of terms—inexplicably equals just what you get from dividing one leg of a triangle by another.

Just this sort of seemingly unexpected connection shows up all the time with infinite series, which is what has made them so attractive to mathematicians, Ramanujan most particularly. Bernoulli numbers, the subject of his first published paper, were defined in terms of infinite series. And every page of his paper was riddled with more of them.

Jacob Bernoulli was among the first in a line of eminent seventeenth- and eighteenth-century mathematicians derived from a merchant family that had fled anti-Protestant massacres in Antwerp and settled in Switzerland. He helped extend calculus, the powerful set of mathematical tools for dealing with continuously varying quantities, beyond the point that Germany's Gottfried von Leibniz, along with England's Sir Isaac Newton, had taken it two decades before. Along the way, he derived the numbers that have since borne his name.

Bernoulli numbers are intimately tied to the quantity *e* which, like π, is a number whose special properties make it ubiquitous in mathematics. It is defined as

$$e = 1 + \frac{1}{1!} + \frac{1}{2!} + \frac{1}{3!} + \ldots$$

Now, when a particular algebraic expression involving *e* is expressed as an infinite series, the coefficients of each term turn out to have special significance. (Coefficients are just the ordinary numbers by which the algebraic parts are multiplied; in the equation $3x + \frac{1}{2}x^2 = 12$, 3 and ½ are the coefficients.) These coefficients were the Bernoulli numbers, which first appeared in his book *Ars Conjectandi*, published after his death in 1713. Notational inconsistencies confuse matters some, but in one system the first few Bernoulli numbers, generically B_n, are $B_1 = -\frac{1}{2}$, $B_2 = \frac{1}{6}$, $B_4 = -\frac{1}{30}$, $B_6 = \frac{1}{42}$. (The odd-numbered ones, except for the first, are all zero.)

Ramanujan had stumbled on Bernoulli numbers for the first time about eight years before, though probably without having ever heard of them as such. The second volume of Carr's *Synopsis* contained references to them in various guises, but Ramanujan may not have seen it until 1904, when he was at Government College—a year after he apparently began working with them. In any case, he'd worked with them ever since, using them repeatedly, through the Euler-Maclaurin summation formula, to approximate the values of mathematical entities known as "definite integrals" (unrelated to "integers") in calculus. Pages 30 and 31 of the first notebook cited them. So did much of chapter 5 of his second.

Now Ramanujan was making them the subject of his first formal contribution to the mathematical literature. Children take the Salk or the Sabin vaccine to protect them against polio. Supersonic aircraft exceed Mach 1, the speed of sound in the measurement system named for Austrian physicist Ernst Mach. In science and medicine, immortality is having something—a treatment, a unit of measurement, a theory—named after you. So, too, in mathematics. Bernoulli numbers bear his name because they appear again and again in a wide variety of mathematical applications. They weren't just flukes of mathematical nature, meaningless chains of digits; there were relationships among them, and Ramanujan had discovered—or, in some cases, rediscovered—what some of them were.

"Some Properties of Bernoulli's Numbers," he called his paper, and it was an apt title. The physical properties of a metal, like its melting point or specific gravity, appear in any chemical handbook; Ramanujan was discovering *mathematical* properties of these numbers. Bernoulli numbers were expressed as fractions; for example, $B_{32} = \frac{7709321041217}{510}$. Well, Ramanujan found that the denominators (the bottom parts) of those fractions were always divisible by 6. He found alternative ways of calculating B_n based on earlier Bernoulli numbers. The sixth of eighteen numbered sections began:

> 6. It will be observed that if n is even but not equal to zero,
>
> (i) B_n is a fraction and the numerator of $\frac{B_n}{n}$ in its lowest terms is a prime number,
>
> (ii) the denominator of B_n contains each of the factors 2 and 3 once and only once,
>
> (iii) $2^n(2^n-1)\frac{B_n}{n}$ is an integer and consequently $2(2^n-1)B_n$ is an *odd* integer.

On and on Ramanujan's paper went like that, filling seventeen pages of the *Journal*. By one reckoning, it stated eight theorems, offering proofs, of a sort, for three of them; two were stated as corollaries of two other theorems, three more as mere conjectures.

Ramanujan's manuscript had problems when it first reached the editor's desk. "Mr. Ramanujan's methods were so terse and novel and his presentation so lacking in clearness and precision," it would later be observed, "that the ordinary [mathematical] reader, unaccustomed to such intellectual gymnastics, could hardly follow him." In other words, his paper was a mess. And this was written by a *champion* of Ramanujan's work. M. T. Narayana Iyengar, a math professor at Central College, Bangalore and the *Journal*'s editor during the early years, confessed later "that the editor's work in connection with Ramanujan's contributions was by no means light," and that the manuscript went back and forth between him and its author three times.

In this first paper, as all through his work, Ramanujan found connections between things that seemed unconnected. Other mathematicians would later prove most of them true; Ramanujan, though, either couldn't be bothered, or didn't see the need to. What proofs he did offer were sketchy or incomplete.

A testament to the influence on him of George Shoobridge Carr? That's what most scholars later concluded. Carr, writing a synopsis of results rather than making original contributions, had given proofs, where he did so at all, only in bare outline. Now Ramanujan, who *was* making original contributions, clung to the pattern. He had asserted, for example, that the numerator of the *n*th Bernoulli number divided by *n* was always a prime number. Proof? Not a shred. Another mathematician later observed, "He takes the numerical evidence as sufficient, and there is no trace of any suggestion that there is need of other proof." Whether Ramanujan cared about proof is debatable; that normally he didn't furnish it, sometimes offering the most provocative results without a scintilla of evidence to support them, is not.

In this particular case, as it happens, Ramanujan *was* wrong. For example, the numerator of $B_{20} / 20 = 174611$, which is not a prime number at all, as he claimed, but equal to 283×617.

This, though, was the exception to prove the rule; much more often, Ramanujan's trust in himself was wholly justified. In his paper on Bernoulli numbers, in the notebooks, in his mathematical correspondence, in his other published papers—he was, with astounding consistency, right.

5. THE PORT TRUST

Appearing in the *Journal of the Indian Mathematical Society*, Ramanujan was on the world's mathematical map at last, if tucked into an obscure corner of it. He was starting to be noticed.

Early the following year, K. S. Srinivasan, a student at Madras Christian College who'd known Ramanujan back in Kumbakonam, dropped by to see him at Summer House.

"Ramanju," he said, "they call you a genius."

Hardly a genius, replied Ramanujan, "Look at my elbow. That will tell you the story." It was rough, dirty, and black. Working from his large slate, he found the quick flip between writing hand and erasing elbow a lot faster, when he was caught up in the throes of his work, than reaching for a rag. "My elbow is making a genius of me," he said.

Why, Srinivasan asked, didn't he just use paper? Can't afford it, replied Ramanujan. He was getting money from Ramachandra Rao. But that only went so far. Paper? He'd need four reams of it a month.

Another friend from the Summer House days, N. Ramaswami Iyer [no relationship to the "Professor"] also recalled Ramanujan's "huge appetite" for paper. Ramaswami pictured him lying on a mat, his shirt torn, "his long hair carelessly bound up with a piece of thin string," working feverishly, notebooks and loose sheets of plain white paper piled up around him. A friend from Pachaiyappa's who met him in Madras a little later, T. Srinivasacharya, recalled that, for want of paper, Ramanujan would sometimes write in red ink on paper already written upon.

It was during this period that, apparently worried something might happen to Ramanujan's notebook, Ramachandra Rao prevailed on him to copy it over. Ramanujan did so, though not without revising and expanding it as he did, incorporating the notes appended to it into appropriate sections of the new one—which comes down to us today as the "second" *Notebook*.

Half a century after Ramanujan was dead, in one of the many memorial books honoring him, one sponsor would be a manufacturer of writing and printing papers in Erode, Ramanujan's place of birth. "Paper, The Great Immortalizer," its one-page ad was headed. "Good Paper," it went on, "has helped preserve and propagate the great thoughts of Man." It would be a fitting tribute.

For about a year, Ramanujan lived on Ramachandra Rao's generosity. He was mathematically productive, peppering the Mathematical Society *Journal* with one interesting new problem after another, and completing a second paper. But he was, after all, unemployed, and this grew to bother him. Not long before, through one of his patrons, Ramanujan got a temporary job in the Madras Accountant General's Office, making twenty rupees per month, but held it only a few weeks. Now, early in 1912, Ramachandra Rao had turned to others among his influential friends, and Ramanujan was applying for a new job:

> Sir,
>
> I understand there is a clerkship vacant in your office, and I beg to apply for the same. I have passed the Matriculation Examination and studied up to the F.A. but was prevented from pursuing my studies further owing to several untoward circumstances. I have, however, been devoting all my time to Mathematics and developing the subject. I can say I am quite

confident I can do justice to my work if I am appointed to the post. I therefore beg to request that you will be good enough to confer the appointment on me.

I beg to remain,
Sir,
Your most obedient servant,
S. Ramanujan

Ramanujan's letter was written in a neat, schoolboy hand unremarkable in every way save for *t*'s whose horizontal arms rarely intersected the vertical stems they were meant to cross, but floated off instead to their right. Appended to it was a hand-copied version of a recommendation by a mathematics professor at Presidency College, E. W. Middlemast, who described him as "a young man of quite exceptional capacity in Mathematics." In fact, recommendation and letter were probably both a matter of form, the job for which he applied doubtless his all along, thanks to Ramachandra Rao.

The letter was dated 9th February 1912. It listed Ramanujan's return address as 7, Summer House, Triplicane. It was addressed to The Chief Accountant, Madras Port Trust.

From its beginnings, Madras had been a trading settlement and port, though nature had equipped it poorly for the job. An uncommonly rough surf crashed relentlessly. There were insidiously tricky ocean currents and peculiar sand buildups. It had no natural harbor at which cargo vessels might unload. Instead, ships would anchor a quarter mile offshore, and *masula* boats—flat-bottomed craft about twenty-five feet long made from thin planks stitched together with coconut fiber—piloted by daring men expert in reading the surf, would row out to the ships, load cargo a few tons at a time, and return with it to the beach. Ninety percent of all losses suffered by ships calling at Madras came in that treacherous last quarter mile.

Yet somehow they'd made a port out of it. In 1796, a lighthouse had been built; it burned coconut oil in lamps visible from seventeen miles away. In 1861, a pier extending eleven hundred feet into the Bay of Bengal was completed, and 1876 saw construction begin on a rectangular artificial harbor, twelve hundred yards on a side, built up from twenty-seven-ton concrete blocks. The new harbor improved matters little. Unloading losses remained high, and the entrance rapidly began silting up, the high-water line on the south side of the harbor advancing seventy feet a year into the bay.

The port of Madras carried more than 60 percent of the Madras Presidency's imports and exports to Britain. Each year, twelve hundred ships called there, bringing in iron and steel, machinery, and railway equipment, and leaving with hides, piece goods, indigo, and raw cotton. Still, by early in the century, it was a troubled operation needing major changes. Placed in charge of the Port Trust in 1904, and as its chief engineer charged with overseeing those changes, was Sir Francis Spring.

A bald man with sleepy eyes, white mustache, and goatee, known in Madras as among the first in South India to own his own motorcar, Sir Francis was in his second career. Born in Ireland in 1849, a graduate of Trinity College in Dublin, he had joined the India Government engineering service in 1870 and for more than thirty years had played a key role in the development of the South Indian Railways System, where he had, among other feats, spanned the Godavari River with a big railroad bridge. For these accomplishments, he had been named knight commander of the Indian Empire in 1911. Seven years before, he'd come to the Port Trust, and with him he'd brought S. Narayana Iyer.

Narayana Iyer was not an engineer by training; the British had set up Indian colleges to train bright clerks to administer the bureaucracy, not equip them to get along without European technical expertise. The son of a Brahmin priest, he held an M.A. from St. Joseph's College, in Trichinopoly, where he'd stayed on to become a lecturer in mathematics. There he'd met Sir Francis. At the Port Trust, he was, as office manager and then as chief accountant, the highest-ranking Indian. Sir Francis relied on him heavily.

Narayana Iyer never succumbed to Western dress but wore the traditional dhoti and turban until his death in 1937; years later, his family would point to that as testimony to his personal strength. All during these years, winds of change from the West influenced even matters of dress. While some ridiculed Indians who adopted European trousers, coat, collared tie, and boots or shoes, more common was the attitude reflected in a *Hindu* editorial in the late 1890s: "There can be no doubt that boots and trousers with the European coat constitute the most convenient dress for moving about quickly. The oriental dress is suited to a life of leisure, indolence, and slow locomotion, whereas the Western costume indicates an active and self-confident life." By the 1910s, educated, upwardly mobile Indians had gotten the message. In a formal photo taken at an

Indian Mathematical Society conference in 1919, Ramachandra Rao, for example, wore Western garb, as did more than half the others. Narayana Iyer, also in the photo, sitting on the ground in the first row, in full turban and flowing robes, did not.

His family would tell of a scrupulously honest, restrained, and dignified man, not a little forbidding in aspect, who championed Indian independence—but quietly. Who, as the patriarch of a house where more than two dozen cousins, brothers, sisters, and assorted hangers-on depended on him, always helped those who came to him for money—but quietly. Who harbored a searchingly independent mind—behind outward behavior conforming in every respect to traditional Hinduism.

These two men, Sir Francis Spring and Narayana Iyer, were to play an important role in Ramanujan's life over the coming years. But now, on March 1, 1912, three weeks after he'd applied for the job, Ramanujan knew only that he worked under them as a Class III, Grade IV clerk in the accounts section, earning thirty rupees per month.

During all this time, Janaki had been far from her husband's side, shuttling back and forth between her parents in Rajendram and her mother-in-law's house in Kumbakonam for "training" in the wifely arts. Now, late in 1912, past puberty and with Ramanujan in a steady job, the two finally became man and wife in something more than name.

Summer House in Triplicane was about three miles from Ramanujan's new job at the offices of the Port Trust opposite the harbor complex north of Fort St. George. So a few months after starting the job, Ramanujan had moved much closer, joining his grandmother in a little house on Saiva Muthiah Mudali Street, off Broadway, in the district known as Georgetown. And it was there that, three years after their marriage, Janaki—along with Ramanujan's mother, Komalatammal—joined him.

Until the visit to Madras in 1906 of the Prince of Wales, the future King George V, Georgetown was still Black Town, the original area set aside for the native, or "black," population; the area within the Fort, set aside for Europeans, was White Town. Its teeming streets held a third of the city's population on 9 percent of its land. Cows and bullocks, chickens and goats, roamed freely. In one street, metalworkers would squat in front of their tiny stalls, hammering out shapes, or tossing scraps of tin into little buckets to be melted. The next street would be clogged with bullocks, shouldering huge sacks of grain. Then streets of jewelry stalls, of textile shops, oilmongers, basketweavers, fruit and vegetable wholesalers . . . And everywhere, driving it all, was muscle power, black-haired men, shoeless and shirtless, clad only in their dhotis, ribs and muscles pushing out against glistening dark brown skin, straining as they pulled carts, or bent low under heavy loads upon their backs, or whipping their animals through the dusty streets.

On Saiva Muthiah Mudali Street, in the tiny house for which they paid three rupees per month, Ramanujan and his family were right on top of each other. Yet he and Janaki, not even a teenager yet, had little contact. They scarcely spoke. During the day, he might ask her to fetch him soap or an article of clothing. At nights, she'd recall, she mostly slept beside Komalatammal—at her mother-in-law's insistence. They were never alone. If Komalatammal had to go to Kumbakonam, his grandmother, Rangammal, remained behind, monitoring their contact.

None of this was uncommon. Until she had children of her own, a new wife's position in the family of her husband verged on that of a slave. She was there only to serve, to do her mother-in-law's bidding.

Later, Ramachandra Rao would say he helped Ramanujan get "a sinecure post" at the Port Trust. And sinecure it was, though whether intended as such from the beginning or becoming that later is not clear. Still, even just putting in the hours made for a life more hectic than Ramanujan was used to. "I used to see him many times running to his office via the Beach Road," recalled a friend from Summer House days, referring to the road that ran right up beside the Port Trust offices. "With his coat, tail and all, flying in the breeze, and his long hair coming undone, a bright *namam* [his trident-shaped caste mark] adorning his forehead, the young genius had no time to waste; he was always in a hurry."

Janaki would later recall how before going to work in the morning he worked on mathematics; and how when he came home he worked on mathematics. Sometimes, he'd stay up till six the next morning, then sleep for two or three hours before heading in to work. At the office, his job probably included verifying accounts and establishing cash balances. At one point, some months after he started, he replaced another clerk, on leave for a month, as "pilotage fund clerk." In any event, the work was hardly taxing, and soon he was being left alone to work on mathematics, being tolerated in this, if not explicitly encouraged, by both Narayana Iyer and Sir Francis.

Once, the story goes, a friend found him around the docks during working hours, prowling for packing paper on which to work calculations. Another time, Sir Francis called Narayana Iyer into his office. How, he demanded to know, sternly regarding his aide, had these pages of mathematical results gotten mixed into this important file? Was he, perhaps, using office time to dabble in mathematics? Narayana Iyer pleaded innocent, claimed the math wasn't in his handwriting at all, that perhaps it was Ramanujan's work. Sir Francis laughed. *Of course* it was Ramanujan's work. He'd known as much all along.

Narayana Iyer, a member of the Mathematical Society and long its treasurer, was not just Ramanujan's immediate superior, but his colleague. In the evenings, they would retire to the elder man's house on Pycroft's Road in Triplicane. There, they'd sit out on the porch upstairs overlooking the street, slates propped on their knees, sometimes until midnight, the interminable scraping of their slate-pencils often keeping others up. Sometimes, after they had gone to sleep, Ramanujan would wake and, in the feeble light of a hurricane lamp, record something that had come to him, he'd explain, in a dream.

Narayana Iyer was no mean mathematician. But in working with him he found that Ramanujan's penchant for collapsing many steps into one left him as lost as a dazed Watson in the wake of a run of Sherlock Holmes logic. How, Narayana Iyer would ask, could he expect others to understand and accept him? "You must descend to my level of understanding and write at least ten steps between the two steps of yours." What for? Ramanujan would ask. Wasn't it obvious? No, Narayana Iyer would reply, it was not obvious. Patiently, he would persist, cajoling him, in the end sometimes getting him to expand a little on his thinking.

It wasn't long before Narayana Iyer was not just a boss to Ramanujan, nor even just a colleague, but advisor, mentor, and friend. "Some people," Janaki later recalled him saying, "look upon him [Ramanujan] as ordinary glass, but they will remain to see him soon to be a diamond." He brought Sir Francis around to his view, too, making him Ramanujan's champion as well.

And it was in coming to the attention of Sir Francis and to the web of contacts radiating out from him that, sometime around the middle of 1912, Ramanujan stepped into *British* India. He had grown up and lived almost his entire life with only the barest contact with the British. Now that was about to change.

6. THE BRITISH RAJ

West of the Madras Presidency, high in the Nilgiri Hills, was Ootacomund, known as "Ooty," the Presidency's summer home, where Englishmen and their families, more steeled to frigid winds blowing in off the North Sea than to the tropical heat, fled to escape the lowland summer. On the far side of the hills lay a narrow north-to-south strip of wet, rainy country, the Malabar coast, dense with tropical vegetation and rich with the pepper, nutmeg, and other spices that had drawn the eyes of Europe to India in the first place. In nearby Mysore, thick forests were home to exotic sandal and rosewood, and animals like the tiger and elephant that had so inflamed the English imagination.

A British presence in India went back to 1600, when the East India Company was formed. For two centuries Britain clashed with the French, Dutch, Portuguese, and others for control of the subcontinent and, in 1876, made India part of the empire. Now, in 1912, more than half a century had elapsed since the Sepoy Mutiny—Indian nationalists called it the First War of Independence—had last challenged British rule. The year before, King George had passed through the Gateway of India erected to his honor in Bombay, then traveled to the new capital, Delhi, to assume his throne. The very notion of India without Britain was, in the words of Lord Curzon, the viceroy, "treason to our trust." Who could imagine that India, led by Gandhi and Nehru, would wrest independence from the Crown in 1947? And that the fair-skinned few now in power were presiding over the last days of the British Raj?

The marvel of British rule was that so few administered it. The Indian Civil Service, the legendary ICS, numbered barely a thousand men. It was India's central nervous system, quietly controlling mechanical arms wielded by Indian clerks and British engineers, physicians, and police. The viceroy and the governors of Madras and the other presidencies were not ICS, but everyone else who mattered was. Evolving out of the East India Company's staff of commercial agents, the ICS was in 1853 thrown open by competitive exam to Indians. But they never held more than a few dozen positions throughout the country. And their salaries were limited to two-thirds those of the British.

Only lexical accident links the ICS to the bureaucratic inefficiency and mediocrity today conjured up by "civil service." In fact, the Indian Civil Service attracted many of Britain's best. Its members were culled from the upper classes and intellectual elite. They were products of the finest public schools, graduates of Cambridge and Oxford. They had passed arduous examinations. In their spare time, they translated works from Sanskrit, deciphered temple inscriptions, wrote grammars, compiled dictionaries. They were men who, as one account later had it, "were fond of thinking of themselves as Plato's ideal rulers." Reared with patrician values, imbued with a sense of responsibility and public trust, they established a reputation for dedication and fair-mindedness.

But there was another side to them—an insufferable smugness, a towering sense of moral superiority. "The members of this service," wrote a retired member of it about this time,

> have generally shown the capacity which is awakened by responsibility in men of British race: with ample salaries they have hardly been tempted by dishonesty, and their detached impartiality has not been disturbed by the importunity of relations or friends. To the credit of their nation they have established and maintained a government, which, for its resources, is exceedingly efficient.

At the district level, the government representative was called a collector, and he wielded the power of a prince. But, as one late nineteenth-century account noted, "The collector and his English staff hardly ever know the vernacular. By the natives they are regarded with awe, not affection." Observed another: "The collector is separated by an impassable gulf from the people of the country. . . . To the eyes of a native, the English official is an incomprehensible being, inaccessible, selfish, overbearing, irresistible."

The central fact of the British presence in India, then, was *distance*. Prints and engravings today on view at Fort St. George in Madras show British life walled off from India, marked by a well-ordered calm reminiscent of nothing so much as an English garden. One view shows turbaned Indians surrounding a snake charmer, as an Englishman, attended by a native servant, looks on from the safe remove of a second-floor porch. In these scenes, natives work—bearing palanquins, or balancing loads atop their heads, or urging *masula* boats out into the angry surf. The English, invariably at their ease, stand shielded from the sun under an umbrella or, in top hat, lean languidly against a column.

An Englishman typically had his own washerman, who got four or five rupees a month, lived on his premises, and functioned as something like his private property. After long enough in India, the Englishman forgot how to so much as brush and fold his clothes. When he finally took the steamer home, he'd be taken aback as an English steward stooped to serve him tea. (Indeed, years later, asked as part of a survey what most struck them about England, students from former Asian and African colonies invariably mentioned the sight of white men doing manual labor.)

That there was an ineradicable split between Englishman and Indian the British themselves were eager to acknowledge. "East is East and West is West and never the twain shall meet," wrote Kipling. Indians might work, even live with you in your bungalow, noted one old India hand, Herbert Compton; but in the end, "there is no assimilation between black and white. They are, and always must remain, races foreign to one another in sentiment, sympathies, feelings, and habits. Between you and a native friend there is a great gulf which no intimacy can bridge—the gulf of caste and custom. Amalgamation is utterly impossible in any but the most superficial sense, and affinity out of the question."

It was this "great gulf" that, in the succeeding months and years, Ramanujan would, of necessity, confront. He had grown up during the reign of Queen Victoria. Coins in his pocket bore likenesses of the British sovereign; until 1902, they'd said VICTORIA EMPRESS, while after her death, they bore the profile of Edward VII, KING AND EMPEROR. In high school, Ramanujan's scholastic prizes included an anthology of patriotic English verse, a collection of Lord Macaulay's essays, and a volume of Wordsworth's poetry—certainly nothing Indian. Later, while he attended Government College in Kumbakonam, the sixty thousand rupees that construction of a new student hostel required were raised in memory not of some Indian notable but of Queen Victoria.

Yet all this had left England for Ramanujan no more than symbol, image, and abstraction; English people he had scarcely known. Now, however, that was changing. His friend Narasimha had introduced him to E. B. Ross, of Madras Christian College. He had met E. W. Middlemast of Presidency College and secured from him a recommendation. Now, at the Port Trust, he had met Sir Francis; soon he would meet Spring's friends.

Whatever prejudices may have distorted the British view of Ramanujan did not likely extend to his intellect. Of the Indian, one English writer noted around this time, "he is cunning and contentious in argument, and his intellectual powers, when educated, are capable of considerable development. . . . In this respect, he puts the Englishman to shame, and were all posts in the Indian Government thrown open to examination in

India, we should probably see the administration filled with Bengali Baboos and Mahratta Brahmins."

But that's as far as British esteem for the Indian temperament went. Herbert Compton, who had run a plantation for years, observed in a book published in 1904 that "whilst you can polish the Hindu intellect to a very high pitch, you cannot temper the Hindu character with those moral and manly qualities that are essential for the positions he seeks to fill." A retired member of the Indian Civil Service, Sir Bampfylde Fuller, marveled at how Hindu boys could flock to classrooms and libraries, and pursue Western literature and science, yet unaccountably cling to . . . well, Indian ways. Indians, he said, were unduly sentimental, wildly inconsistent. "An Englishman is constantly disconcerted by the extraordinary contradictions which he observes between the words and the actions of an educated Indian, who seems untouched by inconsistencies which to him appear scandalous. . . . They give eager intellectual assent to [European] ideals, yet live their lives unchanged."

Other laudable traits the British discerned in the Indian personality verged on damning with faint praise. Of the South's Dravidian stock, one Briton wrote, they were "hardworking, docile and enduring. They are more sober, self-denying and less brutish in their habits than Europeans. They show greater respect for animal life, they have more natural courtesy of manner, and, as servants, attach themselves to those who treat them well with far greater affection than English servants." However seemingly admirable, these were hardly traits that left Indians the equals of their British masters.

It was with some such blend of due regard for his intellect, coupled with a lingering dubiety about his character and temperament, all seen across a vast social divide, that the British would tend to view Ramanujan.

7. THE LETTER

Through Narayana Iyer, Ramachandra Rao, Presidency College mathematics professor Middlemast, and others testifying to his mathematical gifts, it had become clear around Port Trust offices by late 1912 that Ramanujan was something special. The question facing Sir Francis and other British officials was *how* special? And in just *what way* special? And what, in any case, were they to *do* with him? Were his gifts trivial, or profound? Were they the gifts of the genius or, intellectually speaking, the shaman? Was Ramanujan a minor oddity who could be safely dismissed, or a prodigy demanding nurture and guidance?

No one would go out on a limb. The harsh explanation is that, despite firm opinions, they were afraid to, on the chance that events might prove them wrong and history judge them harshly. The kinder and simpler explanation—and the more likely—is that they just didn't know, and *knew* they didn't know.

Narayana Iyer, of course, thought he *did* know; he worked with Ramanujan every day and saw his abilities up close. If Ramanujan was to achieve his promise, it was clear, he needed the British solidly in his corner. Narayana Iyer lobbied on his behalf with Sir Francis.

Drawing on his connections, Ramachandra Rao also tried to gain Spring's ear. "Dear Sir Francis," wrote C. L. T. Griffith, a forty-year-old civil engineering professor at Madras Engineering College, on November 12, 1912, apparently at Ramachandra Rao's behest. "You have in your office an Accountant, on Rs 25, a young man named S. Ramanujan, who is a most remarkable mathematician. He may be a very poor accountant, but I hope you will see that he is left happily employed until something can be done to make use of his extraordinary gifts." Since few could follow, much less meaningfully critique, Ramanujan's work, he went on, he was writing another mathematician (M. J. M. Hill, in London) for advice, and sending him some of Ramanujan's papers. "If there is any real genius in him," wrote Griffith, "he will have to be provided with money for books and with leisure, but until I hear from home," he added, hedging his bets, "I don't feel sure that it is worthwhile spending much time or money on him."

Among those to whom Spring turned for advice was A. G. Bourne, Madras's director of public instruction, who advised that Ramanujan be sent to see one or both of two Madras mathematicians he indicated by name. Then, he added: "If his genius is so elusive or mysterious that good mathematicians, possessed besides of much common sense, cannot recognize and appreciate it even if it carries them beyond their scope, I should doubt its existence."

Two weeks later Ramanujan went to see W. Graham, Madras's accountant general, one of those Bourne had suggested. "Whether [Ramanujan] has the stuff of great mathematicians or not I do not know," Graham wrote after seeing him. "He gives me the impression of having brains." *Gives me the impression* . . . With such care did he word his assessment, no one could possibly fault him should he prove wrong. Confusing matters more, he suggested that "it is possible his brains are akin to those of the calculating boy."

Graham was referring to those freaks of nature, some of them today described as idiot savants, who though lacking real understanding of higher mathematics, possess a peculiar ability to perform extremely rapid calculations—to unerringly multiply and divide long strings of ten-digit numbers, or give the day of the week on which a thousand-year-old battle occurred, or perform similarly trivial computing tasks.

In fact, just as some artists of surpassing brilliance are no good at drawing straight lines or representing the human figure, so does mere facility in arithmetic—whether extracting square roots, or balancing books, or working out tricky word problems—have nothing to do with real mathematics. A mathematician *may* be adept at such skills, just as the artist may be adept at routine draftsmanship or figure drawing. But possession of such skills does not predict mathematical talent.

Ramanujan was more than ordinarily good in arithmetic calculation; on the other hand, his wasn't a skill developed to freakish proportions. And certainly in no other way did he resemble the "calculating boy" model. Nonetheless, it was one more among various possibilities as, during late 1912, British officialdom in Madras groped with the question of what to do with Ramanujan.

Griffith, to whom Graham had also written, wrote Spring the next day: "I think I was right in writing to Prof. Hill," said he, "and we must wait his opinion."

Micaiah John Muller Hill was Griffith's professor from twenty years before, at University College in London, and a teacher known more for the patience and care he lavished on his students than for his mathematical researches. Around mid-December Griffith heard from him at last. He could not look through all Griffith had sent him just now, Hill apologized, but a glance was sufficient to show that Ramanujan had fallen into some pitfalls; some of his results were simply absurd. Should he want to overcome his evident lacks, Bromwich's *Theory of Infinite Series* was the text to consult. If still interested in publication, he ought to write the secretary of the London Mathematical Society. But, Hill warned, "He should be very careful with his [manuscripts. They] should be very clearly written, and should be free from errors; and he should not use symbols which he does not explain"—as he had in the published paper on Bernoulli numbers Griffith had sent him.

But Hill's letter didn't answer the question: Had Ramanujan something extraordinary to offer the world? What was the nature and extent of his genius, if genius it was? "What you say about him personally is very interesting"—presumably a reference to his unusual intellectual history—"and I hope something may come of his work," was about all Hill would add.

A few days later, Hill wrote his former student again. It was a curious letter, still not definitive, but this time more encouraging. On the one hand, Ramanujan's paper on Bernoulli numbers, he said, was riddled with holes. "He has in fact observed certain properties of the earlier Bernoulli numbers and assumed them to be true of them all without proof. For [this and other] reasons, I feel sure that the London Mathematical Society would not have accepted the paper for their Proceedings." On the other hand, he said, "Mr. Ramanujan is evidently a man with a taste for mathematics, and with some ability." His educational deficit was hurting him. He needed to get that Bromwich book, he said again, this time citing the specific chapter that would clear up Ramanujan's misunderstandings.

And then, in a personal aside, Hill said perhaps the most revealing thing of all. "When I was a student in Cambridge, 1876–9, these things were not properly understood," he wrote, referring to the subtle but crucial points undermining Ramanujan's work, "and the modern theory has only recently been established on a firm basis. Many illustrious mathematicians of earlier days stumbled over these difficulties, so it is not surprising that Mr. Ramanujan, working by himself, has obtained erroneous results. I hope he will not be discouraged."

1876–9. Hill's Cambridge years, as it happened, coincided exactly with those of George Shoobridge Carr, author of the book so important to Ramanujan. Here, then, was the first hint of the price Ramanujan had paid in finding no more recent inspiration: he had missed out on all that had been learned in the mathematical capitals of Europe over the past forty years. Ramanujan's mathematics, in effect, was trapped in a time warp. *No wonder* he had gone astray.

Hill, who scarcely remembered his old student Griffith, had been sufficiently intrigued by Ramanujan's work to write two long letters in response to it. Of course, it came to nothing more than that. He was not offering to take him on as a student. Why, if anything, he had judged Ramanujan's first paper unfit for publication. Still, though he did not fully understand all Ramanujan had done, his reply probably contained more serious, reasoned, professional advice than Ramanujan had gotten all his life. And it was encouraging enough to quell most lingering suspicion among the British in Madras that maybe Ramanujan was more crank than genius.

For some time now, many had advised Ramanujan that no one in India properly understood him, that he'd not be able to find there the expertise and encouragement he needed, that he should instead write to Cambridge, or elsewhere in the West, for help. One who did was Singaravelu Mudaliar, his old professor at Pachaiyappa's College, to whom he had drawn close during his brief time there. Another was Bhavaniswami Rao, one of Ramanujan's professors at Kumbakonam College. A third was his friend Narasimha, with whom he had lived in Park Town a couple of years before. More recently, Narayana Iyer probably gave him similar advice.

If Ramanujan needed convincing to look toward the West, now, in the wake of Hill's letter from England, he needed no more. Events had conspired to tell him that he was, in effect, too big for Indian mathematics, and that he was apt to get a more sympathetic hearing from European mathematicians.

India was a quarter of the way around the globe from Europe, but the mail was cheap, reliable, and—long before airmail shrank the world—surprisingly fast; people grumbled if letters to England took as long as two weeks. And so, in late 1912 and early 1913, it was to the international mails that Ramanujan turned. In letters drafted with the help of Narayana Iyer, Sir Francis Spring, and perhaps P. V. Seshu Iyer, he began to write leading mathematicians at Cambridge University, including with his letters samples of his work.

He wrote to H. F. Baker, who held a long string of high honors as a mathematician, including a fellowship of the Royal Society, and had been president of the London Mathematical Society until two years before. Could Baker offer him help or advice?

Either through the kind of formulaic letter of polite discouragement that important men learn to write, or by returning his unsolicited material without comment, or by ignoring his letter altogether, Baker said no.

Ramanujan wrote to E. W. Hobson, an equally distinguished mathematician, also a Fellow of the Royal Society, and holder of Cambridge's Sadleirian chair in pure mathematics.

Hobson, too, said no.

On January 16, 1913, Ramanujan wrote to still another Cambridge mathematician, G. H. Hardy, who at thirty-five, a generation younger than the other men, was already setting the mathematical world of England on its ear. Could Hardy help him?

And Hardy said yes.

CHAPTER FOUR

Hardy

[G. H. Hardy to 1913]

1. FOREVER YOUNG

He was a study in perpetual youth.

One day in the spring of 1901, Hardy took his friend Lytton Strachey to the private green behind Trinity College, to which as a fellow of the college he enjoyed access, for a game of bowls. "He is *the* mathematical genius," Strachey wrote his mother, "and looks a babe of three." Even into his thirties, Hardy was sometimes refused beer and at least once, while at lunch with other Trinity dons, he was mistaken for an undergraduate.

He had ice-clear eyes, a finely chiseled face, and in 1913, straight, close-trimmed hair. He was beautiful. *He* didn't think so, of course, and could scarcely bear to look at himself. His college rooms had no mirrors, and in a hotel room he would cover any with towels, shaving by touch. But he alone was deceived. Even when past fifty, his looks were arresting. His skin, wrote a friend from those years, the novelist C. P. Snow, was tanned to "a kind of Red Indian bronze. His face was beautiful—with high cheek bones, thin nose, spiritual and austere. . . . [Cambridge] was full of unusual and distinguished faces, but even there Hardy's could not help but stand out." He was not, by every yardstick, handsome, at least not "ruggedly" so; his features were too delicate for that. And pursed, ungenerously thin lips, turned down a little at the corners, hinted at a judgmental streak in him.

Hardy was forever judging, weighing, comparing. He rated mathematicians, the work they did, the books and papers they wrote. He held firm opinions on everything, and expressed them. When a Cambridge club to which he'd belonged moved to change its official colors, Hardy took six pages to attack the plan. He faulted a sacrosanct academic tradition of almost two centuries' standing, and condemned it, unrelentingly, for more than twenty years. All his enthusiasms, peeves, and idiosyncrasies were like that—sharp, unwavering, vehement. He hated war, politicians as a class, and the English climate. He loved the sun. He loved cats, hated dogs. He hated watches and fountain pens, loved *The Times* of London crossword puzzles.

In *The Case of the Philosophers' Ring*, a Sherlock Holmes mystery written half a century after the death of Arthur Conan Doyle, the characters include Ramanujan and Hardy. In it, author Randall Collins pictures Hardy as a sort of White Rabbit hopping around the Fellows Garden at Trinity in white flannels and cap, cricket bat in hand, frantically searching for his cricket gloves, crying, "There's a match due to begin, and I can't find them. I'm late, I'm late!" In a prefatory note, Collins abjures all claim to historical accuracy. But in Hardy, he's close to the mark.

Hardy was a cricket aficionado of almost pathological proportion. He played it, watched it, studied it, lived it. He analyzed its tactics, rated its champions. He included cricket metaphors in his math papers. "The problem is most easily grasped in the language of cricket," he would write in a Swedish mathematical journal; foreigners failed to grasp it at all. His highest accolade was to rate a mathematical proof, say, as being "in the Hobbs class"—leaving the benighted to imagine the philosopher Thomas Hobbes, not the legendary Surrey cricketer Jack Hobbs. Hardy would play the game into his sixties. His sister would be reading to him about cricket when he died.

Hardy judged God, and found Him wanting. He was not just an atheist; he was a devout one. As an undergraduate, he was told that to be excused from chapel he had to inform his devout parents; he agonized over what to do—but ultimately wrote them with the crushing news. God, it would be said of him, was his personal enemy. Yet his friends included clerics, and some of his infidel posturing was just that—another of the harmless games he never tired of playing. "It's rather unfortunate," he once grumbled to a friend as a church's six o'clock chimes sounded the end of a sunny day of cricket at Fenner's cricket ground in Cambridge, "that some of the happiest hours of my life should have been spent within sound of a Roman Catholic Church."

Shy and self-conscious, he disliked small talk; cricket, of course, was *not* small talk. He abhorred formal introductions, would not shake hands, would walk, face down, along the street, ignoring those who might expect him to exchange how-do-you-dos. He was "one of the most strange and charming of men," wrote Leonard Woolf, who knew him at Cambridge long before Woolf married Virginia Stephen and, with Strachey and others, launched the Bloomsbury literary movement. His eccentricities would ossify with age, become caricatures of themselves, the stuff of story. But his personality, temperament, and values were already largely formed when he heard from Ramanujan.

Ramanujan knew nothing of this side of Hardy, of course. He knew him only as a mathematician. And in 1913, at the age of thirty-five, Hardy was already a famous one. He had appeared in the mathematical literature for fifteen years, counted more than a hundred papers, and three books, to his credit. He was a Fellow of Trinity College, the mecca of Cambridge mathematics, and hence English mathematics. He had been named to the Royal Society, Britain's most elite body of scientists, in 1910. Indeed, more than sniping at God, or delighting in cricket, or fashioning sly conversational gambits over dinner, Hardy cared about discovering mathematical truth. A brilliant mathematician, he was also a major influence on other mathematicians. A whole school had begun to form around him. He had served on the Council of the London Mathematical Society for three years, would later occupy numerous other posts within the mathematical community. "My devotion to mathematics is indeed of the most extravagant and fanatical kind," he would write. "I believe in it, and love it, and should be utterly miserable without it." His mathematical research, he would say, was "the one great permanent happiness of my life."

Hardy spoke beautifully. He batted out sparkling bons mots the way he did cricket balls from the popping crease—provoking, challenging, asserting. He was scrupulously honest, fastidious about giving others their due, once even admitted that the pro-God position in a debate had been better argued. He was endlessly amusing—but it was all like the gauzy silken shimmer of a woman's dress, meant to distract and disguise more than reveal. Conversation, one of his research students would say, was to him "one of the games which he loved to play, and it was not always easy to make out what his real opinions were."

C. P. Snow once reported that the longer you spent in Einstein's company, the more extraordinary he seemed; whereas Snow found that the longer you spent with Hardy, the more familiar a figure he seemed to become—more like most people, only "more delicate, less padded, finer-nerved"; that his formidable wall of charm and wit shielded an immensely fragile ego; that within lay someone simple, caring, and kind.

There is a picture of Hardy from middle age that shows him slouched in an upholstered wicker chair, one leg crossed over the other, right hand cocked at the wrist, lightly gripping a cigarette, left arm suspended at an unlikely angle across the back of the chair. A wisp of hair slips down over his forehead. He does not look relaxed; in no photograph of Hardy does he ever look relaxed. Always there's that haunted look in his eyes, like "a slightly startled fawn," as Leonard Woolf once said of him. There he sits, brows knitted, lips pursed, peering out over the tops of his reading glasses, imperious and forbidding. Someone spying this picture once said, "To sit that way you have to have been educated in a public school" (or what to Americans is a private boarding school).

Hardy *was* the product of the finest British public school education. But he hadn't come by it in the usual way. There were no viscounts in the Hardy line, no country squires. His family was neither rich nor wellborn—was of humbler lineage, in a sense, than Ramanujan's: in caste-bound India, Ramanujan was a Brahmin, while in England, where social class counted, Hardy came from schoolteacher stock. Indeed, Hardy would later be offered as an "example of how far the English educational system can bring out the personal powers and capabilities of a man." His intellect was so luminous, he was marked from the start. His success implied a blurring of the traditional British class system, a filtering down into the middle classes of opportunities once largely limited to a thin stratum of society at the top.

2. HORSESHOE LANE

In 1896, when Hardy and his new classmates took turns signing the great leather-bound book, with quarter-inch-thick covers, that had been used since 1882 to register each new class at Trinity College, they noted the schools they'd attended previously—Eton, Harrow, Marlborough, and in Hardy's case, Winchester. But a few students, one or two per page of twenty-six names, wrote in no school at all but only "private tutor," or just "private," in the space provided. It was in this sense, then, that Eton, Harrow, and Winchester were "public" schools, not private. Though many went back centuries, public schools as a powerful social institution had blossomed only in the wake of reforms achieved by Thomas Arnold at the Rugby School in the early part of the century. By Hardy's time they were the de rigueur means for fashioning the bodies, minds, characters (and accents) of upper-class boys.

But England, all during Queen Victoria's sixty-four-year reign, was changing. Newly prosperous farmers and tradesmen could hardly send their sons to Eton and Harrow yet were not content, either, with the bare-bones grammar school education afforded the poor. The 1850s and 1860s saw much debate about how best to accommodate this new middle class and the establishment of new schools to serve them.

One was Cranleigh School, located in the county southwest of London known as Surrey. "While the upper class [enjoy] the great public schools, and much has been done and still continues to be done to improve the education of the lower orders, the provision for the sons of farmers, and others engaged in commercial pursuits is so inadequate that the labourer's son often receives a better education than the son of his employer." So advised the school's prospectus at its founding in 1863. Then still called Surrey County School, it aimed to redress this imbalance and, in its early years, did: of 113 boys entering around 1880, for example, 55 were the sons of tradesmen, 20 of clerks, 14 of farmers.

In 1871, taking up the post of assistant master, teaching geography and drawing, was twenty-nine-year-old Isaac Hardy, who had earlier taught at a grammar school in Lincolnshire. Three years later, the school awarded him an extra fifty pounds per year—probably more than half again as much as he then earned; he was getting married, and would be living off school premises. In January 1875, Sophia Hall, three years younger than he, and then senior mistress at the Lincoln Diocesan Training College, became his wife. Little more than a year later, she was pregnant, and on February 7, 1877 she gave birth to Godfrey Harold Hardy. Two years later, he was joined by a sister, Gertrude Edith. There, across the road from the school, on the outskirts of Cranleigh, a village of two thousand souls, they grew up.

Surrey's northern border was formed by the River Thames which, further east, meandered through London. In the 1840s, the railroad had begun pushing out from the great metropolis and in forty years had doubled the county's population, to 342,000; by the yardstick of the centuries, that was breakneck growth. But in Cranleigh, at the other end of the county, things changed more slowly. True, the railroad had in 1865 left Cranleigh just a quick forty miles, by the Guildford and Horsham branch of the Brighton Railway, from London. And rich industrialists had begun to buy farms here and build new homes on them. But during

Hardy's youth, the rolling countryside around Cranleigh remained largely unspoiled, a peaceful tableau of dirt roads, windmills, old manors, and thatched-roof cottages.

Hardy's parents lived midst the old Surrey charm, but were not of it, having moved there from the other side of London, 150 miles away. They had had no money for a university education. Isaac Hardy's father had been a laborer and foundryman. Sophia's, once a turnkey at the county jail, was a baker at the time of her marriage. But both were bright, sought something better in life—and, as schoolteachers, had found a niche in the humbler reaches of the academic world.

Isaac Hardy, as C. P. Snow pictured him, was "a gentle, indulgent, somewhat ineffectual man with more than a touch of the White Knight about him." He was probably a happy man as well, with a refined aesthetic sense and a buoyant attitude toward life. He was the leading tenor in the school choir and, soon after coming to Cranleigh, was giving twice-a-week singing lessons. He edited the school magazine, played football (soccer, to Americans), was active in fraternal organizations, was a member of the Royal Geographic Society. When he died, at age fifty-nine, he was earnestly and sincerely mourned, the shops along High Street closed, their blinds drawn. He was "one of those rare and precious souls," the headmaster would say, who "never uttered an unkind word, never pained any living being, never had an enemy." You might chalk it up to inflated Victorian sentimentalism except for a photograph of him, taken when Harold was a child. It shows a man with sunny eyes, thinning hair, and the kind of thick beard that can make a man seem forbidding; on Isaac Hardy, though, it looks as welcoming as it does on Santa Claus.

The grim formality Victorian photographers more normally found in their subjects returns, however, in a portrait of his wife taken in Cranleigh at about the same time. In it, Sophia Hall Hardy wears an ornate, embroidered dress, her hair combed and piled back onto her head. The small lips, turned down a little at the corners, the startled look in the eyes, would reappear in her son. She was, like Ramanujan's mother, a pious woman. Sundays, she would drag Harold and sister Gertrude to church two or three times. When she left the Lincoln Diocesan College in December 1874 to get married, the school thanked her for the "high religious tone, and consistent Christian conduct" with which she had influenced her students over the past four years; for the good performance of her students in arithmetic; and for "the wise combination of firmness and kindness displayed . . . in the management of the Students, together with the quietness with which she has maintained her authority over them." Altogether, it is hard not to think that Sophia was a stern, upright, surpassingly competent woman.

Like her husband, she had a taste for cultural pursuits. She taught piano. She attended concerts—like a performance of Handel's *Messiah,* held at the school, when Harold was one. As schoolteachers, embracing a world of Art and Learning, both she and Isaac had transcended their roots. And now as parents, proud of the station in life they'd reached, they approached their own children's education with the utmost earnestness and care.

And the sensibilities they hoped to bequeath, at least the intellectual ones, *took.*

In May 1891, Gertrude would enter St. Catherine's School, Cranleigh's sister school, about five miles up the road in Bramley. There, she excelled, earning prize certificates in drawing and Latin. In 1903, after earning a B.A. from the University of London, she returned to her alma mater as art mistress, remaining there most of the rest of her life. In 1926, she became editor of the school magazine, which she filled with poetry, stories, and essays better by far than anything that small had a right to expect. Her editorials were graceful. Her own poems, many of them rich with literary and scholarly allusion, were suffused with an extravagant comfort in, and love of, the whole world of learning. In this uncharacteristic attempt at doggerel, she swiped at the self-satisfied ignorance she found among some students:

There is a girl I can't abide.
Her name? I'll be discreet.
I feel I'd need some savoir dire
Should I her parents meet!

. . .

She says "I never could do Maths.
When Daddy was at school
He could not add!" I'd love to say
"Then Daddy was a fool!"

"In dictée I got minus two;
There's not a verb I know;
I always write the future tense
Of 'rego,' 'regĕbo.'

"But then my Mother cannot write
Or speak a foreign tongue."
Sweet maid, how much the world had gained
If they had both died young!"

Whatever Mr. and Mrs. Hardy laid on their children, it must have come in massive doses, relentlessly administered, to have sired a poem like that. Indeed, the Hardys held firm theories about education, quite definite ideas about raising children that moved Snow to call them "a little obsessive." Harold and his sister had no governess; a nurse taught them to read and write. They had relatively few books, but those they did have were always "good" ones; as a boy, Harold would read *Don Quixote* and *Gulliver's Travels* to his sister. They were never allowed to play with broken toys. It was as if there were an invisible standard of integrity and excellence to which they were invited to compare everything—and to sternly discard or reject anything that failed to measure up.

When Hardy was two, he was writing down numbers into the millions, a common marker of mathematical talent. At church, he'd busy himself seeking the factors in the hymn numbers: *Hymn 84? That's* $2 \times 2 \times 3 \times 7$. At school, young Hardy apparently never sat in a regular math class, but was rather coached privately by Eustace Thomas Clarke, who presided over the school's mathematics instruction. Clarke came to Cranleigh fresh from St. John's College, Cambridge, where, as a "Wrangler," he'd ranked among the top mathematics students. Coupled to his mathematical abilities was uncommon energy and drive, and he won a reputation for instilling those traits in his students.

It was not just mathematics in which Hardy was gifted, but all to which he took his hand. Yet a fragility, a diffidence, sometimes undermined it. He was painfully shy and self-conscious and could scarcely bear going up before the whole school to accept a prize. Sometimes he'd give wrong answers to ensure he wouldn't have to do so. "Over-delicate," his friend Snow described him later. "He seems to have been born with three skins too few."

So was his sister, whom one of her students would describe as "shy and diffident." When she entered St. Catherine's School, Gertrude would recall, "I was very shy, and the atmosphere in which I had been brought up was a poor preparation for a girls' school of that age." Whatever made for that "atmosphere," it gripped both children. Neither ever married. Both spent their lives in academic settings. Both emerged as delicate, enchanted by intellect, and—apparently reacting to their mother's zealotry—contemptuous of religion.

Hardy was an atheist even as a boy. Once, as he and a clergyman walked in the fog, they saw a boy with a string and stick. The clergyman likened God's presence to a kite, felt but unseen. In the fog, he told young Hardy, "you cannot see the kite flying, but you feel the pull on the string." But in fog, Hardy thought, there is no wind and no kite can fly. Gertrude felt much the same. Once, as an old woman confined to a nursing home, she was asked her religious preference. She replied "Mohammedan," bewailed the want of a mosque close by, even set about trying to locate a prayer rug to enhance the deception.

When in the 1920s a photo was taken of the faculty at St. Catherine's School, all twenty or so stared straight ahead into the camera. All, that is, save for "Gertie," as she was known, who gazed off camera to her left, leaving her almost in profile, her left side hidden. Gertrude lost her eye as a child, when Harold, playing carelessly with a cricket bat, struck her; she had to wear a glass eye for the rest of her life. The incident, however, did nothing to disrupt their sibling closeness, and may even have enhanced it. They were devoted to one another all their lives, kept in close touch almost as twins are said to, and for many years shared an apartment in London.

In 1880, when Hardy was three, the board of Cranleigh School approved taking twenty-four of the youngest students and boarding them separately in what had been the sick house across the road, thus freeing up places in the main school and increasing its income. Running it would be Mr. and Mrs. Hardy, helped by a governess. This preparatory school (which may have existed in some form earlier) was a distinct operation, kept separate in the school's account books. In 1881, Mrs. Hardy was paid 281 pounds, 12 shillings, from which, presumably, she met the House's operating expenses.

The House, as the preparatory school came to be known, was a sprawling, barrackslike affair with a double-gabled roof that stood up the slope from Horseshoe Lane opposite the school. At least around 1881, when

Hardy was four, the family lived a few steps down from it toward the road, in a small, two-story brick semidetached house, trimmed with the black and red scalloped clay tiling ubiquitous in Surrey since the seventeenth century. "Mt. Pleasant" the little house was grandly called. It had two small bedrooms on the second floor, and a low-ceilinged sitting room and kitchen, dominated by a big brick fireplace, on the first. From the sitting room, you could see the clock tower of the school across the road and the rose window of its chapel.

Around 1881, census records tell, the house was home not just to Harold and his mother, father, and sister, but also to Eliza Denton, thirty years old, who helped Mrs. Hardy at the House; twenty-two-year-old Catherine Maynard, who may have been the children's nurse; Alice Lee, an eighteen-year-old servant; and another servant, Laura Chandler, a widowed thirty-eight-year-old—eight people, in all, packed into what amounted to a cottage. By today's standards, it must have made for a tight squeeze. But on the other side of the fireplace, in a mirror image of the Hardy house, lived a still larger family—an agricultural laborer and his wife, her father, two sons already at work as laborer and ploughboy, and six other children ranging from eleven down to four months.

Indeed, with only two children, the house might have seemed to Mr. and Mrs. Hardy scandalously empty—making it natural, with the children still young and the family's finances precarious, to rent out sleeping space, probably in the basement, to local servants, some of whom may have worked at the school. During the day, presumably, the servants would scatter to their jobs, Mr. Hardy to school, and Mrs. Hardy to minister to her flock at the House, leaving Harold and his sister with Miss Maynard.

The Hardys were better off than many others who lived along Horseshoe Lane—farmers and laborers with five or six children, most of whom had lived their whole lives in Cranleigh or nearby towns like Alfold, Woking, or Dorking. But Isaac's modest position left them by no means flush. The school's headmaster (principal) made a thousand pounds a year. But at a time when workmen made sixty or seventy pounds a year and an upper-middle-class person three hundred, the school's second master made only a hundred. Assistant master Hardy, it is safe to say, made even less.

The Hardy children, someone later wrote, were brought up in "a typical Victorian nursery"; *typical* is the key word. When he was six, Harold was photographed in a standard-issue Victorian sailor suit, with a bow tie. He grew up on a road that a few years before had been little more than a mud track. His hometown, spreading out haphazardly from its little High Street of gabled, half-timbered shops, was unprepossessing. Meanwhile, the school to which his family had such close ties sent a kindred message: its modest red brick buildings, situated at the end of a sloping driveway up from Horseshoe Lane, were handsome enough, but bore none of the weight of history and tradition boasted by older public schools.

Indeed, the atmosphere at Cranleigh School differed from anything Hardy would meet later. It was not there to groom England's elite. It was ordinary and unpretentious, shot through with a kind of youthful freshness. One visitor in 1875 noted that "Cranleigh boys may wander where they please, and this freedom is characteristic of the establishment throughout." Teachers played on school teams until 1888. Absent were the rigid hierarchies of the older public schools, the tight restrictions on student behavior. Boys went off on their own, smoked pipes, hung out behind the gym.

Snow wrote that Hardy enjoyed a fortunate childhood, "enlightened, cultivated, highly literate. . . . He knew what privilege meant, and he knew that he had possessed it." Indeed, at home, he grew up with an emphasis on intellect and learning that, in ages past, only the aristocracy enjoyed. But at school, his teachers were mostly not public school boys and everyone was aware of being *not* Eton, *not* Harrow, *not* Winchester. And so, if "privilege" it was, it was of a rare and perhaps ideal sort—in circumstances at once economically modest and culturally enriched, like the immigrant Jews of two generations ago, say, or the immigrant Asians of today. And if Hardy grew up, as he did, insistent on intellectual excellence, yet sensitive to those socially scorned or economically unlucky, Cranleigh may have helped make him that way.

In July 1889, J. T. Ward, a fellow of St. John's College, Cambridge, reported to the headmaster at Cranleigh the results of his yearly examinations of students in the school's upper forms. Students at Cranleigh advanced through each "form," or grade, based largely on merit, not age; indeed, one year among the school's 347 boys was a sixth-former who was twenty. Hardy reached the sixth form at twelve. But though five years younger than most other boys, reported Ward, he surpassed most of them in algebra, geometry, and trigonometry. A much older boy came in first that year. But Hardy came next, "far ahead of the third boy, although he has read [studied] no Conic Sections yet, and a very little Mechanics [the science, not the trade]; taking into account his present age, I am confident," wrote Ward, "that he ought to distinguish himself greatly in the future."

Distinguish himself he did. Cranleigh had nourished him but now offered him, at age twelve, nothing more. His parents, who had the highest aspirations for their prodigiously intelligent son, kept tabs on scholarships for which he might be eligible. Around this time, about the scarcest and most coveted scholarship of all was to Winchester, one of England's most hallowed public schools and a traditional proving ground for the mathematically gifted. Its scholarship exam was among the toughest of its kind; one year in the 1860s, 137 boys competed for seven scholarships. To prepare for it, you normally went to a coach or special school. If you won it, you turned down anything else you might also have won. It was the ultimate feather in a boy's academic cap.

Hardy applied for it and in 1890, among a field of 102 candidates, placed first.

By the early 1890s, Cranleigh School was beginning to shed the easygoing openness of its early years, Hardy's father had been named bursar, or treasurer, and the family was living in what was probably a more spacious home, "Connel." But by then, G. H. Hardy was gone from Cranleigh, headed off to the wider world.

3. FLINT AND STONE

If originally endowed, as were many venerable public schools, to educate poor scholars, Winchester had long ago evolved into a preserve for the gentry. A few hours by rail from London's Waterloo Station, in the cathedral city bearing the same name, Winchester was the real thing—the sort of place you conjured up, along with Eton, Harrow, and Rugby, when you thought of *public school,* and which occupied heights to which Cranleigh's founders could only dream of aspiring.

Over the years, the school graduated more than its share of those whom one school historian could aptly class as "gentlemanly rebels and intellectual reformers." And twenty years later, it would be fair to describe Hardy as one of them. Winchester didn't *try* to, of course; its more usual products were reserved, patrician, conservative social and political leaders. But those who did rebel often became distinguished rebels.

At Winchester, Hardy found much against which to rebel. From outside the ancient complex, he confronted the original college wall, all flint and stone, pierced by tiny slotted windows; these formidable architectural details stemmed from the Peasants' Revolt, and from bloody town-gown battles at Oxford, both still of recent memory in the fourteenth century when the school was founded. Scholarship students like Hardy lived in a sort of intellectual ghetto within the college, a fortresslike complex of medieval, gray stone buildings, worlds apart from the sunny openness of Cranleigh.

As a new student, Hardy was grilled on "notions"—a vast lexicon of jargon and slang peculiar to Winchester. Some of them went back to the Latin, some were submerged in the mists of the school's medieval past. Collectively, they defined good form and bad, as Winchester, across the span of centuries, had come to see them. "Tugs" was stale news. To "brock" was to bully. A "remedy," derived from the Latin *remedium,* meant "a holiday." A "tunding" was a flogging at the hands of a prefect, or senior student officer. Learning your notions made for no idle study. There were thousands of words, with whole published glossaries given over to them, some graced with exquisite drawings and illuminated capitals. All had to be memorized.

Making sure they learned their notions and otherwise conformed were the prefects. It was Student Power run amok. Even after Matthew Arnold's reforms, expressly aimed at curbing the power of the older boys at English public schools, Winchester lay in the grip of these prefects, who lorded it over the younger and weaker students. Conditions were better than a quarter century before, when a particularly cruel incident of tunding—thirty strokes, with a ground-ash stick, across the back and shoulders of a student who bristled at taking his notions exam—had outraged a parent and brought in the press. But Winchester was still reckoned among the most brutal of the public schools. Beatings were the rule, not the exception. And student prefects were still left free by school officials to exercise a *Lord of the Flies*–like savagery. The place was a vast adolescent hierarchy, a tribal society built on power, privilege, and force.

And tradition. If anything, the current headmaster, W. A. Fearon, was strengthening tradition's grip on the place. While Hardy was there, he began "Morning Hills," a twice-yearly schoolwide trek to the top of St. Catherine's Hill culminating in prayers and a calling of the school roll. He also revived the ancient processions around the Cloisters, where students sang the morning hymn, *Iam lucis orto sidere,* as they solemnly circled through the stone-arched walkways. After the sweet, soft air of Cranleigh,

Winchester was like a work farm. The amiable wife of Hardy's mathematics teacher, Sarah Richardson, known by everyone as "Mrs. Dick," who held open house on Sunday with cakes and back issues of the *Illustrated London News*, could scarcely do much to ease the grayness. Neither could Fearon's humanizing little trips to London. Neither, amazingly, could cricket.

Back when Hardy was eight, even then the budding writer, he put out his own little newspaper, complete with editorial, advertisements, a speech by Prime Minister Gladstone—and a full report of a cricket match.

As a child growing up in Cranleigh, all through his Winchester years, and beyond, Hardy's world resonated to the sound of cricket bats and the whirl of white flannel under the summer sun. In Britain's more gentlemanly circles, cricket was as ubiquitous and *important* as basketball is in American inner-city neighborhoods, or as baseball once was in Brooklyn. It was the golden age of county cricket in Surrey, and as a youth Hardy would go to the Kennington Oval in London to see cricket greats Richardson and Abel, then in their prime. Back at school, practices were sacrosanct; even detention didn't interfere. The school magazine was heavy with accounts of matches—like this one between school and village in 1888, when Hardy was eleven:

> Played at Cranleigh, July 21st, on a very slow wicket. The village batted first, and were opposed by Robinson and Blaker. Robinson bowled extremely well, and no one but Street could offer any resistance, the whole of the side being dismissed for 53. In the School's innings, Douglas was bowled in the first over and this was only the commencement of a series of disasters, our total only reaching 45, Warner being the only one to obtain double figures. In the 2nd innings, the village did much better. . . .

In cricket, one man hurls a ball, another bats it, and fielders rush to catch it. Sounds like baseball, as does its penchant for statistics. But cricket is a more leisurely game, one inspired by quite a different spirit. The ball is delivered—"bowled"—with a straight arm, at the end of a curious loping run, normally first striking the ground in front of the batter. The "batsman," as he is properly called, uses a paddlelike wooden implement to "defend" the "wicket," a two-foot-high tridentlike affair, stuck in the ground, atop which lie lightly balanced rods. If the bowler dislodges these rods or induces the batsman to do so, the wicket is lost.

The batsman's object is to get as many runs as possible before he is "dismissed."

Such an account scarcely does justice to a game whose roots, by Hardy's time, went back six hundred years, that in organized form had dominated English summers for two hundred, that had its own rich lore, its own etiquette, its own arcane language. (A "sticky wicket," for example, means that the ground in front of the batsman is wet, making the ball's bounce more difficult to estimate.) Cricket, one connoisseur of the game, Neville Cardus, once wrote, "is a thing of personal art and skill; it depends not mainly on results, but on the amount of genius and character which is put into it." *Genius and character*: the English took the game seriously.

And no Englishman more so than Hardy, to whose days cricket gave almost as much meaning as did mathematics. He devotedly studied his *Wisden,* the cricket annual crammed with bowling averages, test-match results, and other arcania of the game. In 1910, the minutes of a Cambridge club would alliteratively cite his command of "the University Constitution, the methods of Canvassing, Clarendon type, and professional cricket." As a young Fellow of Trinity College, he'd play a bastardized form of it in his rooms, with walking stick and tennis ball. He'd play it more seriously into his sixties. He reveled in the batsman's backswing, in tactical nuance, in the logic behind changing bowlers or positioning fielders . . .

But now at Winchester, even cricket gave him little pleasure. Winchester was as obsessed by the game as Cranleigh. Indeed, some had grown alarmed by the rampant athleticism embodied in cricket's hold on students. Come summer, one writer for the school paper lamented in June 1893, three years into Hardy's tenure, "we proceed as if life were one long game of cricket."

But though Hardy was a natural athlete, his cricket skills atrophied at Winchester. He was, as someone later described him, "small, taut, and wiry," and played other sports well, especially soccer. "Hardy made an excellent rush . . . Hardy moved neatly . . . Hardy was magnificent," accounts from the period record. But as for cricket, he never played on the team, and left Winchester feeling slighted. Denied the coaching to which he felt his talents entitled him, Snow remembers him grumbling, his defects as a batsman persisted—thus frustrating, at least in his imaginings, a brilliant cricket career.

That was one grudge Hardy bore against Winchester; given his fixation with the game, that was probably enough for him to forever hate the place. But there were others. Poor teaching crushed any artistic ability he inherited from his father. He was shy and sometimes sickly, in a setting with scant tolerance for such frailties. One winter he got so sick he almost died. Later he would feel a shock of envy for the happier experience of a friend who attended school from home, as a day student. After Hardy left Winchester, he couldn't eat mutton, which was served there, by statute, five days a week. He never returned to visit. He never attended a reunion.

At Winchester, in those days, the classics were still lopsidedly represented. Of twenty-six class hours each week, five each went to Greek, Latin, and history, three to French, two to divinity, two to science, and four to mathematics. But Hardy probably never attended mathematics classes as such, instead working alone with the second master, George ("Dick") Richardson. Something of an anomaly at Winchester, Richardson had attended Cambridge, ranking high in the mathematical Tripos exam there, but was not himself a public school graduate. He was no model teacher, hardly bothering with any but the better students. These, of course, included Hardy, who in 1893 walked off with the school's Duncan Prize in mathematics.

But Hardy was not just interested in mathematics. He studied physics on his own, read Tyndall and Huxley, was a devotee of Ruskin, enjoyed headmaster Fearon's brand of history. So midway through his six years at Winchester, it was not yet clear where his future lay.

During later life, Hardy had a weakness for detective stories, and over the course of one boring London weekend reputedly consumed several dozen of them; so while he enjoyed good literature, he was not above escapist fare. One day when he was about fifteen, he came upon a book written a few years before under the name Alan St. Aubyn, a pseudonym for a certain Mrs. Frances Marshall who, displaying able judgment, chose not to associate her name with it. The book was called *A Fellow of Trinity*. It was about university life at Trinity College, Cambridge. And it was awful, its characters silly, its prose insipid. Yet, somehow, it touched a chord in young Hardy, especially its concluding scene, which shows Herbert Flowers, the earnest hero, rejoicing in academic victory.

> Herbert was back again at Trinity. He wore a long B.A. gown now, with ribbons on it, that he made no attempt to hide, and a fur hood over

> his surplice, and sat in great dignity in the Bachelors' seats in chapel and at the Fellows' table in Hall.
>
> He was a Fellow of Trinity!
>
> He sat at the high table now, and the grave portraits of the founders and the illustrious dead looked down upon him approvingly.
>
> The ardours, the sorrows, the struggles of the race, were all over; only the brilliant achievement remained. The great cloud of witnesses that looked down from those old rafters overhead upon those who feasted there had never approved a more nobly earned success in the rich intellectual history of the past of Trinity.
>
> He wore his honours as he had worn his misfortunes, with becoming modesty, and was warmly welcomed by the grave, scholarly old Fellows who sat around the great horseshoe table in the Combination Room.
>
> Perhaps he never quite realized until he sat there, on that first night of his Fellowship after Hall, mute and wondering, enjoying the walnuts and the wine—and all that the walnuts and the wine round that horseshoe table represented of scholarly and philosophical learning and culture—how great had been his success!

Hardy was bewitched. "Flowers was a decent enough fellow (so far as 'Alan St. Aubyn' could draw one)," Hardy reminisced later, "but even my unsophisticated mind refused to accept him as clever. If he could do these things, why not I?" *He,* Hardy, would become a Fellow of Trinity, and mathematics would be his ticket to it.

Omitted from Hardy's account was that Trinity *was,* after all, Cambridge's crown jewel, enjoying the richest mathematical tradition of any Cambridge or Oxford college. Still, for a Wykehamist, as Winchester men were known, New College, Oxford, was the more orthodox choice. Both Winchester and New College had been founded by William of Wykeham, bishop of Winchester and twice lord chancellor, who conceived the two institutions as one, encompassing all the years of school and university. So of the two great universities, Wykehamists normally picked Oxford, in particular New College. In Hardy's year, 1896, forty-six Wykehamists went to various Oxford colleges, fifteen of them to New College. Only eight went to all the Cambridge colleges combined.

Hardy was one of them, and on a scholarship to boot. "Congratulations are due to Smyth and Hardy for their brilliant success," noted *The Wykehamist*, the school newspaper, with uncharacteristic fervor. "It is an unique honor to gain two of the four best scholarships of the year; Winchester has certainly never accomplished it before, and we doubt if any other school can boast such a performance."

Back in Cranleigh, it was a fine time to be a Hardy. Isaac Hardy's students had done so well in the big art examination, gushed headmaster Rev. G. C. Allen at the school's annual Speech Day, that the examiner had compared them favorably to the best in the country. *Hear hear*, and warm applause erupted from the audience. And now, Allen went on, Mr. Hardy's son had won a Major Scholarship to Trinity—an achievement which, the local paper wrote, "was one of the highest things even Winchester could expect to get."

4. A FELLOW OF TRINITY

It would be impossible to exaggerate the role in British life of Cambridge and Oxford. When Havelock Ellis in 1904 prepared his *Study of British Genius,* based on a thousand or so eminent Britons, of the half who had been to university, 74 percent had been to one or the other of them.

Cambridge, like Oxford eighty miles to the southwest, was first a town—a midsized trading settlement that grew out of a Roman encampment on a hill north of the River Cam. Beginning in the thirteenth century, scholars congregated there and formed colleges. What ultimately emerged was a federation of colleges more like America under the Articles of Confederation—states unto themselves exercising a jealous grip on their rights and powers—than America under the strong central government of the Constitution; you were first a student of your college, only then of the university. Until the Senate House was built in 1720, the university had no central meeting place.

Each college had its own history, its own endowment, its own faculty with its own academic strengths, its own predominant architectural style. Each attracted its own mix of students, who wore their own distinctive undergraduate gowns. Each had its own roster of distinguished graduates. There was Peterhouse, oldest of the Cambridge colleges, established in 1282. And King's College, with its storied chapel—Wordsworth's "immense and glorious work of fine intelligence!" And Magdalene, pronounced "maudlin," at the north end of town across the River Cam, founded by Benedictine monks in 1428 and whose Samuel Pepys Library contained his famous *Diary*. There were St. John's, Jesus, Gonville and Caius, Pembroke . . . by Hardy's time close to twenty in all.

Some had just one or two hundred undergraduates and fellows; some, like Trinity, closer to a thousand.

Architecturally, the colleges could be imagined as variations on a powerful and persistent theme: facing the street was a more or less continuous wall of two- or three-story buildings in stone or brick. Entering the college through a great arched gate, you found yourself in a dark passage, to one side of which stood the Porter's Lodge. There a man in a bowler hat, whose accent instantly marked him as porter and not student, saw to the daily workings of the college—admitting visitors, delivering messages, keeping track of keys. Past the Porter's Lodge was the enclosed court of the college, whose smartly coiffed grass plot would, even on the grayest of English days, leap at you with green. Around the court, typically on all four sides, were walls of buildings. These included the chapel, the library, and the dining hall, typically a grand, high-ceilinged affair, richly paneled in wood, and studded with framed portraits of college luminaries.

A student might attend university lectures, but he mostly learned through "supervisions," at the elbow of a college tutor, in his rooms within the college, two or three students at a time. He also had a tutor who watched over his "moral" development, and a director of studies, who monitored his scholarly progress; these, too, were members of the college, not drawn from the university at large. Likely as not, the undergraduate lived "in college"—in a set of rooms off a court of the college reached through one of the staircases spaced every dozen or so paces around it.

"Rooms" at Cambridge did not mean the bare furnished room conjured up by Americans today in "room for rent." In fact, they were often small apartments of their own, some of them well furnished and ornate, with fine wallpaper, rugs, framed paintings, sometimes even formal dining-room tables. Cambridge men were deemed wholly unable to cook, clean, or otherwise care for themselves. They were served by bedmakers and "gyps," who brought coal up to feed the fireplace, took in the mail, fetched lunch, set out bed linen, towels, and tea cloths, served tea. Students were invariably addressed as Mister and treated with no little respect. But perks at Cambridge did not extend to the modern plumbing and conveniences that ordinary middle-class Americans were enjoying by this time. The few baths were clustered together, often across a windswept court from the student's room. Hot and cold running water were unknown. The gyp's first job, upon showing up at seven in the morning, was to fill a tin saucer with cold water for washing.

Trinity was among the oldest of the Cambridge colleges; it was certainly the largest, and the most famous, formed from two smaller colleges each of which went back to the fourteenth century. It had taken its present form under Henry VIII, whose plump, strutting sculpted figure faced Hardy each time he walked past the bicycles massed outside the gate to Great Court.

Dominated by a picturesque domed fountain that went back to 1602, Great Court was the largest among Trinity's five courts and, indeed, the largest enclosed quadrangle in Europe, more than four hundred feet diagonally across; you could drop a baseball diamond within it and its walls would make for respectably distant outfield fences.

Hardy was assigned a room in Whewell's Court, across cobblestoned Trinity Street from Great Court, in a complex of buildings completed just thirty years earlier. More intimately scaled than most of the rest of Trinity, and a little isolated, Whewell's Court was its own quiet preserve of stone and lawn. From his room on the second floor of Staircase M, Hardy could look out through arched Gothic windows to the busy sidewalk, or to the stone ramparts of Sidney Sussex College across the street. Yet within the inner sanctum of the court itself, the sound of bicycles clattering down Bridge Street or of excited conversation among students in Jesus Lane did not penetrate.

After Winchester, of course, Trinity was not primarily an aesthetic delight for Hardy. Winchester was even older than Trinity, or almost anyplace else at Cambridge for that matter, its chapel as stately, its courts as tradition bound, its cloisters as imposing. So for nineteen-year-old Hardy, as he signed the Trinity Admissions Book in 1896, it was what Cambridge and Trinity *represented* that mattered. And what it represented was being at the top of the intellectual heap.

Hardy, however, had a shock coming—a disillusioning that would lead him to weigh dropping mathematics. It came in the form of a venerable Cambridge institution called the Tripos.

The word is pronounced *try*-poss, and it originally referred to a three-legged stool. On it, in olden days at Cambridge, sat a man whose job was to dispute, sometimes humorously, sometimes aggressively, the candidate for the degree in mathematics. The word subsequently came to refer generally to the examinations mathematical candidates took to earn their degrees. Still later, the word was extended to examinations in other fields—a Classical Tripos, a Natural Sciences Tripos, and so on.

The mathematical Tripos was impossibly arduous. You sat for four days of problems, often late into the evening, took a week's break, then came back for four more days. The first half, which stressed quickness and counted even mere arithmetical facility, covered the easy stuff; showing even modest aptitude was enough to earn a degree. The second half weighed doubly and encompassed more difficult material. Here, sometimes on problem after problem, the brightest students, destined for distinguished mathematical careers, would not even know where to begin. It was a frightful ordeal. Recollections invariably lapse into awed hyperbole. The Tripos, wrote one English-born mathematician years later, "became far and away the most difficult mathematical test that the world has ever known, one to which no university of the present day can show any parallel."

But the Tripos was more than an examination. It was an institution. By the time Hardy arrived in Cambridge in 1896, it comprised the academic rituals surrounding it, the esteem accorded it, the system built to support it, even the style of mathematics it encouraged, all rolled into one. The Tripos went back, in earliest form, to 1730 and had always been difficult. But over the years, as it grew more demanding and more important, it also, in inimitable English fashion, took on the luster—and deadweight—of Tradition.

Tripos candidates were ranked on the basis of their performance, and a whole ritual surrounded the reading of the Order of Merit at the Senate House. Those ranking in the first of three classes were deemed "Wranglers," the topmost among them being named Senior Wrangler; in the early days of the Tripos, the disputants "wrangled" over points of logic. To learn who the Senior Wrangler was, everyone flocked to the Senate House. Even women, normally banished to the fringes of Cambridge life, showed up. Mingled reverence and sex appeal, if it can be imagined, accrued to the Wranglers, who "usually expected," a Cambridge vice-chancellor was advised back in 1751, "that all the young Ladies of their Acquaintance . . . should wish them Joy of their Honour."

If anything, the Joy and Honour were greater by Hardy's time, the Senior Wrangler and those just behind him in the order of merit earning applause and hurrahs from friends and college mates. At graduation, the vice-chancellor sat on the dais at one end of the Senate House, and college tutors presented to him those taking their degrees. The recipient kneeled before him, and the vice-chancellor took his hand between his own and repeated in Latin the awarding of the degree.

Standing below the Wranglers were the Senior Optimes, or second class, and the Junior Optimes, or third. When the lowest scoring of the Junior Optimes—the bottom-ranking man in the class—went to receive his degree, his friends solemnly lowered from the Senate House gallery the Wooden Spoon. In fact it was a huge malting shovel, as large as the man receiving it, inscribed in Greek, and flamboyantly decorated. Upon rising to his feet, he'd take the ungainly thing and, shouldering it triumphantly, stride from the hall with his friends.

Of course, the Wooden Spoon was strictly a consolation prize. The honor accorded the Senior Wrangler, on the other hand, was no laughing matter, but clung to him, like an aura, for a lifetime. "If one person were consensus All-American, a Rhodes scholar, and Bachelor of the Year," observed one account aimed at American readers, "he would not come close to commanding the lasting distinction that came to the Senior Wrangler." To a lesser extent, Wranglers among the first ten or so shared some of the glory and were virtually guaranteed distinguished careers. Obituaries of English mathematicians half a century later invariably noted that you had once been Senior Wrangler, or Second Wrangler, or Fourth. As a history of the Cambridge Philosophical Society accurately noted, "the Senior Wrangler was not invariably a great mathematician . . . [but] it was virtually certain that he could, if he wished, be an influential one."

Around Cambridge and to an extent elsewhere in Britain, the Senior Wrangler was a celebrity. People related to him much as they do the winner of the Kentucky Derby—without knowing the first thing about horseflesh. He was a star. *The Times* of London invariably recorded his elevation. Picture postcards bearing his photographic likeness were sold around town. One from around the turn of the century shows the victor sitting outside, posed on an ornate chair, polished shoes gleaming, hands demurely clasped together—the Derby winner shown off to the press.

All this was very nice in its way, very British. No one questioned that Tripos success reflected, more or less, mathematical ability. Nor that, if anyone was to become a great mathematician, it would be the Wrangler and not the Wooden Spoon. On the other hand, precise ranking in the Order of Merit, everyone knew, meant far less than it seemed. Indeed, it was an open secret that Second Wranglers seemed to distinguish themselves more than Senior Wranglers. Maxwell, the great mathematical physicist who united electricity and magnetism, had been Second Wrangler. So had J. J. Thomson, discoverer of the electron. Then there was thermodynamicist Lord Kelvin, then still William Thomson, surely the best mathematician of his year. Everyone, including himself, thought he was a shoo-in for Senior Wrangler. "Oh, just run down to the Senate House, will you, and see who is Second Wrangler," he asked his servant. The servant returned and said, "*You,* sir." Someone else, his name today forgotten, had proved better able to master Tripos mathematics.

And that was the problem: there *was* such a thing as "Tripos mathematics," and it bore little kinship to anything of interest to serious, working mathematicians. The Tripos was tricky and challenging, and certainly separated the Wranglers from the Wooden Spoon; in 1881, for example, the Senior Wrangler got a score of 16,368 "marks," out of a possible 33,541, the Wooden Spoon 247. But there was a datedness to the problems, a preoccupation with Euclid, and Newton, and exercises in mathematical physics—a sphere spinning on a cylinder with the candidate asked to establish the equations governing its motion, or a problem based on Carnot's Cycle in thermodynamics, and so on. They demanded accuracy and speed in the manipulation of mathematical formulas, a shallow cleverness, perhaps, but not real insight.

And not even stubborn persistence; a proof demanded by a Tripos question couldn't be too long or too involved; so you learned to look for that hidden Tripos twist. During one Tripos exam, a top student—that year's Senior Wrangler—observed a less capable candidate making short work of a problem over which he agonized. *Must be a trick,* he realized—and went back and found it himself. The personal qualities encouraged by the Tripos, J. J. Thomson would make so bold as to suggest, made it excellent training—for the bar.

In homage to the Tripos, an alternative educational system had sprung up. In the nineteenth century, university lecturers at Oxford and Cambridge inhabited a world of their own with scant contact with students. Teaching was left to tutors in the individual colleges, who were hardly up to the task of preparing students for the Tripos. So a Third Force had emerged to fill the vacuum—private coaches.

These coaches were not out to teach you mathematics for its own sake, but to school you, for handsome fees, in Tripos mathematics. You were trained, one future Senior Wrangler wrote, like a racehorse. In groups of four or five, you and your coach went over problems. Maybe three times a week, he lectured to you. He pored over old examinations, wrote little private tracts, codifying mathematical knowledge into neat bundles. Only rarely did a coach himself produce important new mathematical results.

But "produce" he did—Senior Wranglers. At one point, coach E. J. Routh churned out some two dozen of them in as many years.

For students, the work load was prodigious. One distinguished mathematician, J. E. Littlewood, wrote that "to be in the running for Senior Wrangler one had to spend two-thirds of the time practicing how to solve difficult problems against time." Attending lectures became a luxury. Of his professors in the late 1870s, A. R. Forsyth has written that they

> did not teach us; we did not give them the chance. We did not read their work: it was asserted, and was believed, to be of no help in the Tripos. Probably, many of the students did not know the professors by sight. Such an odd situation, for mathematical students in a University famed for mathematics, was due mainly, if not entirely, to the Tripos and its surroundings which, as undefined as is the British constitution, had settled into a position beyond the pale of accessible criticism.

That was how things stood in Forsyth's time, and that's how they remained during Hardy's. The Tripos system discouraged exploration of any area of mathematics, however personally satisfying, not apt to show up on the examination. It granted professional success—a fellowship at a good college, say—to those doing well on the exam, not those demonstrating a bent for research, or boldness in pursuing it. Tripos success became, like marriage for the prototypical Southern belle, not a happy prelude to one's life but its culmination.

Hardy was drawn up into the system, in his first term being handed over to R. R. Webb, the day's foremost producer of Senior Wranglers. At Cambridge, the academic year consisted of three terms of seven or eight weeks. Typically, for ten straight terms the coaches did their work, and their students did theirs. Until finally, on a cold January in the unheated Senate House, the student would sit down to begin the Tripos ordeal. This was what Hardy had to look forward to.

He was not alone among the mathematically gifted revolted by the banality of the Tripos. Bertrand Russell, who ranked as Seventh Wrangler when he took the Tripos in 1893, and who would make numerous contributions to mathematical philosophy in the years ahead, wrote later how preparing for it "led me to think of mathematics as consisting of artful dodges and ingenious devices and as altogether too much like a crossword puzzle." The Tripos over, he swore he'd never look at mathematics again, and at one point sold all his math books.

But Russell *did* go through it. And so did most others. Hardy's future colleague, Littlewood, later confided that he, too, had seen Tripos mathematics as empty. But he simply gritted his teeth and, as Hardy told the story years later, "decided deliberately to postpone his mathematical education, and to devote two years to the acquisition of a complete mastery of all the Tripos technique, resuming his studies later with the Senior Wranglership to his credit and, he hoped, without serious prejudice to his career." And that, Hardy declared "in hopeless admiration," was precisely what he did.

But the young Hardy, new to Cambridge and deeply disenchanted, began to think he just couldn't go through with it. It was all too stupid. Maybe he would quit mathematics altogether.

"I cannot remember ever having wanted to be anything but a mathematician," Hardy would later write. "I suppose that it was always clear that my specific abilities lay that way, and it never occurred to me to question the verdict of my elders." Modestly—and falsely—he would disclaim all artistic ability. As a philosopher, he decided, he would have been insufficiently original. Only as a journalist might he have made a go of it. But these ruminations lay almost half a century in the future. At the time, it was none of these fields that he weighed entering, but history . . .

In about 1064, two years before being killed by an arrow in the eye at the Battle of Hastings, Harold, son of Godwin, swore to Prince William of Normandy not to contest the English throne, and in return was crowned Earl of Wessex. The scene was depicted in the famous Bayeux tapestries, and as a boy, Harold Hardy had taken that masterpiece as his model for a brightly colored illumination of surpassing grace, onto which he dutifully calligraphed the caption "Ye Crowninge of Ye Earl Harold." Later, Winston Churchill, among others, would also tell the story of Harold, in *The Birth of Britain*, the first of his epic four-volume *History of the English-Speaking Peoples*. And it was a comprehensive history something like that which Hardy, at the age of about ten, dared undertake. Unfortunately, he approached it in overmuch detail and so never got much past the good Earl Harold.

An historical essay Hardy had written on entering Trinity impressed his examiners so much he could as well have gotten his scholarship in history as in math. Now, recalling Headmaster Fearon's treatment of the subject as a bright spot of the drab Winchester years, he flirted with changing fields. He might well have gone ahead with it. But in the midst of his confusion, he went to his director of studies, who brought another influence to bear on his malleable young mind in the person of Augustus Edward Hough Love.

A thirty-three-year-old man with a huge, bushy mustache, mutton-chop sideburns, and a vast bald oval of a head, Love had been named, a few years before, a Fellow of the Royal Society, Britain's most distinguished scientific body. In 1893, he'd finished his two-volume *Treatise on the Mathematical Theory of Elasticity,* summarizing what was then known of how materials deform under impact, twisting, and heavy loads. But Love did not push Hardy into his own field. Though an applied mathematician, he had a bent for fundamentals, basic principles, abstract formulations. Once, talking with a friend who was explaining something geometrically, Love shook his head, and said he didn't follow. "You see it is all x, y, z for me, and not your pictures at all." And so, perhaps attuned to Hardy's natural bent, he suggested Hardy read a mathematical text alien to the libraries of most applied mathematicians, the Frenchman Camille Jordan's *Cours d'analyse de l'Ecole Polytechnique.*

When Jordan died almost three decades later, it would be Hardy who wrote his obituary in *Proceedings of the Royal Society* and took the opportunity to comment on the book that changed his life: "To have read it and mastered it is a mathematical education in itself," Hardy wrote, "and it is hardly possible to overstate the influence which it has had on those who, coming to it as I did from the elaborate futilities of 'Tripos' mathematics, have found themselves at last in [the] presence of the real thing."

Jordan and other Continental mathematicians were taking seemingly obvious mathematical concepts and subjecting them to the most searching scrutiny. For example, mathematicians spoke of "continuous functions"—relationships between variables unmarked by weird and abrupt lurches. And in just such vague, intuitive ways did they tend to think of them. But what, exactly, *was* a function? And what did it mean to say it was continuous? Among "analysts," as this breed of mathematician was known, just such questions were their meat.

Say you draw a circle on a piece of paper; obviously, the circle divides the paper into two regions—within the circle, and outside it. Now, say you've got two points on the paper, both of which lie outside the circle: surely you can connect them (not necessarily with a straight line) without cutting the circle, right? And just as surely, if one point lies outside the circle and the other inside, then any continuous line linking them has to cut the circle . . .

"Surely?" "Just as surely?" For English mathematicians untouched by the precision of the Continent, such notions *were* obvious, scarcely worthy of another thought. But Jordan actually stated these seemingly self-evident truths as theorems and set about trying to prove them rigorously. In fact, he couldn't do it, or at least not completely; his proofs were laced with flaws, and his successors had later to correct them. But they invoked just the kind of close, sophisticated reasoning that Hardy, coming upon Jordan now, at the age of barely twenty, found beguiling. "I shall never forget," Hardy later wrote of Jordan's book, whose second, much-improved edition had just appeared in 1896, "the astonishment with which I read that remarkable work, the first inspiration for so many mathematicians of my generation, and learnt for the first time as I read it what mathematics really meant."

Of course none of this had anything in the least to do with the Tripos, which—now a mathematician again more than ever—Hardy would have to take. And *that,* realistically speaking, meant studying with a coach. So Hardy—yes, his friend Littlewood later seemed to gloat, "even the rebel Hardy"—surrendered. It was a bitter pill; "real" mathematics was put aside, Tripos mathematics reluctantly embraced. "When I look back upon those two years of intensive study," he would recall later, "it seems to me almost incredible that anyone not destitute of ability or enthusiasm should have found it possible to take so much trouble and to learn no more."

Inaugurating something of a trend among the better students, Hardy took the Tripos at the end of his second year, rather than his third, thus freeing himself from its grip that much sooner. Still he managed to come in fourth—by any measure a superb performance. "Hurrah," one friend, the future historian G. M. Trevelyan, wrote when Hardy notified him. "It is a great triumph, not only for you but the good cause of taking the triposes in the second year." And yet for all his contempt for the Tripos system, Hardy told his friend Snow later, it rankled that he wasn't Senior Wrangler. "He was enough of a natural competitor," wrote Snow, "to feel that, though the race was ridiculous, he ought to have won it."

From then on, though, Hardy's stature among his generation of English mathematicians steadily rose. In 1898, when he was twenty-one, his first mathematical publication appeared; like Ramanujan's, it was in the form of a question, followed three issues later by its solution. Hardy graduated in 1899 and the following year took Part II of the Tripos; Part I was the one for which the coaches groomed their students and which determined Wrangler standing, Part II the more provocative and more challenging. On it, Hardy scored first and was promptly named a Fellow of Trinity.

5. "THE MAGIC AIR"

At its two-hundred-fiftieth meeting, on May 23, 1901, the Cambridge Shakespeare Society held a reading of the first three acts of *Twelfth Night*. Hardy was there—not to perform but, as his friends deemed his role, as "critic." With him was his close friend R. K. Gaye, a classical scholar, who played Malvolio. Others lending retrospective luster to the evening were Lytton Strachey, playing Maria, Leonard Woolf in the dual role of Valentine and the Captain, and J. T. Stephen, the future Virginia Woolf's brother, who played Sir Toby Belch. James Hopwood Jeans, who came up to Trinity the same year as Hardy and would enjoy almost as distinguished a career in applied mathematics as Hardy would in pure, took photographs by magnesium light. What a pity they don't survive! The men sported lavish costumes, sparing nothing in the detail they expended on the ladies' gowns some of them wore. Gaye wore the yellow stockings, cross-gartered of Shakespeare's text. Stephen wore a stomacher, an elaborately embroidered chest piece. Woolf wore his old bowler hat. It was a madcap evening. The reading went well enough but, as the secretary was delighted to record, "elementary stage rules, such as facing the audience, were neglected by those who should have known better. Let us hope it was modesty!"

A few months before, on January 22, Queen Victoria had died; the century that bore her name, with its straitlaced proprieties and its moral fervor, was over, and the first stirrings of the new, more relaxed era heralded by Edward's reign could be felt.

On September 3, Hardy was in Cranleigh for the funeral of his father, who'd died four days before, at the age of fifty-nine. There he joined his sister and mother in the procession which bore his father's body, in a wheeled bier covered with wreaths and laced with ivy, from the House, to the school chapel, and thence through the village to the parish church, in the shadow of whose squat, stone tower it was laid to rest.

Back at Trinity, where he now occupied rooms in Great Court, twenty-four-year-old Hardy took up his old life again. Shy and sensitive though he was, Hardy had, since taking the Tripos in 1898, become something of a social animal. He was now part of a circle to which the sons of schoolteachers normally did not belong and whose values his family *might* scarcely have been able to imagine. He attended Shakespeare

Society meetings. He belonged to Decemviri and Magpie & Stump, two Cambridge debating societies, which took up such questions as "An enlightened selfishness is the highest virtue," and "The complete degradation of the Latin races is final and irrevocable." But he also belonged to another group, which neither advertised its meetings, nor openly recruited members, nor even proclaimed its existence.

Woolf and Strachey were both present or future members of a secret intellectual society known as the Apostles. So were two others in the *Twelfth Night* cast. And so was Hardy.

It had been started in 1820 as the Conversazione Society, one of many Cambridge student groups devoted to debate and good fellowship. Founded as it was by twelve men, it soon became known as the Apostles, and in time as simply the Society. By the time Hardy joined, its ranks had included some of the most brilliant men Cambridge had ever produced. There was Tennyson, the poet; Whitehead, the philosopher; James Clerk Maxwell, the physicist; Bertrand Russell; and many others whose names are less recognizable today. Of the ten whom Leonard Woolf would later count as the nucleus of "Old Bloomsbury," the iconoclastic intellectual community that would reach full flower in the 1920s and 1930s, no fewer than seven were Apostles.

A scientist, as one Apostle noted round this time, was elected only if "he was a very nice scientist." Hardy, apparently, was nice enough; his crystalline intellect, his sly charm, his aesthetic sensibilities, his good looks, his love of good conversation—all these would have endeared him to such men. Actually, for reasons unknown, he first declined membership. But finally, in a gala, well-attended meeting of the Society on the evening of February 19, 1898, he was inducted.

His "father," the man putting him up for membership, was G. E. Moore. Leonard Woolf would later write of Moore that he pursued truth "with the tenacity of a bulldog and the integrity of a saint." An Apostle since 1894, he was a charismatic philosopher who brooked no imprecision in thought, feeling, or expression; always he was asking, What do you exactly mean to say? His philosophy, set out in *Principia Ethica* in 1903, coupled with what Woolf would call "his peculiar passion for truth, for clarity and common sense," made him the dominant influence on his whole generation of Cambridge intellectuals, and ultimately on Bloomsbury as a whole.

Each Saturday night the Apostles met. (The normal fare was "whales," the name given to sardines or anchovies on toast.) One of them read a paper on some intellectual topic, such as, "Is Any Event Necessary?" Or, "Can Moral Philosophy Provide Any Antidote for Unhappiness?" Or, "Does Youth Approve of Age?" The question would then be debated and put to the floor for a vote. (On that last one, Hardy voted no.) The group wanted for nothing in arrogance and preciousness, as its peculiar jargon suggested: An "embryo" was a candidate for membership, a "birth" his induction ceremony. "Phenomena" was the rest of the world—anything or anyone not an Apostle; "reality" was the Society, its members and activities.

"There were to be no taboos, no limitations, nothing considered shocking, no barriers to absolute freedom of speculation," wrote Bertrand Russell, who was inducted six years before Hardy. "We discussed all manner of things, no doubt with a certain immaturity, but with a detachment and interest scarcely possible in later life." Meetings lasted till about one in the morning, followed by informal discussion up and down the cloisters of Nevile's Court, at Trinity, which along with King's supplied most of the group's members during this period. "The soul of the thing, as I felt it," wrote Goldsworthy Lowes Dickinson, who had joined in 1885, "is incommunicable. When young men are growing in mind and soul, when speculation is a passion, when discussion is made profound by love, there happens something [that would be unbelievable] to any but those who . . . breathe the magic air."

A classical scholar by way of King's, a short, vigorous man of formidable charm who had briefly swerved into medicine and later would become a peace activist, Dickinson was also a homosexual. Not all the Apostles were homosexual, perhaps not even most; and yet a homosexual current ran through the Society and, in the years after Hardy joined, intensified. In 1901, E. M. Forster, who was to write *A Passage to India*, joined; he was homosexual. In 1902, Lytton Strachey joined, and in 1903 John Maynard Keynes, the economist (who once advised Hardy that had he followed stocks as avidly as he did cricket, he would have become rich). Both were homosexual, too, and their election, as one chronicler of the Society later put it, transmuted "its naughty verbal mannerisms and Walt Whitmanesque feelings of comradeship into overt full-blooded—almost aggressive—homosexuality." Homosexuality was elevated almost to the status of an art form or aesthetic doctrine. "The Higher Sodomy," Apostles termed it—the assertion that the love between man and man could be higher and finer than that of man for woman, thus raising homosexual relationships to an almost spiritual plane.

At the time Hardy joined, the Apostles had not yet reached the point where, as Duncan Grant would put it, "even the womanisers pretend to be sods, lest they shouldn't be thought respectable." But while he formally "took wings" from the Society just before Strachey and Keynes joined, Hardy maintained friendships with "brethren," many of them homosexual, for years thereafter.

That Hardy himself was at least of homosexual disposition is scarcely in doubt. No woman, aside from his mother and sister, played the slightest substantive role in his life. And he had numerous male friends of whom he was passionately fond.

In 1903, for example, he shared a double suite of rooms at Trinity with R. K. Gaye, who was also a fellow of the college. Gaye, who had joined in the *Twelfth Night* revelry two years before, entered Trinity the same year as Hardy and showed evidence of being as brilliant a classical scholar as Hardy was a mathematician. Already, at twenty-five, he had received a host of medals and prizes, soon to include the Hare Prize for his *Platonic Conception of Immortality and Its Connexion with the Theory of Ideas*. Much later, he would commit suicide. But for now, Leonard Woolf would write, he and Hardy "were absolutely inseparable; they were never seen apart and rarely talked to other people." Woolf records them in their rooms sitting in domestic tranquility beside the fire, quiet and dejected after returning from the veterinarian with their emaciated, worm-ridden cat.

A younger mathematician who knew Hardy much later, when during the 1920s he was at Oxford, says that there was indeed a "rumor of a young man" then. Later, when Hardy visited America in the 1930s, he would impress the mathematician Alan Turing, himself homosexual, as, in the words of his biographer, Andrew Hodges, "just another English intellectual homosexual atheist." And during this period, too, he would meet an Oxford man many years his junior to whom he would later dedicate a book and whom one account simply refers to as "his beloved John Lomas."

But all that came later, and if Hardy was a practicing homosexual during his early days as a Trinity undergraduate and fellow, he was remarkably discreet about it. Several of those who knew him well, while willing to see in him homosexual leanings, note that he never displayed any of the stereotyped mannerisms of dress and behavior imputed to homosexuals. Littlewood, who worked with him for almost forty years, called him "a non-practicing homosexual." Then, too, no explicit record of homosexual activity comes down to us, none of the kind of blunt, gossipy X-was-lovers-with-Y tales that, in recent years, surround other venerable Cambridge figures. Indeed, the only sure knowledge we have of Hardy as sexual being is that one Saturday night in 1899, when the Apostles debated whether masturbation (they called it "self-abuse") was bad as an end, as opposed to a means, Hardy voted with the ayes.

Hardy was a product of the English public schools, a monastic environment that served as a crucible for homosexual relations among the boys. Everyone knew it, everyone accepted it, no one took it seriously, and for most, such adolescent flings didn't carry into adult life. In their pioneering look at homosexuality, *Sexual Inversion,* published in 1897, the year after Hardy left Winchester, Havelock Ellis and John Addington Symonds took note of "these school-boy affections and passions," noting that most passed with time.

If Hardy did have homosexual experiences during this time, he would scarcely have been likely to bruit the news about. A suggestion that homosexuality might be instinctual, and was no proper matter for the exercise of moral control, came through Edward Carpenter in *Homogenic Love,* published in 1896, and *The Intermediate Sex,* published in 1908. But these were no more than bubbles of tolerance in a sea of homophobia. Indeed, the year 1895, while Hardy was in his teens, brought the anguish of the Oscar Wilde trials, in which the great dramatist was sentenced to two years' hard labor for homosexual acts; his wife changed her name and those of her two sons as well, and Wilde's plays stopped being produced. *That* was the norm; Carpenter and Havelock Ellis, who also wrote of homosexuality as natural, were the barest breath in the cultural wind. "The last five years of the century were much less open to discussions of love, sex, and marriage," one chronicler of the era, Samuel Hynes, has written in *The Edwardian Turn of Mind.* "It was as though the Victorian age, in its last years, had determined to be relentlessly Victorian while it could."

Hardy reached young adulthood in the early 1900s, with its freer Edwardian sensibilities. But his attitudes had been formed earlier, in the Victorian era, a time preoccupied with public morality. It was a time when the face you put on things, as Gertrude Himmelfarb has stressed in *Manners and Morals Among the Victorians,* was taken with the utmost seriousness. Whatever people did on the sly, they were careful to keep it quiet, and led whole lives that way; it wasn't "hypocrisy," but simple propriety. Vastly less play was given "natural" drives. If anything, the idea was to bottle them up, or channel them, but in any case to control them; whatever your private fantasies, you didn't have to act on them, for goodness sake, and shouldn't.

Hardy's Cambridge in the early 1900s was a curious blend of the old Victorianism and the new, freer Edwardianism. If within the Apostles and in Cambridge generally the homosexual undercurrent surged toward the surface, it never broke through. Even among these, the most avant of the avant-garde, Victorian echoes sounded. As Himmelfarb has pointed out, referring to the Bloomsbury movement that largely grew out of the Apostles:

> It is ironic that people who prided themselves on their honesty and candor, especially in regard to their much-vaunted "personal affections"—in contrast, as they thought, to Victorian hypocrisy and duplicity—should have succeeded for so long in concealing the truth about those personal affections. Even so perceptive and psychoanalytic-minded a critic as Lionel Trilling was able to write a full-length study of Forster in 1943 without realizing that he was a homosexual. Nor did Roy Harrod, in his biography of Keynes published in 1951 (the definitive biography, as it seemed at the time and as it remained for more than thirty years), see fit to mention Keynes' homosexuality—a deliberate suppression, since Harrod was a friend of Keynes and was perfectly well aware of his sexual proclivities and activities. Nor did Leonard Woolf, in a five-volume autobiography that was entirely candid about his wife's mental breakdowns, give any indication of the frenetic sexual affairs of everyone around him.

Given the times out of which they came, then, that Hardy's long-time friend, Littlewood, called him "a non-practicing homosexual" may mean little. Littlewood was himself known to have fathered a child by a married woman but would not publicly acknowledge her until near his death in 1977. (When he did idly mention it one day in the Combination Room at Trinity, he was stunned that no one seemed to care.) Homosexuality, of course, defied a yet more rigid taboo. So if Hardy did lead a homosexual life about which Littlewood knew, Littlewood might well have denied it.

Then again, Littlewood may have known nothing, Hardy keeping the various parts of his life scrupulously walled off from one another. Hardy knew all the Apostles, went to meetings. And his name shows up in diaries, memoirs, and biographies of Strachey, Forster, Woolf, Russell, and the others. But it doesn't show up *much;* one senses him at the edge of their world, not its center. Being a mathematician, and a pure mathematician at that, may have isolated him; within the Shakespeare Society, for example, he was gently ribbed for putting "his knowledge of higher mathematics" to use in calculating the tab for a recent dinner at five shillings, one penny—an "alarming sum." In the avant-garde world of which he was part, Hardy was tolerated, respected, appreciated for his inimitable personal charm; but the core of his life, one gathers, lay elsewhere.

In mathematics? Certainly.

In a homosexual underworld? Only perhaps.

That Hardy's life was spent almost exclusively in the company of other men, that he scarcely ever saw a woman, was, in those days, not uncommon. After all, among Havelock Ellis's thousand or so British "geniuses," 26 percent never married. In the academic and intellectual circles of which Hardy was a part, such a monastic sort of life actually represented one pole of common practice.

Thus, at Cranleigh School, all the teachers, except for the House staff, were men, most of them bachelors; dormitory masters *had* to be bachelors. Winchester was the same way. So was Cambridge. "In my day we were a society of bachelors," wrote Leslie Stephen in *Some Early Impressions* of his time at Cambridge during the early 1860s. "I do not remember during my career to have spoken to a single woman at Cambridge except my bed-maker and the wives of one or two heads of houses."

Not much had changed by the time Hardy reached Cambridge a generation later. Among the twenty or so colleges, two—Girton and Newnham—had been established for women in the previous two decades. But though women, with the lecturer's consent and chaperoned by a woman don, could attend university lectures, by 1913 they still kept mostly to themselves and played little part in undergraduate life. Until 1882, college fellows *couldn't* marry, but even after that most fellows remained bachelors. In 1887, a proposal was made to offer degrees to women; it was soundly defeated. Ten years later, on a May day in 1897, a straw-hatted mob thronged outside the Senate House, where the matter was again being taken up, demonstrating against the measure. A woman was hanged in effigy. A large banner advised (after Act II, Scene I of *Much Ado About Nothing*) "Get you to Girton, Beatrice. Get you to Newnham. Here's no place for you maids."

It was an almost laughably artificial environment, with dons left woefully ignorant of domestic life. One time at St. John's College, the story goes, an elderly bachelor at High Table congratulated someone on the birth of his son. "How old is the little man?" he asked.

"Six weeks," came the reply.

"Ah," said the bachelor don, "just beginning to string little sentences together, I suppose."

About the only time Hardy and other fellows encountered women was among the bedmakers who tidied up college rooms—and they were said to be selected for their plainness, age, and safely married status, presumably so as to minimize the distraction they represented to students and fellows of the colleges.

Within so exclusively male a setting, steeped in a sterner sense of public morality, and free of today's pervasive sexual drumbeat, even passionate and devoted friendships took physical form less frequently. David Newsome, writing of late Victorian Cambridge, alludes to

> romantic friendship within exclusively male communities, a phenomenon so normal and respected throughout the period . . . that the greatest care must be taken to avoid slick and dismissive judgements. . . . In the nineteenth century the normality of both men and women forming highly emotional relationships with those of their own sex, of the same age or sometimes older or younger . . . was neither questioned as necessarily unwholesome, nor felt to inhibit the same relationship with the opposite sex leading to perfectly happy marriage.

One physician in the late 1890s wrote Ellis and Symonds of several such cases of passionate, yet presumably nonphysical relationships:

> In all these, I imagine, the physical impulse of sex is less imperative than in the average man. The emotional impulse, on the other hand, is very strong. It has given birth to friendships of which I find no adequate description anywhere but in the dialogues of Plato; and beyond a certain feeling of strangeness at the gradual discovery of a temperament apparently different to that of most men, it has provoked no kind of self-reproach or shame. On the contrary, the feeling has been rather one of elation in the consciousness of a capacity of affection which appears to be finer and more spiritual than that which commonly subsists between persons of different sexes. . . . In all these cases, a physical sexual attraction is recognized as the basis of the relation, but as a matter of feeling, and partly also of theory, the ascetic ideal is adopted.

It was just such kinds of relationships, remote to American life today, that C. P. Snow, who knew Hardy as well as anyone, imputes to him. Hardy, he wrote, did not normally form close, demonstrative bonds among even those he called his friends.

> But he had, scattered through his life, two or three other relationships, different in kind. These were intense affections, absorbing, non-physical but exalted. The one I knew about was for a young man whose nature was as spiritually delicate as his own. I believe, though I only picked this up from chance remarks, that the same was true of the others. To many people of my generation, such relationships would seem either unsatisfactory or impossible. They were neither the one nor the other; and unless one takes them for granted, one doesn't begin to understand the temperament of men like Hardy . . . nor the Cambridge society of his time.

Despite suggestive evidence, then, one cannot conclude that Hardy was a practicing homosexual. And yet, in one sense, it doesn't matter. *Either* he led an almost wholly asexual life, scarcely knowing what he was as a sexual being, and submerging any sexual desires behind a screen of Victorian propriety, *or* he led a secret sexual life so elaborately and successfully hidden that even friends knew nothing of it and those who did kept quiet. In either case, he would have required a vast architecture of personal defenseworks to pull it off, heroic acts of will performed day in and day out over the years. And though made somewhat easier and more ordinary by the times in which he lived, it would, in the end, have had to exact its toll.

And it did. There was a hauntedness to Hardy that you could see in his eyes. "I suspect," remembered an Oxford economist, Lionel Charles Robbins, who knew him later, that "Hardy found many forms of contact with life very painful and that, from a very early stage, he had taken extensive measures to guard himself against them. Certainly in his friendlier moments—and he could be very friendly indeed—one was conscious of immense reserves." Always, he kept the world at bay. The obsession with cricket, the bright conversation, the studied eccentricity, the fierce devotion to mathematics—all of these made for a beguiling public persona; but none encouraged real closeness. He was a friend of many in Cambridge, an intimate of few.

In the years after 1913, Hardy would befriend a poor Indian clerk. Their friendship, too, would never ripen into intimacy.

6. THE HARDY SCHOOL

In 1900, Hardy became a Fellow of Trinity College. In 1901, he won one of two Smith's Prizes, named after a former master of Trinity College, and since 1769 the blue ribbon of Cambridge mathematics.

In 1903, he was named an M.A., which at English universities was normally the highest academic degree. (Cambridge didn't offer the doctorate, a German innovation, until after World War I, hoping to lure Americans otherwise drawn to Germany.)

In 1906, be became a Trinity lecturer. He gave six hours a week of lectures, usually in two courses, elementary analysis and the theory of functions. He occasionally gave informal classes during this period, but he was never actually a college tutor. He was there to do research.

Hardy would later say he blossomed slowly. In a sense, that was true; most of his more important mathematical contributions lay in the future. But already in the first decade of the twentieth century he was batting out papers at a prodigious clip, ten or a dozen a year, most of them on integrals and series. Like "Research in the Theory of Divergent Series and Divergent Integrals," which appeared in *Quarterly Journal of Mathematics* in 1904; and "On the Zeros of Certain Classes of Integral Taylor Series," in the *Proceedings of the London Mathematical Society* in 1905. In much of this work, he was refining and enhancing ideas suggested by Camille Jordan in the book that had so inspired Hardy as an undergraduate.

In later years, Hardy himself would set little stock in the work he did during this period. "I wrote a great deal during the next ten years, but very little of importance," he would say of the period before he met Littlewood and Ramanujan; "there are not more than four or five papers which I can still remember with some satisfaction." Still, by 1907, they added up to a corpus substantial enough that, on October 31 of that year, he was put up for membership in the Royal Society.

Many of the top names in the Cambridge mathematical establishment went to bat for him. A. E. H. Love, who had introduced Hardy to Camille Jordan when he was still an undergraduate, did. So did E. W. Hobson, one of those who would, in a few years, hear from Ramanujan. So did T. J. I'A. Bromwich, whose book on infinite series Ramanujan had been urged to consult just before he wrote Hardy. Like most who would place the coveted F.R.S. after their names, Hardy didn't get it first time out. But in 1910, he was elected, at the age of thirty-three. A London photographer took his picture for the occasion and did a little retouching around the eyes and mouth. But it was probably a reflex action; Hardy still looked boyish.

Many years later, Hardy would insist that none of the mathematics he had done during his career was ever in the least "useful." But around now came one exception. During the previous century an Austrian monk named Gregor Mendel had done experiments in crossing tall pea plants with dwarfs, found that each generation of the progeny bore fixed, predictable proportions of dwarf and tall, and so laid the basis for the science of genetics. At the time, of course, nobody cared, and Mendel's experiments lay forgotten. But the publication of his work in 1900, sixteen years after his death, sparked a flurry of interest in his insights, and the next few years saw them the subject of much active debate.

One controversy surrounded the fate of recessive and dominant traits in succeeding generations. A recessive trait was normally "silent"; it needed to be represented in both parents to show up in the children. A dominant trait, on the other hand, needed only one copy of the gene. An article in the *Proceedings of the Royal Society of Medicine* argued that Mendelian genetics predicted that a dominant trait, like the stunted finger growth known as brachydactylism, would tend to proliferate in the population. That this ran flatly counter to the evidence, the author asserted, undermined Mendel.

But no, showed Hardy, in a brief letter to the American scholarly journal *Science* in 1908, a dominant trait would not proliferate in the population. Assign symbols to the probabilities of a particular gene type, work through the simple algebra, and you'd find that the proportion of each gene would tend to stay fixed generation after generation. In other words, if matings took place randomly—that is, not skewed one way or the other by Darwinian natural selection—dominant traits would not take over and recessive traits would not die out. A German physician, Wilhelm Weinberg, showed something similar the same year, and the principle became known as the Hardy-Weinberg Law. It exerted a marked impact on population genetics. It was applied to the study of the genetic transmission of blood groups and rare diseases. It appears today in any scientific dictionary or genetics textbook.

Hardy himself, of course, set no great store by it. As for the mathematics, it was trivial. Besides, it was not, well, *useless*. And in Hardy's mind that made it verge on the execrable.

Havelock Ellis once wrote that "by inborn temperament, I was, and have remained, an English amateur; I have never been able to pursue any aim that no passionate instinct has drawn me towards." There was, as in Ellis, a streak of disdain in the English character for mere necessity; the amateur, bless his heart, did what he did for love, for the sake of beauty or truth, not because necessity compelled it. This streak had gained more reasoned form in the philosophy of G. E. Moore, Hardy's Apostolic "father." His *Principia Ethica* represented, in the words of Gertrude Himmelfarb, a "manifesto of liberation" stressing love, beauty, and truth. "And even love, beauty, and truth were carefully delineated as to remove any taint of utility or morality. Useless knowledge was deemed preferable to useful, corporeal beauty to mental qualities, present and immediately realizable goods to remote or indirect ones."

G. H. Hardy's mathematics would emerge as the consummate manifestation, within his own field, of Moore's credo.

"I have never done anything 'useful,'" is how he would put it years later. "No discovery of mine has made or is likely to make, directly or indirectly, for good or ill, the least difference to the amenity of the world." He would never say so, perhaps he did not even see it, but he had taken Moore's sensibilities and applied them to mathematics. "Hardyism," someone would later dignify this doctrine, so hostile to practical applications; and Hardy's *Mathematician's Apology,* written almost half a century later, would embody it on every page.

That mathematics might aid the design of bridges or enhance the material comfort of millions, he wrote, was scarcely to say anything in its defense. For such mathematics, he bore only contempt.

> It is undeniable that a good deal of elementary mathematics . . . has considerable practical utility. [But] these parts of mathematics are, on the whole, rather dull; they are just the parts which have least aesthetic value. The "real" mathematics of the "real" mathematicians, the mathematics of Fermat and Euler and Gauss and Abel and Riemann, is almost wholly "useless."

Hardy went on to pity the mathematical physicist who might use mathematical tools to understand the workings of the universe: was not his lot in life a little pathetic?

> If he wants to be useful, he must work in a humdrum way, and he cannot give full play to his fancy even when he wishes to rise to the heights. "Imaginary" universes are so much more beautiful than this stupidly con-

> structed "real" one; and most of the finest products of an applied mathematician's fancy must be rejected, as soon as they have been created, for the brutal but sufficient reason that they do not fit the facts.

Not so in real mathematics, Hardy continued, which "must be justified as art if it can be justified at all." It was in this light that the Hardy-Weinberg Law, hopelessly useful as it was, and barren of anything in the least mathematically beautiful, violated his principles.

While these aesthetic principles, as it were, may not yet have been so fully formed or so well articulated at the time Hardy heard from Ramanujan in early 1913, they had certainly by then jelled—and had begun to find an outlet in the school of English mathematics forming around him. It was a school with scant interest in anything so dull and pragmatic as, say, genetics; or even, for that matter, with the mathematical physics that had for years been the special English strength. It was a school, rather, that embraced the Continental purity of Camille Jordan.

Alone in their island kingdom, cut off from the Continent, the British were an insular people, suspicious and intolerant of all things foreign. At the end of the nineteenth century and beginning of the twentieth, however, they were probably even more so, basking in the glow of empire, fat and happy. "It must have seemed like a long garden party on a golden afternoon—to those who were inside the garden," Samuel Hynes has written, in *The Edwardian Turn of Mind,* of British attitudes during this period. "But a great deal that was important was going on outside the garden: it was there that the twentieth-century world was being made"—in mathematics, he might have added, as elsewhere.

Since the seventeenth century, Britain had stood, mathematically, with its back toward Europe, scarcely deigning to glance over its shoulder at it. Back then, Isaac Newton and the German mathematician Gottfried Wilhelm von Leibniz had each, more or less independently, discovered calculus. Controversy over who deserved the credit erupted even while both men lived, then mushroomed after their deaths, with mathematicians in England and on the Continent each championing their compatriots. Newton was the premier genius of his age, the most fertile mind, with the possible exception of Shakespeare's, ever to issue from English soil. And yet he would later be called "the greatest disaster that ever befell not merely Cambridge mathematics in particular but British mathematical science as a whole." For to defend his intellectual honor, as it were, generations of English mathematicians boycotted Europe—steadfastly clung to Newton's awkward notational system, ignored mathematical trails blazed abroad, professed disregard for the Continent's achievements. "The Great Sulk," one chronicler of these events would call it.

In calculus as in mathematics generally, the effects were felt all through the eighteenth and nineteenth centuries and on into the twentieth. Continental mathematics laid stress on what mathematicians call "rigor," the kind to which Hardy had first been exposed through Jordan's *Cours d'analyse* and which insisted on refining mathematical concepts intuitively "obvious" but often littered with hidden intellectual pitfalls. Perhaps reinforced by a strain in their national character that sniffed at Germanic theorizing and hairsplitting, the English had largely spurned this new rigor. Looking back on his Cambridge preparation, Bertrand Russell, who ranked as Seventh Wrangler in the Tripos of 1893, noted that "those who taught me the infinitesimal Calculus did not know the valid proofs of its fundamental theorems and tried to persuade me to accept the official sophistries as an act of faith. I realized that the Calculus works in practice but I was at a loss to understand why it should do so." So, it is safe to say, were most other Cambridge undergraduates.

Calculus rests on a strategy of dividing quantities into smaller and smaller pieces that are said to "approach," yet never quite reach, zero. Taking a "limit," the process is called, and it's fundamental to an understanding of calculus—but also, typically, alien and slippery territory to students raised on the firm ground of algebra and geometry. And yet, it is possible to blithely sail on past these intellectual perils, concentrate on the many practical applications that fairly erupt out of calculus, and never look back.

In textbooks even today you can see vestiges of the split—which neatly parallels that between Britain and the Continent in the nineteenth century: the author briefly introduces the limit, assumes a hazy intuitive understanding, then spends six chapters charging ahead with standard differentiation techniques, maxima-minima problems, and all the other mainstays of Calc 101 . . . until finally, come chapter 7 or so, he steps back and reintroduces the elusive concept, this time covering mine-strewn terrain previously sidestepped, tackling conceptual difficulties—and stretching the student's mind beyond anything he's used to.

Well, the first six chapters of this generic calculus text, it could be said, were English mathematics without the Continental influence. Chapter 7 was the new rigor supplied by French, German, and Swiss mathematicians. "Analysis" was the generic name for this precise, fine-grained approach. It was a world of Greek letters, of epsilons and deltas representing infinitesimally small quantities that nonetheless the mathematicians found a way to work with. It was a world in which mathematics, logic, and Talmudic hairsplitting merged.

First Gauss, Abel, and Cauchy had risen above the looser, intuitive nostrums of the past; later in the century, Weierstrass and Dedekind went further yet. None of them were English. And the English professed not to care. Why, before the turn of the century, Cauchy—*the* Cauchy, Augustin Louis Cauchy, the Cauchy who had launched the French school of analysis, the Cauchy of the Cauchy integral formula—was commonly referred to around Cambridge as "Corky."

Since Newton's time, British mathematics had diverged off on a decidedly applied road. Mathematical physics had become the British specialty, dominated by such names as Kelvin, Maxwell, Rayleigh, and J. J. Thomson. Pure math, though, had stultified, with the whole nineteenth century leaving England with few figures of note. "Rigor in argument," J. E. Littlewood would recall, "was generally regarded—there *were* rare exceptions—with what it is no exaggeration to call contempt; niggling over trifles instead of getting on with the real job." Newton had said it all; why resurrect these arcane fine points? Calculus, and the whole architecture of mathematical physics that emanated from it, *worked*.

And so, England slept in the dead calm of its Tripos system, where Newton was enshrined as God, his *Principia Mathematica* the Bible. "In my own Tripos in 1881, we were expected to know any lemma [a theorem needed to prove another theorem] in that great work by its number alone," wrote one prominent mathematician later, "as if it were one of the commandments or the 100th Psalm. . . . Cambridge became a school that was self-satisfied, self-supporting, self-content, almost marooned in its limitations." Replied a distinguished European mathematician when asked whether he had seen recent work by an Englishman: "Oh, we never read anything the English mathematicians do."

The first winds of change came in the person of Andrew Russell Forsyth, whose *Theory of Functions* had begun, in 1893, to introduce some of the new thinking—though by this time it wasn't so new anymore—from Paris, Göttingen, and Berlin. Written in a magisterial style, it burst on Cambridge, as E. H. Neville once wrote, "with the splendour of a revelation"; some would argue it had as great an influence on British math-

The passport photo. Ramanujan in 1919, on his way back to India. "He looks rather ill," G. H. Hardy wrote when he first saw the photo in 1937, "but he looks all over the genius he was." *Master and Fellows of Trinity College, Cambridge*

Komalatammal, Ramanujan's mother, and the decisive influence on him in his youth. No photo of K. Srinivasa Iyengar, Ramanujan's father, is known to exist. *Ragami's Collections, Madras, South India*

Ramanujan's house, on Sarangapani Sannidhi Street, Kumbakonam, South India. Once, while in high school, he found that a formula he had thought original with him actually went back 150 years. Mortified, he hid the paper on which he had written it in the roof of the house. *Ragami's Collections, Madras, South India*

A recent photo of the *pial*, or front porch, of Ramanujan's house in Kumbakonam. Here he would sit for hours and work on mathematics while his friends played in the street.

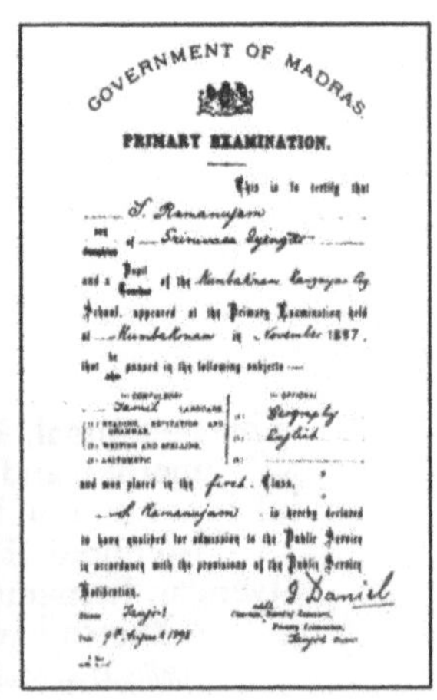

GOVERNMENT OF MADRAS

PRIMARY EXAMINATION.

This is to certify that S. Ramanujam son of Srinivasa Iyengar and a pupil of the Kumbakonam [illegible] School, appeared at the Primary Examination held at Kumbakonam in November 1897, that he passed in the following subjects :—

COMPULSORY: Tamil Language (1) Reading, Recitation and Grammar. (2) Writing and Spelling. (3) Arithmetic

OPTIONAL: (1) Geography (2) English (3)

and was placed in the first Class.

S. Ramanujam is hereby declared to have qualified for admission to the Public Service in accordance with the provisions of the Public Service Notification.

I. Daniel

Ramanujan scored high on this examination, which he took when he was nine. But later, once he discovered mathematics and lost interest in all else, he regularly failed his exams. (Here, the English spelling of his Tamil name was rendered as Ramanujam.) *Ragami's Collections, Madras, South India*

A recent photo of the Sarangapani Temple, just up the street from Ramanujan's house, which is visible on the right.

At Town High School, in his hometown of Kumbakonam, Ramanujan was still a conventionally good student, earning prizes and winning praise from his elders. Here, the school's campus in a recent photograph.

They call it Ramanujan Hall today, in honor of Town High School's most distinguished alumnus. The "m" at the end reflects how Ramanujan's Tamil name often gets transliterated into English. It was under this spelling that it first appeared in Indian mathematical journals.

S. Tirunarayanan, born in 1905 when Ramanujan was 17. His older brother would tease him, carry him on his shoulders, tell him stories. *Ragami's Collections, Madras, South India*

Janaki, Ramanujan's wife, in a picture taken after Ramanujan's death. As a widow, and mostly cut off from Ramanujan's family, she supported herself as a seamstress. *Ragami's Collections, Madras, South India*

Narayana Iyer, Ramanujan's immediate boss at the Madras Port Trust, one of his most ardent champions, and himself a fine mathematician. When the two worked on mathematics, often until late into the night, Ramanujan's penchant for collapsing many steps into one left him dazed. "You must descend to my level of understanding," he would complain. *Ragami's Collections, Madras, South India*

P. V. Seshu Iyer. One of Ramanujan's mathematics teachers at Government College in Kumbakonam and later one of his supporters. But one friend recalled Ramanujan complaining that Seshu Iyer, like everyone else, had at first been "indifferent" to him and his work. *Ragami's Collections, Madras, South India*

After Ramanujan married, he scoured South India for a job or a patron, his friends often putting him up for a while. One night, while staying with a friend at Victoria Hostel in Madras, shown here in a recent photo, he compared his impoverished lot to that of Galileo, persecuted by the Inquisition and misunderstood in his own time.

Soon after Ramanujan received his first encouraging letter from Hardy in Cambridge, he was named research scholar at the University of Madras. Freed from money worries for the first time in his life, he would come here, to the Connemara Library, seen here in a recent photo, and lose himself in mathematics.

Shrine to the goddess Namagiri, Ramanujan's family deity, in a recent photo. It was here in the South Indian town of Namakkal that Ramanujan came in late 1913 with Narayana Iyer. He stayed on the temple grounds for three nights and by the end had resolved to go to England, in defiance of Hindu tradition.

E. W. Hobson and H. F. Baker. Both were eminent Cambridge mathematicians. Both received Ramanujan's appeals for help. Both dismissed them. *Master and Fellows of Trinity College, Cambridge*

The tank, a large ritual pool, opposite the Parthasarathy Temple, the central religious shrine of the Triplicane district of Madras. It was in Triplicane, down the street from the tank, that Ramanujan lived in the period before he left for England.

(1)

In page 36 it is stated that "the no of prime nos less than $x = \int_2^x \frac{dt}{\log t} + \rho(x)$ where the precise order of $\rho(x)$ has not yet been determined."

The precise order itself is not sufficient to find the value of $\rho(x)$. Even if it is known that $\frac{\rho(x)}{\phi(x)} = 1$ when x becomes infinite, $\phi(x)$ being a known function of x, $\rho(x)$ cannot be supposed to have been found with sufficient accuracy. for example $(x + \frac{x}{\log\log x})/x = 1$ when x becomes infinite, yet the difference between $x + \frac{x}{\log\log x}$ and x is very great.

From the forms of $\rho(x)$ given in page 53, viz.

$O\left\{\frac{x}{(\log x)^a}\right\}$, $O(xe^{-a\sqrt{\log x}})$, $O(\sqrt{x})$, &c it appears that from particular numerical values the forms have been guessed

Even in regular functions it is difficult to have an idea of the form from the numerical values. In such a complicated function as $\rho(x)$ it is difficult to have an idea even for large values of $\log x$; for example even if we give billions for x, $\rho(x)$ is very difficult to be found.

I have observed that $\rho(e^{2\pi x})$ is of such a nature that its value is very small when x lies between 0 and 3 (its value is less than a few hundreds when $x = 3$) and rapidly increases when x is greater than 3

I I have found a function which exactly represents the no. of prime nos less than x, 'exactly' in the sense that the difference between the function and the actual no of primes is generally 0 or some small finite value even when x becomes infinite

The first of nine pages of mathematical results Ramanujan sent G. H. Hardy from India in 1913. *Syndics of Cambridge University Library*

(9)

IX Theorems on Continued Fractions.

a few examples are :—

(1) $\frac{4}{x+}\frac{1^2}{2x+}\frac{3^2}{2x+}\frac{5^2}{2x+}\frac{7^2}{2x+\&c} = \left\{\frac{\Gamma(\frac{x+1}{4})}{\Gamma(\frac{x+3}{4})}\right\}^2$

(2) If $P = \frac{\Gamma(\frac{x+m+n+1}{4})}{\Gamma(\frac{x-m+n+1}{4})} \cdot \frac{\Gamma(\frac{x+m-n+1}{4})}{\Gamma(\frac{x-m-n+1}{4})} \times \frac{\Gamma(\frac{x-m+n+3}{4})}{\Gamma(\frac{x+m+n+3}{4})} \cdot \frac{\Gamma(\frac{x-m-n+3}{4})}{\Gamma(\frac{x+m-n+3}{4})}$, then

$\frac{1-P}{1+P} = \frac{m}{x+}\frac{1^2-n^2}{x+}\frac{2^2-m^2}{x+}\frac{3^2-n^2}{x+}\frac{4^2-m^2}{x+\&c}$

(3) If $z = 1 + (\frac{1}{2})^2 x + (\frac{1\cdot3}{2\cdot4})^2 x^2 + \&c$

and $y = \pi \cdot \frac{1 + (\frac{1}{2})^2(1-x) + (\frac{1\cdot3}{2\cdot4})^2(1-x)^2 + \&c}{1 + (\frac{1}{2})^2 x + (\frac{1\cdot3}{2\cdot4})^2 x^2 + \&c}$, then

$\frac{1}{(1+a^2)\cosh y} + \frac{1}{(1+9a^2)\cosh 3y} + \frac{1}{(1+25a^2)\cosh 5y} + \&c$

$= \frac{1}{2} \frac{2\sqrt{z}}{1+}\frac{(az)^2}{1+}\frac{(a^2z)^2 x}{1+}\frac{(3az)^2}{1+}\frac{(4az)^2 x}{1+\&c}$

a being any quantity

(4) If $u = \frac{x}{1+}\frac{x^5}{1+}\frac{x^{10}}{1+}\frac{x^{15}}{1+}\frac{x^{20}}{1+\&c}$

and $v = \frac{\sqrt[5]{x}}{1+}\frac{x}{1+}\frac{x^2}{1+}\frac{x^3}{1+\&c}$

then $v^5 = u \cdot \frac{1-2u+4u^2-3u^3+u^4}{1+3u+4u^2+2u^3+u^4}$

(5) $\frac{1}{1+}\frac{e^{-2\pi}}{1+}\frac{e^{-4\pi}}{1+}\frac{e^{-6\pi}}{1+\&c} = \left(\sqrt{\frac{5+\sqrt{5}}{2}} - \frac{\sqrt{5}+1}{2}\right)\sqrt[5]{e^{2\pi}}$

(6) $\frac{1}{1-}\frac{e^{-\pi}}{1+}\frac{e^{-2\pi}}{1-}\frac{e^{-3\pi}}{1+\&c} = \left(\sqrt{\frac{5-\sqrt{5}}{2}} - \frac{\sqrt{5}-1}{2}\right)\sqrt[5]{e^{\pi}}$

(7) $\frac{1}{1+}\frac{e^{-\pi\sqrt{n}}}{1+}\frac{e^{-2\pi\sqrt{n}}}{1+}\frac{e^{-3\pi\sqrt{n}}}{1+\&c}$ can be exactly found if n be any positive rational quantity.

A more typical page of Ramanujan's first letter to Hardy. *Syndics of Cambridge University Library* ematics as any work since Newton's *Principia*. By the standards of the Continent, however, it was hopelessly sloppy and was soundly condemned there. "Forsyth was not very good at delta and epsilon," Littlewood once said of him, referring to the Greek letters normally used for dealing with infinitesimally small quantities. Still, it helped redirect the gaze of English mathematicians toward the Continent. It charted a course to the future, but did not actually follow it.

That was left to Hardy.

As a spokesman for the new rigor, Hardy exerted his impact not alone by what he had to say, but through the force, grace, and elegance with which he said it, both in print and in person.

In lectures, his enthusiasm and delight in the subject fairly spilled over. "One felt," wrote one of his later students, E. C. Titchmarsh, "that nothing else in the world but the proof of these theorems really mattered." Norbert Wiener, the American mathematical prodigy who would later create the field known as "cybernetics," attended Hardy's lectures. "In all my years of listening to lectures in mathematics," he would write, "I have never heard the equal of Hardy for clarity, for interest, or for intellectual power." Around this time, a pupil of E. W. Barnes, director of mathematical studies at Trinity, sought Barnes's advice about what lectures to attend. Go to Hardy's, he recommended. The pupil hesitated. "Well," replied Barnes, "you need not go to Hardy's lectures if you don't want, but you will regret it—as indeed," recalled the pupil many years later, "I have." Others who missed his lectures may not, in retrospect, have felt such regret: so great was Hardy's personal magnetism and enthusiasm, it was said, that he sometimes diverted to mathematics those without the necessary ability and temperament.

But as lucid as were his lectures, it was his writing that probably had more impact. Later, speculating about what career he might have chosen other than mathematics, Hardy noted that "Journalism is the only profession, outside academic life, in which I should have felt really confident of my chances." Indeed, no field demanding literary craftsmanship could fail to have profited from his attention. "He wrote, in his own clear and unadorned fashion, some of the most perfect English of his time," C. P. Snow once said of him. That Hardy's impressions of Ramanujan would be so relentlessly quoted, and would go so far toward fixing Ramanujan's place in history, owes not alone to his close relationship with Ramanujan but to the sheer grace with which he wrote about him.

Hardy didn't much like his early style, he decided later, terming it "vulgar." Of course, he didn't much like his glorious good looks, either. "Everything Hardy did," Snow once wrote of him, "was light with grace, order, a sense of style." And his writing exemplified that. He wrote, for the *Cambridge Review,* about the philosophy of Bertrand Russell. His obituaries of famous mathematicians were rounded, gracious, and wise. He could write about geometry and number theory for lay audiences. And his *Mathematician's Apology*, which became a classic, is almost mesmerizing in its language's hold on the reader.

He applied his gifts even to the most densely mathematical of his work. In collaborations, it was almost always he who wrote up the joint paper and shepherded it through publication. "He supplied the gas," recalled Littlewood, who was content if what he had to say was simply correct. Hardy wanted more; the "gas," as he once defined it, was the "rhetorical flourishes," the equivalent of "pictures on the board in the lecture, devices to stimulate the imagination of pupils." A reviewer would say of one of his mathematical texts that Hardy had "shown in this book and elsewhere a power of being interesting, which is to my mind unequalled."

Thought, Hardy used to say, was for him impossible without words. The very act of writing out his lecture notes and mathematical papers gave him pleasure, merged his aesthetic and purely intellectual sides. Why, if you didn't know math was supposed to be dry and cold, and had only a page from one of his manuscripts to go on, you might think you'd stumbled on a specimen of some new art form beholden to Chinese calligraphy. Here were inequality symbols that slashed across the page, sweeping integral signs an inch and a quarter high, sigmas that resonated like the key signatures on a musical staff. There was a spaciousness about how he wrote out mathematics, a lightness, as if rejecting the cramped, ungenerous formalities of the printed notation. He was like a French impressionist, intimating worlds with a few splashes of color, not a maker of austere English miniatures.

All through the first decade of the twentieth century Hardy used his pen to seduce a generation of young English mathematical students into taking seriously the new Continental rigor. When a review of Bertrand Russell's *Principles of Mathematics* ran in the *Times Literary Supplement* in 1903, it was Hardy who wrote it. While English mathematics often turned a deaf ear to events across the Channel, Hardy used the pages of the *Mathematical Gazette* to comment on foreign books. In 1903, he reviewed

Einleitung in die Funkionentheorie, by Stolz and Gmeiner; in 1905 *Leçons sur les fonctions des variables réeles*, by Borel.

Meanwhile, he held English and American texts to strict account for their lapses. In one *Mathematical Gazette* review in 1907, for example, he wrote of an American calculus text by W. Woolsey Johnson. Oh, it wasn't bad of its type, he allowed. But its type was a breed of English book that, while forgivable thirty years before, was no longer. Perhaps it was all right to pass over theoretical difficulties in laying the foundations of calculus, he wrote.

> But there are different ways of passing over difficulties. We may simply and absolutely ignore them: that is a course for which there is often much to be said. We may point them out and avowedly pass them by; or we may expand a little about them and endeavour to make our conclusions plausible without professing to make our reasoning exact.

But, Hardy went on:

> There is only one course for which no good defence can ever be found. This course is to give what profess to be proofs and are not proofs, reasoning which is ostensibly exact, but which really misses all the essential difficulties of the problem. This was Todhunter's [the author of a kindred text] method, and it is one which Prof. Johnson too often adopts.

Whereupon Hardy launched into a mathematical example to show how the author employed arguments "entirely destitute of validity."

Hardy felt he could do better, and in September 1908 completed *A Course of Pure Mathematics*, the first rigorous exposition in English of mathematical concepts other texts sloughed over in their rush to get to practical applications or cover broad expanses of mathematical ground. Such rigor was sorely needed, said Hardy in his preface. "I have [very rarely] encountered a pupil who could face the simplest problem involving the ideas of infinity, limit, or continuity with a vestige of the confidence with which he could deal with questions of a different character and of far greater intrinsic difficulty."

Like everything else Hardy ever wrote, his textbook was *readable*. This was not simply page after gray page of formula. His were real explanations of difficult ideas presented in clear, cogent English prose. What in other hands would be buried in a sea of abstractions, in Hardy's fairly jumped out at you, sometimes as the culmination of a passage actually verging on suspenseful.

Early on, for example, he addressed "rational" numbers, numbers like 6, $2/3$, $112/3890$, or 19 that can be expressed as ordinary fractions or integers. Between any two numbers representing points on a line segment, he showed that more rational points can always be squeezed in. Between $1/2$ and $2/3$, you can fit a $3/5$. Between $3/5$ and $2/3$, you can fit $5/8$. And so on, forever, resulting in an infinity of such points. Then he goes on:

> From these considerations the reader might be tempted to infer that these rational points account for all the points of the line, i.e. that *every* point on the line is a rational point. And it is certainly the case that if we imagine the line as being made up solely of the rational points, all other points (if any such there be) being imagined to be eliminated, the figure which remained would possess most of the properties which common sense attributes to the straight line and would, to put the matter roughly, look and behave very much like a line.

Something is coming, the reader rightly suspects, without knowing just what.

Hardy then showed that within the same line segment there was, roughly speaking, *another* infinity of points that could be crowded into the interstices between these rational numbers—the "irrational" numbers, which cannot be expressed as fractions, and whose properties he then proceeded to explore.

This loving attention to fundamentals was just what English mathematics needed. As one review of the book commented, "When Mr. Hardy sets out to prove something, then, unlike the writers of too many widely read textbooks, he really does prove it. . . . If the book is widely read, I for one shall hope to avoid in the future the many weary hours that have usually to be spent in convincing University students that 'proofs' which they have laboriously learned at school are little better than nonsense."

Hardy's book *was* widely read. For the next three-quarters of a century, and through ten editions and numerous reprints, it became the single greatest influence on the teaching of English mathematics at the university level. Through it—and through his lectures at Cambridge, through his papers and reviews, through his relationships with other mathematicians—Hardy made rigor no longer the preserve of a few Teutonic zealots but something that bordered on the mathematically fashionable.

* * *

At the root of Britain's mathematical backwardness, Hardy was sure, lay the Tripos system. Originally the means to a modest end—determining the fitness of candidates for degrees—the Tripos had become an end in itself. As Hardy saw it, English mathematics was being sapped by the very system designed to select its future leaders.

Around 1907, he became secretary of a panel established to reform it. But in fact, he championed its reform only as a first step toward doing away with it altogether, and only because he saw no hope, just then, for more radical change. As he later told a meeting of the Mathematical Association, "I adhere to the view . . . that the system is vicious in principle, and that the vice is too radical for what is usually called reform. I do not want to reform the Tripos but to destroy it."

Hardy did not oppose examinations in general; he saw a place for them, a sharply limited one—as a floor, a *minimum* standard necessary to earn a degree. "An examination," said he, "can do little harm, so long as its standard is low." But the Tripos laid no such meager claims; it meant to appraise, to sift, to grade. Undergraduates, as Hardy pictured them, exhausted "themselves and their tutors in the struggle to turn a comfortable second [class] into a marginal first." *That,* in his view, was the problem: the Tripos distorted teaching and learning alike, and English mathematics was the loser for it.

With others among the younger dons, Hardy succeeded in forcing changes through a reluctant senate, the university's governing arm. Chief among them was abolishment of the Order of Merit; a degree candidate still took the Tripos but, beginning in 1910, was ranked only by broad category—as Wrangler, Senior Optime, or Junior Optime. There would no longer be a Senior Wrangler to which to aspire, no longer the merciless pressure it created, no longer the ambition-driven need for coaches. Overnight, the most notorious abuses of the Tripos system were eliminated.

But Hardy's more ambitious goal was futile; the Tripos, in modified form, exists still—in part because while many pointed out its failings, few did so with Hardy's ferocity. There was a mild-manneredness in the English personality that Hardy, when it came to the Tripos, trespassed. "It is useless to propose anything revolutionary to Englishmen," it would be pointed out to a mathematical audience some few years later. "Existing institutions always have merits, which are as deep-seated as their defects are patent. . . . Our English way is to alter the defective institution a little bit at a time, so that it comes a little nearer to what we desire." Hardy's friend Littlewood, while no fan of the Tripos, was also less heated about it. "I do not claim to have suffered high-souled frustration," he wrote of his experience with it. "I took things as they came; the game we were playing came easily to me, and I even felt a satisfaction of a sort in successful craftsmanship."

Hardy's enmity, then, was something different, almost beyond reason. Plainly, his own experience influenced him. Back in 1896, the prospect of two years of Tripos tedium had nearly deflected him from mathematics altogether. But in the end, he had meekly surrendered; he had acquiesced to a coach, climbed on board the System, put "real" mathematics aside. When he did at last pit himself against the Tripos, he could almost be said to have "failed" it; for someone as competitive as he, that's what being Fourth Wrangler meant. The Tripos, in a sense, had beaten him.

Hardy's vehemence suggests a peculiar rift within his personality. Here was a man—a friend would one day liken him to "an acrobat perpetually testing himself for his next feat"—who set up rating scales at the least provocation, loved competitive games, grilled new acquaintances on what they knew, held up mathematical work to the highest standards—yet swore eternal enmity to the Tripos system which, in a sense, was the ultimate rating scale, the ultimate test.

In Hardy coexisted a stern, demanding streak with an indulgent liberal-mindedness, a formidable and forbidding exterior with a soft and fragile core. He would later claim to have scant interest in his less able students. But this, by all accounts, was nine-tenths bluster; he never failed any of them. "He simply *couldn't* think that way," Mary Cartwright, a former research student, told a friend, "because he was so kind to the weak ones."

Hardy disdained social niceties, ever kept his distance, arrogantly dismissed God. Yet he could be kind and endlessly obliging. Even the obituaries he wrote showed a largeness of spirit that, as someone once put it, "must have made every mathematician wish that he could have seen his own career described in the same generous terms."

So he was demanding, distant, emotionally astringent—*and* large-hearted, caring, and kind. It takes no straining of the facts to lay this split to the respective influences of his mother and father. But whatever their source, these two contrasting strands wound through his personality always. And both would emerge in his relationship with Ramanujan over the next seven fateful years.

* * *

Winter 1913. Europe stirred, armies marshaled. The world was restless with change. Picasso's first cubist drawings had appeared barely a year before. In Paris, Diaghilev, whose Russian Ballet had given its first London performance two years before, prepared for the premiere of Stravinsky's tempestuous *Le Sacre du Printemps,* in which a maiden dances herself to death. In England, George V was King, Edward having died suddenly in 1910. In 1911, the Parliament Bill had stripped the House of Lords of its veto power on acts of Commons. All through Britain, workers struck and militant suffragettes smashed windows. Ireland seethed.

But in Cambridge, things were as they always were. Hardy neared his thirty-sixth birthday with his face bearing scarcely a mark of it. He'd visit Bertrand Russell in Nevile's Court and discuss Bergson and the philosophy of religion; once, Norbert Wiener and his father met him there and took him to be an undergraduate. In 1912, Hardy published nine more papers, including his first collaborative one with Littlewood, "Some Problems of Diophantine Approximation." His first key paper on Fourier series was coming out later in 1913, the revised edition of his popular textbook the following year. Hardy's friend from the Apostles, Leonard Woolf, recently back from Ceylon, found Cambridge much as he'd left it; on the train back to London, he wrote, "I felt the warmth of a kind of reassurance. I had enjoyed my week-end. There was Cambridge and Lytton and Bertie Russell and Goldie, the Society and the Great Court of Trinity, and Hardy and bowls—all the eternal truths and values of my youth—going on just as I had left them seven years ago."

Hardy held to a regular routine. He read the London *Times* over breakfast, especially the cricket scores. He worked for four hours or so in the morning, then had a light lunch in Hall, perhaps played a little tennis in the afternoon. His career was well in place, his life comfortable, his future secure.

Then the letter came from India.

CHAPTER FIVE

"I Beg to Introduce Myself . . ."

[1913 to 1914]

1. THE LETTER

The letter, borne in a large envelope covered with Indian stamps, was dated "Madras, 16th January 1913," and began:

> Dear Sir,
>
> I beg to introduce myself to you as a clerk in the Accounts Department of the Port Trust Office at Madras on a salary of only £20 per annum. I am now about 23 years of age. I have had no University education but I have undergone the ordinary school course. After leaving school I have been employing the spare time at my disposal to work at Mathematics. I have not trodden through the conventional regular course which is followed in a University course, but I am striking out a new path for myself. I have made a special investigation of divergent series in general and the results I get are termed by the local mathematicians as "startling."

Some insignificant clerk in some backwater of an office five thousand miles away apparently sought to incite both pity *and* wonder. There was a nerviness about him: *I have not trodden through the conventional regular course.* By the second paragraph he was insisting he could give meaning to negative values of the gamma function. By the third he was disputing an assertion in a mathematical pamphlet Hardy had written three years before, part of a series called the Cambridge Tracts in Mathematics and Mathematical Physics.

It was called *Orders of Infinity: The 'Infinitarcalcul' of Paul Du Bois-Reymond*, and in it Hardy dealt with how mathematical functions can grow toward infinity more or less rapidly. For example, $f(x) = x^3$ approaches infinity faster than $g(x) = 3x$. Both functions, as x grows larger, grow without bound; both, it can be crudely said, "reach" infinity. But the first does so more quickly than the second. By the time $x = 100$, for example, the first function has exploded to 1,000,000, while the second is still mired at 300. At one point, Hardy had cited a familiar mathematical expression from the theory of prime numbers. This expression consisted of, first, a term involving logarithms and, second, an error term, $\rho(x)$, that simply represented how far wrong the first term was. On page 36, Hardy had asserted that "the precise order of $\rho(x)$ has not been determined."

Well, Ramanujan now wrote Hardy, it *had* been determined; *he* had determined it. "I have found an expression [for the number of prime numbers] which very nearly approximates to the real result, the error being negligible." He was saying that the prime number theorem, as it was known in the mathematical world, and as it had first been given form by Legendre and then more precisely by Gauss, was inadequate and incomplete, and that he, an unknown Indian clerk, had something better.

This was the hook with which Ramanujan set out to snare Hardy's attention. He concluded:

> I would request you to go through the enclosed papers. Being poor, if you are convinced that there is anything of value I would like to have my theorems published. I have not given the actual investigations nor the expressions that I get but I have indicated the lines on which I proceed. Being inexperienced I would very highly value any advice you give me. Requesting to be excused for the trouble I give you.
>
> I remain,
> Dear Sir,
> Yours truly,
> S. Ramanujan

This was not, of course, the end of the matter, but the beginning. The "enclosed papers" to which Ramanujan referred went on for nine pages (and also probably included a copy of his published paper on

Bernoulli numbers). The first page or so read like an inventor's patent claim, and with the same almost rhythmic ring of brash certainty:

> I have found a function which exactly represents the no. of prime nos. less than x, "exactly" in the sense that the difference between the function and the actual no. of primes is generally 0 or some small finite value even when x becomes infinite. I have got the function in the form of infinite series and have expressed it in two ways.
>
> . . .
>
> I have also got expressions to find the actual no. of prime nos. of the form $An + B$, which are less than any given number however large.
>
> . . .
>
> I have found out expressions for finding not only irregularly increasing functions but also irregular functions without increase (e.g. the no. of divisors of natural nos.) not merely the order but the exact form. The following are a few examples from my theorems. . . .

Now ordinary English virtually disappeared, giving way to the language of algebra, trigonometry, and calculus. There were theorems in number theory, theorems devoted to evaluating definite integrals, theorems on summing infinite series, theorems on transforming series and integrals, theorems offering intriguing approximations to series and integrals—perhaps fifty of them in all.

The whole letter, all ten pages of it, was written out in large, legible, rounded schoolboy script distinguished only by crossed *t*'s that didn't cross. His handwriting had always been neat; but here, if possible, it was neater still, as if he realized the gulf of skepticism that divided him from Hardy and dared not let an illegible scrawl widen it.

It was a wise precaution; the gulf was indeed great. For Hardy, Ramanujan's pages of theorems were like an alien forest whose trees were familiar enough to call trees, yet so strange they seemed to have come from another planet; it was the strangeness of Ramanujan's theorems that struck him first, not their brilliance. The Indian, he supposed, was just another crank. He was forever getting bizarre manuscripts from strangers that, as his friend Snow later put it, "pretended to prove the prophetic wisdom of the Great Pyramid, the revelations of the Elders of Zion, or the cryptograms that Bacon had inserted in the plays of the so-called Shakespeare."

And so, after a perfunctory glance, he put the manuscript aside and soon lost himself in the day's London *Times* which, in late January 1913, told of opium abuse in China, port hands in Lisbon gone on strike, the

French battling Arab rebels near Mogador, the House of Lords debating home rule for Ireland. . . . He may have skipped over the account of the Buxton divorce trial, where Mrs. Buxton was accused of adultery with Henry Arthur Mornington Wellesley, Lord Crowley. But he likely didn't miss the news of England's one-goal rugby victory over the French before twelve thousand spectators at Twickenham.

Around nine that morning, he set to work on mathematics, kept at it until about one, then ambled over to Hall for lunch. Then it was off to the university courts on Grange Road for a game of "real" tennis (which is what the English called the indoor variant that antedated the lawn tennis more popular today). But that day, in the corner of his mind normally left serene by vigorous athletics, something was wrong. The Indian manuscript scraped and tugged at his composure with, as Snow wrote, its "wild theorems. Theorems such as he had never seen before, nor imagined."

Were they wild and unimaginable because they were silly, or trivial, or just plain wrong, with nothing to support them? Or because they were the work of some rare flower of exotic genius?

Or maybe they were merely well-known theorems the Indian had found in some book and cleverly disguised by expressing in slightly different form—making it just a matter of time before Hardy found them out?

Or perhaps it was all a practical joke? Hoaxes, after all, were much in vogue just then. Many Englishmen holding high positions in the Indian Civil Service had endured the mathematical Tripos or were otherwise versed in mathematics—well versed enough, perhaps, to pull off such a stunt. And how best to dupe your old Cambridge friend Hardy? Why, you'd garb familiar "theorems" in unfamiliar attire, purposely twist them into weird shapes. But who in India was adept enough to do it? Maybe the hoax had originated in Europe. But would the perpetrator have gone to the trouble of securing a genuine Madras postmark . . . ?

Vagrant, fragmentary thoughts like these bubbled through Hardy's head as, returning from tennis, he walked back across one of the Cam bridges, then over the expanse of lawn that was the Backs, and through the gateway into New Court. Back in his second-floor suite of rooms, which were built over one of the gateways, he again sat down with the letter from India. Outside the Gothic mullioned windows of his room, the winter light began to fade.

Years later, most of the formulas in Ramanujan's letter would become the subjects of papers in the *Journal of the London Mathematical Society* and other mathematical journals. In them, their authors, including Hardy himself, would take two, or five, or ten pages to formally prove those not already known. But now, proving them wasn't Hardy's aim. Now he was content to see if there was anything to them at all. And even that was not apparent—in part because, as Hardy wrote later, "some curious specialization of a constant or a parameter made the real meaning of a formula difficult to grasp." Roughly speaking, it was as if, instead of stating the aphorism "penny-wise, pound-foolish," Ramanujan had for his own reasons expressed it as "two pennies wise, seven-and-a-half pounds foolish"—leaving the listener, distracted by the particulars, harder pressed to extract its meaning. In any case, it only compounded Hardy's perplexity.

Darkness fell. It was almost time for dinner. The formulas grew no more straightforward, the quality of the man who had written them no clearer. Genius or fraud? You couldn't idly riffle through these pages and tell. Yes, Hardy decided, Littlewood would have to see them, too.

John Edensor Littlewood was just two years older than Ramanujan, but while Ramanujan foundered in India, he had been mathematically schooled by England's best.

He came from old English yeoman stock; Littlewood archers, it was said, fought at the Battle of Agincourt in 1415. More recently, his ancestors had been robustly middle-class professionals—ministers, schoolmasters, publishers, doctors and the like. Both his grandfather and father had studied mathematics at Cambridge; his grandfather became a theologian, his father headmaster of a school in South Africa, where John lived for eight years. Back in England when he was fourteen, he attended St. Paul's School. There he caught the mathematics bug.

In his memoir, *A Mathematician's Miscellany*, Littlewood would assert, without artifice or conceit, that he was a "prodigy." Prodigy or not, he profited from an education that was everything Ramanujan's was not. About the time Ramanujan was studying S. L. Loney's *Trigonometry*, so was Littlewood. But whereas Ramanujan's formal exposure to mathematical ideas ended there, Littlewood's had just begun. Over the next three years he made his way through Macaulay's *Geometrical Conics*, Smith's *Analytical Conics*, Edwards's *Differential Calculus*, Williamson's *Integral Calculus*, Casey's *Sequel to Euclid*, Hobson's *Trigonometry*, Routh's *Dynamics of a Particle*, Murray's *Differential Equations*, Smith's *Solid Geometry*, Burnside and Panton's *Theory of Equations*, and more—all before so much as sitting for the Trinity College entrance scholarship exam in December 1902. By this time, Ramanujan had not even encountered Carr.

Two years later, in 1905, Littlewood was Senior Wrangler. But the fellowship normally his almost by right mysteriously went to someone else. For three years, he left to assume a lectureship at the University of Manchester. There, in "exile," he endured an oppressive load of lecturing, conferences, and paperwork—the normal lot of faculty members at provincial universities. After that, beginning in 1910, he was back at Trinity for good.

In the words of two of his biographers, he was "a rough-hewn earthy person with a charm of his own." He was strong, virile, vigorous. He had been a crack gymnast at school, had played cricket, would become an accomplished rock climber and skier and, even into his eighties, could be seen hiking through the East Anglian countryside around Cambridge. He wasn't especially tall, but a photograph of him lecturing in academic robes suggests an enormous, hulking masculinity. Another of him and Hardy together shows Littlewood dominating the picture, hat planted firmly atop his head, slope-shouldered, feet spread as if ready for a fight; the smaller Hardy seems to recede into the background.

Like Hardy, Littlewood remained a bachelor. But unlike Hardy, he thoroughly enjoyed the company of women. While at Manchester, he was the best dancer in his group. He would write of his grandmother that she was "a remarkable woman from an able family, but unfortunately very saintly." *He* was not. His long-term relationship with a married woman, with whose family he long shared a house in Cornwall, and with whom he had a daughter, would become well known in Cambridge.

As a mathematician, Littlewood was "the man most likely to storm and smash a really deep and formidable problem: there is no one else who can command such a combination of insight, technique, and power." So said Hardy himself. The two men first met, at least intellectually, in 1906 when an early Littlewood paper to the London Mathematical Society sparked disagreement as to its merits and it went to Hardy as referee. When Littlewood returned to Trinity in 1910, he worked closely enough with him for Hardy to acknowledge his help in the preface to *Orders of Infinity*, the book that had caught Ramanujan's eye in India.

Their actual collaboration began inauspiciously, when a proof they submitted to the London Mathematical Society in June 1911 turned out to be flawed. But then, the first of their more than one hundred papers appeared in 1912, and after that they were mathematically inseparable. "Nowadays," somebody said later, "there are only three really great English mathematicians: Hardy, Littlewood, and Hardy-Littlewood." Because Littlewood disdained bright, sparkling company and stayed away from mathematics conferences, some—at least in jest—doubted he existed at all. But exist he did, and in 1913, when Ramanujan's letter arrived, it was natural that Hardy thought to show it to him.

Littlewood had recently moved into rooms on D Staircase of Nevile's Court. Pausing in the arched doorway at the staircase's base, he could sight through the portico to the arches across the court framed within it. It was an arresting view—perhaps one reason he remained there for sixty-five years, until his death in 1977. From Hardy's rooms, it was just nineteen steps down the winding stone staircase, then forty paces through the gate into Nevile's Court and around to D Staircase. Yet normally the two men communicated by mail or college messenger and did not, in any case, routinely run off to confer with one another in person.

And so, that winter evening in 1913, to let Littlewood know he wished to meet with him after Hall, Hardy sent word by messenger.

About nine o'clock, as Snow reconstructed the day's events, they met, probably in Littlewood's rooms, and soon the manuscript lay stretched out before them. Some of the formulas were familiar while others, Hardy would write, "seemed scarcely possible to believe." Twenty years later, in a talk at Harvard University, he would invite his audience into the day that had so enriched his life. "I should like you to begin," he said, "by trying to reconstruct the immediate reactions of an ordinary professional mathematician who receives a letter like this from an unknown Hindu clerk." It was a mathematical audience, so Hardy introduced them to some of Ramanujan's theorems. Like this one, on the bottom of page three:

$$\int_0^\infty \frac{1+\left(\frac{x}{b+1}\right)^2}{1+\left(\frac{x}{a}\right)^2}\cdot\frac{1+\left(\frac{x}{b+2}\right)^2}{1+\left(\frac{x}{a+1}\right)^2}\cdots dx = \tfrac{1}{2}\pi^{\frac{1}{2}}\frac{\Gamma(a+\frac{1}{2})\,\Gamma(b+1)\,\Gamma(b-a+\frac{1}{2})}{\Gamma(a)\,\Gamma(b+\frac{1}{2})\,\Gamma(b-a+1)}$$

The elongated *S*-like symbol appearing on the left-hand side of this equation, and in many other equations all through the letter, was an integral sign, a notation originating with Newton's competitor Leibniz. An integral—the idea goes back to the Greeks—is essentially an addition, a sum, but one of a peculiar, precise, and, at first glance, infuriating kind.

Imagine cutting a hot dog into disclike slices. You could wind up with ten sections half an inch thick or a thousand paper-thin slices. But however thin you sliced it, you could, presumably, reassemble the pieces back into a hot dog. Integral calculus, as this branch of mathematics is called, adopts the strategy of taking an infinite number of infinitesimally thin slices and generating mathematical expressions for putting them back together again—for making them whole, or "integral." This powerful additive process can be used to determine the drag force buffeting a wing as it slices through the air, or the gravitational effects of the earth on a man-made satellite, or indeed to solve any problem where the object is to piece together the contributions of many small influences.

You don't need integral calculus to determine the area of a neat rectangular plot of farmland; you just multiply length times width. But you *could* use it. And you could use the same additive methods applicable to wings and satellites to calculate the area of an irregularly shaped plot where length-times-width *won't* work. Furnish the function that mathematically defines its shape, and in principle you can get its area by "integrating" it—that is, by performing the additive process in a particular, precisely defined way.

Calculus books come littered with hundreds of ways to integrate functions. And yet, pick a function at random and chances are it can't be integrated—at least not straightforwardly. With "definite integrals" like those Ramanujan offered in his letter to Hardy, however, you're offered a back-door route to a solution.

A definite integral is "definite" in that you seek to integrate the function over a definite numerical range; the little numbers at top and bottom of the elongated *S*—the ∞ and 0 in Ramanujan's equation—tell what it is. (In other words, you mark off a piece of the farm plot whose area you want reckoned.) When you evaluate a definite integral, you don't wind up with a general algebraic formula (as you do with indefinite integrals) but, in principle, an actual number. And sometimes, by applying the right mathematical tools, you can determine this number without integrating the function first—indeed, without being *able* to integrate it at all.

Broadly, this was what Ramanujan was doing in the theorem on page 3 of his letter to Hardy and all through the section labeled "IV. Theorems on Integrals."

This particular integral, he was saying, could be represented in terms of gamma functions. (The gamma function is like the more familiar "factorial"—4!, read "four factorial," = 4 × 3 × 2 × 1—except that it extends the idea to numbers other than integers.) Hardy figured he could prove this theorem. Later he tried, and succeeded, though it proved harder than he thought. None of Ramanujan's other integrals were trifling exercises, either, and all would wind up, years later, the object of papers devoted to them. Still, Hardy judged, these were among the *least* impressive of Ramanujan's results.

More so were the infinite series, two of which were:

$$1-5\left(\frac{1}{2}\right)^3+9\left(\frac{1\cdot 3}{2\cdot 4}\right)^3-13\left(\frac{1\cdot 3\cdot 5}{2\cdot 4\cdot 6}\right)^3+\ldots=\frac{2}{\pi}$$

and

$$1+9\left(\frac{1}{4}\right)^4+17\left(\frac{1\cdot 5}{4\cdot 8}\right)^4+25\left(\frac{1\cdot 5\cdot 9}{4\cdot 8\cdot 12}\right)^4+\ldots=\frac{2^{3/2}}{\pi^{1/2}\{\Gamma(\frac{3}{4})\}^2}$$

The first wasn't new to Hardy, who recognized it as going back to a mathematician named Bauer. The second seemed little different. To a layman, in fact, it and kindred ones in Ramanujan's letter might seem scarcely intimidating at all; save for pi and the gamma function they were nothing but ordinary numbers. But Hardy and others would show how these series were derived from a class of functions called hypergeometric series first explored by Leonhard Euler and Carl Friedrich Gauss and as algebraically formidable as anybody could want.

Sometime before 1910, Hardy learned later, Ramanujan had come up with a general formula, later to be known as the Dougall-Ramanujan Identity, which under the right conditions could be made to fairly spew out infinite series. Just as an ordinary beer can is made in a huge factory, the ordinary numbers in Ramanujan's series were the deceptively simple end product of complex mathematical machinery. Of course, on the day he got Ramanujan's letter, Hardy knew nothing of this. He knew only that these series formulas weren't what they seemed. Compared to the integrals, they struck him as "much more intriguing, and it soon became obvious that Ramanujan must possess much more general theorems and was keeping a great deal up his sleeve."

Some theorems in Ramanujan's letter, of course, did look comfortably familiar. For example,

If $\alpha\beta = \pi^2$, then

$$\alpha^{-1/4}\left(1 + 4\alpha \int_0^\infty \frac{xe^{-\alpha x^2}}{e^{2\pi x} - 1}\,dx\right) = \beta^{-1/4}\left(1 + 4\beta \int_0^\infty \frac{xe^{-\beta x^2}}{e^{2\pi x} - 1}\,dx\right)$$

Hardy had proved theorems like it, had even offered a similar one as a mathematical question in the *Education Times* fourteen years before. Some of Ramanujan's formulas actually went back to the days of Laplace and Jacobi a century before. Of course, it was quite something that this Indian had rediscovered them.

But now, then, what was Hardy to make of this one, which he found on the last page of Ramanujan's letter?

$$\text{If } u = \frac{x}{1+}\,\frac{x^5}{1+}\,\frac{x^{10}}{1+}\,\frac{x^{15}}{1+\ldots},\quad v = \frac{x^{1/5}}{1+}\,\frac{x}{1+}\,\frac{x^2}{1+}\,\frac{x^3}{1+\ldots}$$

$$\text{then } v^5 = u\,\frac{1 - 2u + 4u^2 - 3u^3 + u^4}{1 + 3u + 4u^2 + 2u^3 + u^4}$$

This was a relationship between continued fractions, in which the compressed notation for, say, the function u actually means this:

$$u = \cfrac{x}{1 + \cfrac{x^5}{1 + \cfrac{x^{10}}{1 + \cfrac{x^{15}}{1 + \ldots}}}}$$

The publication of this result some years hence would set off a flurry of work by English mathematicians. Rogers would furnish one ten-page proof for it in 1921. Darling would explore it, too. In 1929, Watson would approach it from a different angle, trying to steer clear of the tricky mathematical terrain of theta functions. But in 1913, Hardy could make nothing of it, classing it among a group of Ramanujan's theorems which, he would write, "defeated me completely; I had never seen anything in the least like them before. A single look at them is enough to show that they could only be written down by a mathematician of the highest class." And then, in a classic Hardy flourish, he added: "They must be true because, if they were not true, no one would have the imagination to invent them."

As Hardy and Littlewood probed the theorems before them, trying to make out what they said, where they fit into the mathematical canon, and how they might be proved or disproved, they began to reach a judgment. That the Indian's mathematics was strange and individual had been evident from the start. But now they were coming to see his work as something more. It was not "individual" in the way a rebellious teenager tries to be, camouflaging his ordinariness behind bizarre dress or hair. It was much more. "There is always more in one of Ramanujan's formulae than meets the eye, as anyone who sets to work to verify those which look the easiest will soon discover," Hardy would write later. "In some the interest lies very deep, in others comparatively near the surface; but there is not one which is not curious and entertaining."

The more they looked, the more dazzled they became. "Of the theorems sent without demonstration, by this clerk of whom we had never heard," one of their Trinity colleagues, E. H. Neville, would later write, "not one could have been set in the most advanced mathematical examination in the world." Hardy would rank Ramanujan's letter as "certainly the most remarkable I have ever received," its author "a mathematician of the highest quality, a man of altogether exceptional originality and power."

And so, before midnight, Hardy and Littlewood began to appreciate that for the past three hours they had been rummaging through the papers of a mathematical genius.

It wasn't the first time a letter had launched the career of a famous mathematician. Indeed, as the mathematician Louis J. Mordell would later insist, "It is really an easy matter for anyone who has done brilliant mathematical work to bring himself to the attention of the mathematical world, no matter how obscure or unknown he is or how insignificant a position he occupies. All he need do is to send an account of his results to a leading authority," as Jacobi had in writing Legendre on elliptic functions, or as Hermite had in writing Jacobi on number theory.

And yet, if Mordell was right—if "it is really an easy matter"—why had Gauss spurned Abel? Carl Friedrich Gauss was the premier mathematician of his time, and, perhaps, of all time. The Norwegian Niels Henrik Abel, just twenty-two at the time he wrote Gauss, had proved that some equations of the fifth degree (like $x^5 + 3x^4 + \ldots = 0$) could never be solved algebraically. That was a real coup, especially since leading mathematicians had for years sought a general solution that, Abel now showed, didn't exist. Yet when he sent his proof to Gauss, the man history records as "the Prince of Mathematics" tossed it aside without reading it. "Here," one account has him saying, dismissing Abel's paper as the work of a crank, "is another of those monstrosities."

Then, too, if "it is really an easy matter," why had Ramanujan's brilliance failed to cast an equal spell on Baker and Hobson, the other two Cambridge mathematicians to whom he had written?

Certainly Henry Frederick Baker, forty-eight at the time he heard from Ramanujan, qualified as the kind of "leading authority" Mordell had in mind. He held a special Cayley lectureship. He had been elected a Fellow of the Royal Society, at the age of thirty-two, in 1898. He had received the Sylvester Medal in 1910. He had been president of the London Mathematical Society until the year before.

But as one biographer noted after his death, Baker "was little affected by the revolution brought about amongst the Cambridge mathematical analysts by G. H. Hardy in the first decade of the twentieth century. During this period Baker's position was essentially that of one of the leaders of the older generation." That he was immune to Hardy, of course, did not itself explain his indifference to Ramanujan; it did, however, suggest some reticence about embracing new ideas. Indeed, Baker was said to so revere the great mathematicians of the past that it choked his own originality. His upcoming second marriage, which took place in 1913, may also have left him less open to the importunings of an unknown Indian clerk.

The other Cambridge mathematician, a Senior Wrangler, was E. W. Hobson, who was in his late fifties when he heard from Ramanujan and more eminent even than Baker. His high forehead, prominent mustache, and striking eyes helped make him, in Hardy's words, "a distinguished and conspicuous figure" around Cambridge.

But he was remembered, too, as a dull lecturer, and after he died his most important book was described in words like "systematic," "exhaustive," and "comprehensive," never in language suggesting great imagination or flair. "An old stick-in-the-mud," someone once called him. For some years, he was a Tripos coach (one student was John Maynard Keynes), largely ignoring mathematical research. He would take a conventional stand on the coming war, and vehemently opposed granting degrees to women. These were sensibilities, then, hardly primed for unfamiliar theorems coming from an unorthodox source.

Of course, Ramanujan's fate had always hung on a knife edge, and it had never taken more than the slightest want of imagination, the briefest hesitancy, to tip the balance against him. Only the most stubborn persistence on the part of his friend Rajagopalachari had gained him the sympathy of Ramachandra Rao. And Hardy himself was put off by Ramanujan's letter before he was won over by it. The cards are stacked, against any original mind, and perhaps properly so. After all, many who claim the mantle of "new and original" are indeed new, and original—but not better. So, in a sense, it should be neither surprising nor reason for any but the mildest rebuke that Hobson and Baker said no.

Nor should it be surprising that no one in India had made much of Ramanujan's work. Hardy was perhaps England's premier mathematician, the beneficiary of the finest education, in touch with the latest mathematical thought *and*, to boot, an expert in several fields Ramanujan plowed. . . . And yet a day with Ramanujan's theorems had left him bewildered. *I had never seen anything in the least like them before.* Like the Indians, Hardy did not know what to make of Ramanujan's work. Like them, he doubted his own judgment of it. Indeed, it is not just that he discerned genius in Ramanujan that redounds to his credit today; it is that he battered down his own wall of skepticism to do so.

That Ramanujan was Indian probably didn't taint him in Hardy's eyes. True, Hardy's knowledge of India may have been as mired in imperial stereotypes as that of most other Englishmen; when he was ten, back in Cranleigh, the school magazine had featured a day at "An Indian Bazaar," rife with bejeweled maidens, dagger-bearing Ghurkas, and filthy fakirs cursing "English dogs" under their breaths. But by 1913, Hardy had already made mathematical contact with several Indians. A professor of mathematics at Allahabad, Umes Chandra Ghosh, had, in 1899, given the solution to one of his earliest questions in the *Educational Times*. And in 1908, another of Hardy's questions, on infinite series, had drawn a response by V. Ramaswami Iyer, who two years before had founded the Indian Mathematical Society and two years later would befriend Ramanujan. (Then, too, the Indian cricket sensation Ranjitsinjhi had been in his prime when Hardy was an undergraduate, perhaps also helping to overturn any lingering prejudices.)

Growing up along Horseshoe Lane and attending Cranleigh School may have made Hardy readier than other Cambridge dons to see merit dressed in exotic garb. All his life, certainly, he was sympathetic to the underdog. Mary Cartwright, who met him a few years after his Ramanujan period, recalled that, as a woman mathematician, "I was a depressed class"—and so enjoyed Hardy's favor. Snow wrote that Hardy preferred the downtrodden of all types "to the people whom he called the *large bottomed*: the description was more psychological than physiological. . . . [They] were the confident, booming, imperialist bourgeois English. The designation included most bishops, headmasters, judges, and . . . politicians." When Hardy attended a cricket match and knew none of the competitors, he would tap his own favorites on the spot. These "had to be the under-privileged, young men from obscure schools, Indians, the unlucky and diffident. He wished for their success and, alternatively, for the downfall of their opposites."

Further tipping the balance in Ramanujan's favor was Hardy's willingness to stray from safe, familiar paths. Hobson and Baker were both from an earlier generation, more settled, perhaps at a time in their lives when they were less eager to take on something new. Hardy, on the other hand, was a generation younger and had a penchant for the unorthodox and the unexpected. He had left familiar Cranleigh for Winchester. He had allowed a sixth-rate novelist, "Alan St. Aubyn," to deflect him toward Trinity. He had weighed leaving mathematics; then, he had embraced Professor Love's suggestion that he dip into Camille Jordan's *Cours d'analyse*. He had joined the Apostles. He had broken precedent to take the Tripos after his second year instead of his third.

Each time Hardy had opened himself, he had come away enriched. Now, something wildly new and alien had presented itself to him in the form of a long, mathematics-dense letter from India. Once again, he opened his heart and mind to it. Once again, he would be the better for it.

2. "I HAVE FOUND IN YOU A FRIEND . . ."

"No one who was in the mathematical circles in Cambridge at that time can forget the sensation" caused by [Ramanujan's] letter, wrote E. H. Neville years later. Hardy showed it to everyone, sent parts of it to experts in particular fields. (Midst all the excitement, Ramanujan's original cover letter, along with one page of formulas, got lost.) Meanwhile, Hardy had sprung into action, advising the India Office in London of his interest in Ramanujan and of his wish to bring him to Cambridge.

It was not until a windy Saturday, February eighth, the day following his birthday, that Hardy sat down to deliver to Ramanujan the verdict on his gifts that Cambridge already knew. "Trinity College, Cambridge," he wrote at the top, and the date, then began: "Dear Sir, I was exceedingly interested by your letter and by the theorems . . ." With an opening like that, Ramanujan would, at least, have to read on.

But in the very next sentence, Hardy threw out his first caveat: "You will however understand that before I can judge properly of the value of what you have done, it is essential that I should see proofs of some of your assertions."

Proof. It wasn't the first time the word had come up in Ramanujan's mathematical life. But it had never before borne such weight and eminence. Carr's *Synopsis*, Ramanujan's model for presenting mathematical results, had set out no proofs, at least none more involved than a word or two in outline. That had been enough for Carr, and enough for Ramanujan. Now, Hardy was saying, it was *not* enough. The mere assertion of a result, however true it might seem to be, did not suffice. And all through his letter to Ramanujan he would sound the same insistent theme:

> I want particularly to see your proofs of your assertions here. You will understand that, in this theory, *everything* depends on rigorous exactitude of proof.

And again:

> assuming your proofs to be rigorous . . .

And:

> Of course in all these questions everything depends on *absolute* rigour.

On the whole, Hardy's letter was lavish with encouragement. True, some of Ramanujan's theorems were already well known, or were simple extensions of known theorems. But even these, Hardy allowed, represented an achievement. "I need not say that if what you say about your lack of training is to be taken literally, the fact that you should have rediscovered such interesting results is all to your credit."

Then, too, Hardy hazarded, some of Ramanujan's results, while themselves of little note, were perhaps examples of general methods and thus more important than they seemed. "You always state your results in such particular forms that it is difficult to be sure about this."

Littlewood was also intrigued by Ramanujan's work, Hardy mentioned, even adding as a sort of appendix, "Further Notes Suggested by Mr. Littlewood." Most of these dealt with Ramanujan's work on prime numbers, a subject in which Littlewood had recently made a stunning, if not yet published, advance. So Hardy's message from Littlewood carried an especially fervent plea: "Please send the *formula* for the no. of primes & . . ."—here it was again—"as much proof as possible quickly."

Hardy's whole letter was like that—shot through with urgency, with a barely contained excitement, that Ramanujan would have been dull indeed not to sense. At the bottom of page 6, Hardy wrote, "I hope very much that you will send me *as quickly as possible*"—he underlined it with a veritable slash across the page—"a few of your proofs, and follow this more at your leisure by a more detailed account of your work on primes and divergent series."

And he went on: "It seems to me quite likely that you have done a good deal of work worth publication; and if you can produce satisfactory demonstration, I should be very glad to do what I can to secure it."

Hardy's letter probably arrived late in the third week of February. But his endorsement of Ramanujan had reached Madras earlier. Almost a week before writing Ramanujan, Hardy had contacted the India Office and, by February 3, a certain Mr. Mallet had already written Arthur Davies, secretary to the Advisory Committee for Indian Students in Madras. Later in the month, Davies met with Ramanujan and, at Sir Francis Spring's behest, Narayana Iyer, apprising him of Hardy's wish that Ramanujan come to Cambridge.

But as Hardy soon learned, Ramanujan wasn't coming. Religious scruples, or a cultural resistance that verged on it, got in the way; Brahmins and other observant Hindus were enjoined not to cross the seas. And that, it seemed for a long time, was that.

Meanwhile, in Madras, the balance delicately poised since Ramanujan's meeting with Ramachandra Rao in late 1910, teetering inconclusively between success and failure, now came down firmly on Ramanujan's side. All that had been wanting was for a mathematician of unimpeachable credentials to weigh in with a verdict. Now Hardy had delivered it.

On February 25, Gilbert Walker was shown Ramanujan's work. Walker was a former Senior Wrangler, a former fellow and mathematical lecturer at Trinity, and was now, at the age of forty-five, head of the Indian Meteorological Department in Simla. At the time of his appointment, he had no meteorological background whatever. But so crucial was predicting the onset of the monsoon in India, and so lively the press furor in the wake of several years of bad predictions, that it was thought a professional mathematician of Walker's standing—he had just been named a Fellow of the Royal Society—might help defuse the situation.

Now, as Walker was passing through Madras, Sir Francis prevailed on him to look through Ramanujan's notebooks. The next day, Walker wrote Madras University, asking it to support Ramanujan as a research student. "The character of the work that I saw," he wrote the university registrar,

> impressed me as comparable in originality with that of a Mathematics fellow in a Cambridge College; it appears to lack, however, as might be expected in the circumstances, the completeness and precision necessary before the universal validity of the results could be accepted. I have not specialised in the branches of pure mathematics at which he has worked, and could not therefore form a reliable estimate of his abilities, which might be of an order to bring him a European reputation. But it was perfectly clear to me that the University would be justified in enabling S. Ramanujan for a few years at least to spend the whole of his time on mathematics without any anxiety as to his livelihood.

Walker, his letter as much as said, was as mystified by Ramanujan's work as everyone else was. Heaven knows, he was no pure mathematician. As a young man, he'd shown interest in gyroscopes and electromagnetics. Early prominence had come from his studies of aerodynamic forces on the boomerang, which as an undergraduate he liked to throw on the Cambridge Backs. Now, as India's chief weatherman, his most recent paper was entitled "The Cold Weather Storms of Northern India." In other words, he was a mathematician of applied bent whose work lay as far distant from Ramanujan's as it was possible to be. His statement, *I have not specialised* . . . was gross understatement, his admission that he *could not form a reliable estimate of his abilities* the plain, simple truth.

And yet none of this discouraged him from eagerly recommending Ramanujan for a special scholarship. Because now, in any appraisal of Ramanujan, there was a new factor to consider: Hardy. Spring, who had introduced Walker to Ramanujan's work, and everyone else at the Port Trust knew that Ramanujan had Hardy's imprimatur. If Walker harbored any doubt as to Ramanujan's merits, Hardy's verdict erased it. The wheels of Ramanujan's career, for ten years barely creaking along, now, greased by Hardy's approval, began to whirr and whine like a finely tuned race car engine.

If Ramanujan had doubts about what Hardy's endorsement might mean to him, they had been dispelled by the last two days and Walker's ringing endorsement. On February twenty-seventh, he wrote Hardy a second letter, again packed with theorems. "I am very much gratified on perusing your letter of 8th February 1913," he wrote. "I have found a friend in you who views my labours sympathetically."

In fact, it was not just Walker, Spring, and others in the Madras mathematical community who had been fortified by Hardy's letter. It was Ramanujan himself. For all his confidence in his mathematical prowess, Ramanujan needed outside approval, affirmation. Now he had it. Hardy's letter *took him seriously*. And the pronouncement delivered by this unseen F.R.S., this man reputed to be the finest pure mathematician in England? It was no vague, empty one filled with glowing accolades that Ramanujan might, in an anxious moment, dismiss, but rather nine pages of specific, richly detailed comment: a statement Ramanujan had written on his sixth page of theorems about a series expressible in terms of pi and the Eulerian constant could be deduced from a theorem in Bromwich's *Infinite Series* (the book M. J. M. Hill, in his letter of two months before, had advised that he consult). A theorem on the same page involving hyperbolic cosines Hardy himself had proved in *Quarterly Journal of Mathematics*. Hardy *knew*.

That Ramanujan had every bit the "invincible originality" with which Hardy would later credit him didn't mean he didn't care what others thought of him. He did care. In assessing Ramanujan's work, Hardy had informally broken it down into three broad categories—those results already known or easily derived from known theorems; those curious, and perhaps even difficult, but not terribly important; and those promising to be important indeed, if they could be proved. In Hardy's mind, plainly, it was those in this third category that weighed most heavily.

But not in Ramanujan's.

"What I want at this stage," he wrote Hardy, "is for eminent professors like you to recognize that there is some worth in me." And such "worth," he felt, accrued not alone, or even primarily, through the theorems Hardy saw as novel and important, but those he'd ranked in the first category, as already known. It was these, he wrote, "which encourage me now to proceed onward. For my results are verified to be true even though I may take my stand upon slender basis." Before the world whose judgment mattered so much to him, he could stand tall and declare, *I have been pronounced by competent authority to be just as I said I was*—which was more, for example, than his repeated flunking out of school had seemed to say.

In his second long letter to Hardy, Ramanujan seemed buoyant, pumped up, cocky. Hardy wanted proof? Well, he wrote

> If I had given you my methods of proof I am sure you will follow the London Professor [Hill]. But as a fact, I did not give him any proof but made some assertions as the following under my new theory. I told him that the sum of an infinite no. of terms of the series:- $1 + 2 + 3 + 4 + \ldots = -1/12$ under my theory. If I tell you this you will at once point out to me the lunatic asylum as my goal. I dilate on this simply to convince you that you will not be able to follow my methods of proof if I indicate the lines on which I proceed in a single letter. You may ask how you can accept results based upon wrong premises. What I tell you is this: Verify the results I give and if they agree with your results, got by treading on the groove in which the present day mathematicians move, you should at least grant that there may be some truths in my fundamental basis.

Got by treading on the groove! Ramanujan was flying high. Four years of hawking his mathematical wares had left him neither shy about going after what he needed nor above stooping to self-pity: "I am already a half-starving man," he wrote Hardy now. "To preserve my brains I want food and this is now my first consideration. Any sympathetic letter from you will be helpful to me here to get a scholarship either from the University or from Government."

In this vein Ramanujan continued for two pages, then proceeded to pile up more theorems, expanding on the ideas about prime numbers on which Hardy had challenged him, and going on to new work—in all, nine more theorem-stuffed pages. "I have also given meanings to the fractional and negative no. of terms in a series as well as in a product," he wrote, "and I have got theorems to calculate such values exactly and approximately. Many wonderful results have been got from such theorems. . . ."

Later, many would see in Ramanujan an appealing and genuine humility. But here were hints of another Ramanujan, one already dreaming of a place for himself in mathematical history: "You may judge me hard that I am silent on the methods of proof," he wrote. But his silence was due only to lack of space in which to set them down, not unwillingness to do so. No, he wrote, "I do not mean that the methods should be buried with me."

Humble? E. H. Neville would later describe Ramanujan as "perfect in manners, simple in manner, resigned in trouble and unspoilt by renown, grateful to a fault and devoted beyond measure to his friends." Nowhere, though, did he call him humble, or suggest that Ramanujan shrank from a sanguine assessment of his own gifts. Nor does his later remark to Janaki that, in her words, "his name would live for one hundred years," suggest undue humility. There had been a stubborn, self-confident streak in Ramanujan all along; for him to work, unrecognized and alone for years, fortified only by his delight in the work itself, it *had* to have been there. And here, in his second letter to Hardy, it unabashedly surfaced.

Hardy, we may be sure, was not put off by any of this. "Good work," he once wrote, "is not done by 'humble' men."

On March 13, spurred by Walker's letter, B. Hanumantha Rao, professor of mathematics at the engineering college, invited Narayana Iyer to a meeting of the Board of Studies in Mathematics to discuss "what we can do for S. Ramanujan. . . . You have seen some of his results & can help us to understand them better than the author himself." On the nineteenth, the board met and recommended to the syndicate, the university's governing body, that Ramanujan receive a research scholarship of seventy-five rupees a month—more than twice his Port Trust salary—for the next two years.

But when the syndicate met on April 7, Ramanujan's case encountered a setback. Such scholarships were reserved for those with master's degrees, and Ramanujan lacked even a bachelor's; why, he'd flunked out of every college he'd ever attended. By one Indian account, it was the English—led by Richard Littlehailes, an Oxford-educated professor of mathematics at Presidency College and later portrayed as one of Ramanujan's champions—who invoked this technicality and, "with all their vehement speeches," lined up against him. In any case, the syndicate's vice-chancellor, P. R. Sundaram Iyer, chief justice of the Madras High Court, then rose. Did not the preamble of the act establishing the university, he asked, specify that one of its functions was to *promote research*? And, whatever the lapses of Ramanujan's education, was he not a proven quantity as a mathematical researcher?

That argument won the day. "The regulations of the University do not at present provide for such a special scholarship," the registrar later wrote. "But the Syndicate assumes that Section XV of the Act of Incorporation and Section 3 of the Indian Universities Act, 1904, allow of the grant of such a scholarship, subject to the express consent of the Governor of Fort St. George in Council."

It was a measure of how far, six weeks after receipt of Hardy's letter, Madras opinion had swung behind Ramanujan: now, the authorities were stretching the rules to accommodate him.

By April 12, Ramanujan had learned the good news. The scholarship set him free to do mathematics, to attend lectures at the university, to use its library. So that now, as Neville would write, he "entered the Presidency College in Madras to practice as a virtue that singleminded devotion to mathematics which had been a vice in Kumbakonam nine years earlier."

3. "DOES RAMANUJAN KNOW POLISH?"

It was a great open space that, as you came out of the alleys leading to it, offered a broad blue sweep of sky, a little as the piazza of an Italian hill town does. But it was not a piazza, or a square, or a park. It was a "tank," a sort of religious swimming pool—an expanse of water, normally almost square, with a sandy bottom, granite steps leading down to the water on several sides, and a man-made island of richly detailed religious sculpture at its center.

This one was large, perhaps a hundred yards across, and notably handsome. Its legendary predecessor gave the whole district its name—Triplicane, a corruption of Tiru Alli Keni, meaning "sacred lily tank." Adjacent to it stood the ancient Parthasarathy Temple, a Vaishnavite shrine whose principal deity was Krishna as the "sarathi," or charioteer, in the great battle of the *Bhagavad Gita*, and which still bore the inscription of a Pallava king's gift of land in A.D. 792. Here Muslim, Dutch, and French soldiers had encamped over the centuries. Here the faithful would bathe on ritual occasions—sometimes a few at a time, occasionally hundreds or thousands of them, closely packed together, a sea of bare-chested brown men.

Off one street bordering the tank to the south ran a little lane, Hanumantharayan Koil Street. And a few doors down, almost where the street turned abruptly left, was a little house, set back a courtyard's depth from the street. Here, about a mile and a half from Presidency College, Ramanujan and his family now lived. With research scholarship in hand and on leave from the Port Trust, they no longer had to stay in congested Georgetown, and probably around May had moved back to Triplicane.

Ramanujan had nothing to do now but pursue mathematics and, every three months, submit a progress report. For this he received seventy-five rupees a month. Back in Kumbakonam, the five or ten rupees the family got monthly from boarders made for a sizable chunk of their income. Even at the Port Trust, he'd still only made twenty-five or thirty rupees, possibly as much as fifty at one point. And what with at least Janaki, Komalatammal, and Komalatammal's own mother to provide for, that didn't go far; sometimes he'd had to moonlight, tutoring college students on the side. Now, though, he was almost flush, almost at the hundred rupees a month, for example, that the Indian Mathematical Society set as its threshold for paying full dues. Ramanujan had friends in high places now, was published in mathematical journals, was corresponding with one of the West's top mathematicians.

During the early mornings, and then again at night, he'd work with Narayana Iyer, no longer boss now, but colleague. Sometimes he'd borrow math books from K. B. Madhava, a statistician who lived down the street. Often, he could be found in the alcoves of the Connemara Library, a wing of which also housed the university collection. The Connemara, named for a former governor of Madras, was like a secular church, a soaring chamber of arching stained-glass windows, filled with wood engravings and columns and arches rich with molded plaster ornamentation, an Indo-Saracenic temple of learning. Here Ramanujan would lose himself on some of those deliciously free days after his scholarship had set him free.

It was a little like the period before his marriage. More completely than ever he was able to throw himself into mathematics, leaving day-to-day cares to others. Sometimes, Janaki would recall, "he would ask his mother or grandmother to wake him up after midnight so that he could go on with his work in the silent and cooler hours of the after-night." Sometimes he "had to be reminded about his food. On some occasions his grandmother or mother would serve, in his hand, food made of cooked rice mixed with sambhar, *rasam,* and curd successively. This they did so that his current of thought might not be broken." At other times, she or his mother would prepare brinjal in just the way he liked. Normally this teardrop-shaped vegetable, about the size of a peach, was eaten cooked. But he had learned to love it the special way his mother made it for him—quartered, then quartered again, but with a tab of flesh keeping all eight segments together, the junction steeped in tamarind and *masala* for an hour or so, then eaten raw.

Ramanujan now had his own workroom upstairs. He and Janaki usually slept at separate times, but it seemed that "whenever I opened my eyes," as she would recall, "he would be working," the scratch of stylus on slate sounding through the house. Once, on the spur of the moment, he rigged up a little science experiment for her; from a jug of water and some kind of tubing, he made a siphon and showed her how gravity drew the water to lower points. But otherwise, contact between them was slight. Janaki was barely fourteen. He worked in the heady realms of pure mathematics, whereas she, beyond simple Tamil, had no education at all. Occasionally, when he took a break, he would ask her to later jog his memory with language like, "the one you were working on downstairs," or "the one you were working on before eating." But intellectually, they could share nothing. Nor did he try to force-feed her. She didn't ask him to, and he didn't volunteer.

Ramanujan and Hardy, meanwhile, danced around one another uncertainly. In their correspondence, Hardy was defending the mathematical revolution he had wrought in England, trying to wrest rigorous proofs from Ramanujan. Ramanujan held back, offering excuses. At least that was how his stance could be read, and by mid-March the situation verged on a real row. "How maddening his letter is in the circumstances," Littlewood wrote Hardy. "I rather suspect he's afraid that you'll steal his work." Taking up the issue in his next letter, Hardy took pains to reassure him:

> Let me put the matter quite plainly to you. You have in your possession now 3 [Ramanujan had only two] long letters of mine, in which I speak quite plainly about what you have proved or claim to be able to prove. I have shown your letters to Mr. Littlewood, Dr. Barnes, Mr. Berry, and other mathematicians. Surely it is obvious that, if I were to attempt to make any illegitimate use of your results, nothing would be easier for you than to expose me. You will, I am sure, excuse my stating the case with such bluntness. I should not do so if I were not genuinely anxious to see what can be done to give you a better chance of making the best of your obvious mathematical gifts.

Whether Hardy and Littlewood had misread Ramanujan's letter or not, Ramanujan now denied misgivings and even managed, rather successfully, to seem hurt. "I am a little pained to see what you have written at the suggestion of Mr. Littlewood," he wrote in mid-April.

> I am not in the least apprehensive of my method being utilised by others. On the contrary my method has been in my possession for the last eight

years and I have not found anyone to appreciate the method. As I wrote in my last letter I have found a sympathetic friend in you and I am willing to place unreservedly in your possession what little I have.

Ramanujan the special research student, Ramanujan the friend of Cambridge was now a hot topic in Madrasi circles, and people were forever trooping through his house, as if to do him homage. In August, his name came up at a get-together of professors and their students; everyone marveled how Ramanujan had garnered such intellectual standing "without," as one there that evening put it, "the help of books or teachers."

In September, Narayana Iyer submitted some theorems on the summation of series to the *Journal of the Indian Mathematical Society*, at one point adding: "The following theorem is due to Mr. S. Ramanujan, the Mathematics Research Student of the Madras University."

On October 26, perhaps out to win over one who had before campaigned against Ramanujan's scholarship, Narayana Iyer took Ramanujan to see Richard Littlehailes, professor of mathematics at Presidency College and soon to become Madras's director of public instruction. Ramanujan was never good at explaining his own methods, so Narayana Iyer did the talking. Littlehailes assured them that, after the first of December, he looked forward to studying Ramanujan's results.

In November, the mathematician E. B. Ross of Madras Christian College, whom Ramanujan had met a few years before, stormed into class, his eyes glowing. "Does Ramanujan know Polish?!?" he asked his students. Ramanujan didn't, of course. Yet his most recent quarterly report had anticipated the work of a Polish mathematician whose paper had just arrived by the day's mail.

As a research scholarship holder, Ramanujan's only obligation was to prepare reports every three months detailing his progress. He delivered three of them in late 1913 and early 1914, all dutifully on time. Like much of his work, the theorems he described had their roots in his notebooks; some went back to notes appearing around page 180 of his first notebook, some to chapters 3 and 4 of the second. Most of them dealt with evaluating definite integrals.

"At present," he wrote in his first report, addressed to the Board of Studies in Mathematics and dated August 5, 1913, "there are many definite integrals the values of which we know to be finite but still not possible of evaluation by the present known methods." The theorem he offered—it would later be called "Ramanujan's Master Theorem"—

would provide means of evaluating many of them. "This paper," his cover letter noted, "may be considered the first installment of the results I have got out of the theorem." More, he promised, were coming soon—as indeed they were in the second and third reports.

As in his letters to Hardy earlier that year, Ramanujan was attacking definite integrals that resisted every effort to reduce them to simpler, more useful forms, defeated the whole arsenal of mathematical tools brought to bear on them. Ramanujan was fashioning new tools.

Like a screwdriver, saw, or lathe, a mathematical "tool" is supposed to *do* something; those used to evaluate a definite integral perform mathematical operations on it that, one hopes, get it ready for the next tool—the next theorem or technique—and ultimately lead to a solution. But just as a screwdriver tightens screws but can't saw wood, a mathematical tool may work for evaluating one integral but not others. If you don't know in advance, you try it. If it doesn't work, you try something else.

One tool, which Ramanujan had apparently encountered in an 1896 textbook on integral calculus, was Frullani's integral theorem. Now, in late 1913, Ramanujan was telling of a powerful generalization of it that could defeat a wider range of formerly unyielding integrals. To apply, Frullani's theorem demanded that two particular functions be equal; in Ramanujan's generalization, they didn't have to be, thus expanding its applicability to many more cases. Back in 1902, Hardy had written a paper on the Frullani integral. But he had never seen in it what Ramanujan saw now.

While this time Ramanujan furnished proofs for many of his assertions, more analytical mathematicians would later shoot them full of holes. Yet the results themselves—the theorems Ramanujan offered as true—*were* true. Bruce Berndt, an American Ramanujan scholar, would see in that curious split a message for mathematicians today: "We might allow our thoughts to occasionally escape from the chains of rigor," he advised, "and, in their freedom, to discover new pathways through the forest."

4. A DREAM AT NAMAKKAL

While life was sweet to Ramanujan during this period, his long-distance correspondence with Hardy had soured. In the cover letter to his first quarterly report, in August, he cited Hardy much as the author of a book might quote favorable review comment: "The integral treated in Ex. (v) note Art. 5 in the paper, Mr. G. H. Hardy, M.S., F.R.S., of Trinity

College, Cambridge, considers to be 'new and interesting.' " But by then, in fact, the two of them were scarcely writing at all. Perhaps it was the tartness in their earlier exchanges. Or Hardy's disappointment at Ramanujan's refusal to come to England, and frustration at communicating across so formidable a physical and cultural gap. Or simply the press of other work. In any case, while Ramanujan wrote him at least once during this period, Hardy for months failed to respond.

Finally, though, in early January, Ramanujan found a long letter from Hardy, responding to a proof he had earlier supplied, pointing out its flaws, and showing how they had led him astray. "You will see that, with all these gaps in the proof, it is no wonder that the result is wrong." But, Hardy went on, walking on eggs, "I hope you will not be discouraged by my criticisms. I think your argument a very remarkable and ingenious one. To have proved what you claimed to have proved would have been about the most remarkable mathematical feat in the whole history of mathematics."

And oh yes, there was one more thing. "Try," he added almost carelessly, "to make the acquaintance of Mr. E. H. Neville, who is now in Madras lecturing. He comes from my college and you might find his advice as to reading and study invaluable."

This was not exactly a lie, but it wasn't the whole truth either. Hardy had more in mind than getting Ramanujan the right books to read. He had made Neville the instrument of his plan. He had deputized him to bring Ramanujan to England.

Of course, that had been his intent from the beginning. As Snow put it, "Once Hardy was determined, no human agency could have stopped Ramanujan" from coming. Even before writing Ramanujan, he had contacted the India Office to that effect. But then, the answer had come back from Madras that, for religious reasons, Ramanujan had no intention of coming.

The way Ramanujan told the story about a year later in a letter to Hardy, he had gotten a letter in February 1913 from Arthur Davies, secretary to the Advisory Committee for Indian Students in Madras. Could Ramanujan meet him in his office the following noon? Sir Francis asked Ramanujan's "superior officer"—this, certainly, was Narayana Iyer—to accompany him and answer any questions.

The next day, Davies dropped the big question: Was Ramanujan prepared to go to England?

Ramanujan's mind raced. What did the offer mean? Go to England and study math? Go to England and take examinations, and doubtless fail them? Would it mean . . . ?

But before he could speak, Narayana Iyer—whom Ramanujan depicted as "a very orthodox Brahmin having scruples to go to foreign land"—replied that no, of course Ramanujan could not go to England. Then, according to Ramanujan, "the matter was dropped."

That, at least, was Ramanujan's later version of these events. But almost certainly, it was a face-saving story concocted to explain his past unwillingness to go. What blocked his going was, indeed, Brahminic "scruples to go to foreign land." But it was not Narayana Iyer's scruples, but those of Ramanujan's friends, and his family, and himself.

"His people shunned him for having lost his caste," it was recorded of an early eighteenth-century South Indian who had traveled to Europe in defiance of his caste. "He was practically a dead man in their estimation." While abroad, the man had been granted numerous honors, had been named Chevalier of the Order of St. Michael. But back in India, none of it counted; he was despised. When, generations later, another Indian, T. Ramakrishna, asked his mother whether he might travel to England to study, the story was invoked in warning. Ramakrishna waited for his mother to die before defying her wishes and making the trip.

For an orthodox Hindu—and Ramanujan came from a very orthodox Hindu family—traveling to Europe or America represented a form of pollution. It was in the same category as publicly discarding the sacred thread, eating beef, or marrying a widow. And, traditionally, it had the same outcome—exclusion from caste. That meant your friends and relatives would not have you to their homes. You could find no bride or bridegroom for your child. Your married daughter couldn't visit you without herself risking excommunication. Sometimes, you couldn't go into temples. You couldn't even get the help of a fellow casteman for the funeral of a family member. Here was the grim, day-to-day meaning of the word *outcaste*.

A quarter century before Ramanujan, Gandhi had met similar obstacles in going to England for his education. "Will you disregard the orders of the caste?" its *Sheth*, or headman, had asked him. He did—and was pronounced an outcaste. By Ramanujan's time, scruples against foreign travel had relaxed but slightly; for all but the most adventurous, it was still taboo.

However much he may have yearned for England, Ramanujan found in Madras little reprieve from tradition's hold. Madras was no oil-fed Houston, or railroad-driven Chicago, where ambitions surged and dreams were meant to be lived. Its population was static, barely changing in the past decade. It was flat, built low, spread out over the countryside; there were no Himalayan peaks to draw the imagination upward, no great towers to serve as symbols of human aspiration. More than the other great cities of India, Madras clung in spirit to the villages and towns of the surrounding countryside. Calcutta and Bombay roiled over with unmarried workers from all over India, bore a sort of rude, masculine, Wild West dynamism. But Madras was more socially cohesive, with men from the villages bringing their wives and children with them. Like Ramanujan, the one in sixteen Madrasis who were Brahmins mostly inhabited the area around Triplicane's Parthasarathy Temple. All around Ramanujan they lived—orthodox Brahmins who, like his mother, buttressed tradition and encouraged conformity. All around him the forces of social order were securely in place.

And Ramanujan was no rebel. If he wavered in his acceptance of the bar on foreign travel, he was not about to say so—and was certainly not, on his own, going to defy it. For him to go to England, strong outside forces would have to be brought to bear. External voices he respected would have to sanction it.

Around New Year's Day in 1914, Hardy's man, Eric Harold Neville, arrived in Madras.

An able mathematician who had never done "anything which has aroused any real enthusiasm": that's how Hardy assessed Neville a few years later. "I should *expect* [his work] to be good, but he is not a man like Littlewood of whom one would say it *must* be good." That said almost as much about Hardy as it did Neville, of course, for Neville was already an important young mathematician. Just twenty-five, he had been among the last to take the old-style Tripos—sitting for it a year early in order to have a shot at becoming the last Senior Wrangler before Hardy's reforms took hold. (He came in second.) Two years later, in 1911, Neville won the Smith's Prize, and a year after that was named a Fellow of Trinity College. Now, in the winter of 1913, he'd come to Madras to give a series of lectures on differential geometry at the university.

He had, of course, one additional charge—to convince Ramanujan to come to England.

It was in the Senate House, the university's examination hall and offices, located across the road from Marina Beach, that in early January the two men met. On the outside, the Senate House was an extravagant architectural blend of Italianate, Byzantine, and Indo-Saracenic influences, executed in brick and stone, shot through with rose windows, minarets, parapets, and chiseled stone pillars. On the inside, it was little more than a vast void—a great hall built to seat sixteen hundred people, with only a carved, fifty-four-foot-high ceiling and stained-glass windows to recall its ornate exterior. Here Neville had come to deliver the twenty-one lectures that, over the next month, would draw mathematicians from all over South India.

After one of the first lectures, Ramanujan was introduced to him. This was not "the uncouth, unshaven, unclean figure of Ramachandra Rao's picture," Neville would write later, "but a man at once diffident and eager." As for his English, once so poor it had undermined his school career, "ten years had worked wonders, for a more fluent speaker or one with a wider and better used vocabulary I have seldom met."

At least three times they sat down together with Ramanujan's notebook. Neville was stunned—so much so that when, after their third meeting, Ramanujan asked him whether he might like to take the notebook away to peruse at his leisure, it struck him as "the most astounding compliment ever paid to me. The priceless volume had never before [so Neville assumed] been out of his hands: no Indian could understand it, no Englishman could be trusted with it."

Perhaps, Neville came to think, Ramanujan mistrusted the Englishmen he'd met in Madras. Perhaps he mistrusted Hardy, with whom he had corresponded only at a continent's remove. But he, Neville, a fresh wind from afar, bore none of the old baggage. A pure mathematician, naturally sympathetic to Ramanujan's mathematics, within a year or so of him in age, he had lavished three days on the notebooks that were Ramanujan's life's work. He had gained Ramanujan's trust.

And so now he struck, acting on Hardy's charge to him: would Ramanujan come to Cambridge? Anticipating a negative response, Neville silently marshaled his arguments. Yet now, unaccountably, he didn't need them. "To my delight and surprise," he wrote later, he learned "that Ramanujan needed no converting and that his parents' opposition had been withdrawn."

* * *

What miracle had wrought this transformation?

By one account it was K. Narasimha Iyengar, a family friend with whom Ramanujan had stayed in Madras early in 1911 and with whom he had remained in touch since, who helped get Ramanujan's mother, certainly a key obstacle, to acquiesce to the trip. Seshu Iyer also exerted pressure on her, Ramaswami Iyer and Ramachandra Rao on Ramanujan. "I lent all the weight of my influence to induce him to go," recalled Ramachandra Rao later. So did M. T. Narayana Iyengar, the Bangalore mathematician and editor of the *Journal of the Indian Mathematical Society* who had worked closely with Ramanujan three years before to get his first paper ready for publication; scrupulously orthodox himself, his arguments carried added weight.

But if these influences may be said to have ultimately triumphed, they did not by themselves change Ramanujan's mind. Something more was needed—something, at least for public consumption, beyond his mother's merely human will, or Ramanujan's. Neville learned what it was: "In a vivid dream his mother had seen [Ramanujan] surrounded by Europeans and heard the goddess Namagiri commanding her to stand no longer between her son and the fulfillment of his life's purpose." Details differ, but this and other versions of the story agree in substance—that permission for Ramanujan to go came personally through the intervention of the goddess Namagiri, residing in her shrine at Namakkal.

In the vicinity of Namakkal, eighty miles west of Kumbakonam, hills abruptly push up from the otherwise flat plain. Palm and banana trees can still be seen. So can rice fields. But overall, the vegetation is markedly thinner than along the Cauvery. Mud and brick huts have given way to stone, not much seen to the east, chiseled out of the surrounding hills and from rocky outcroppings jutting out from the fields.

One of those outcroppings, more impressive by far than the rest, gave Namakkal its name. A town of seven thousand people, about twenty miles from the nearest train station (in Karur, near where Ramanujan's mother and grandparents originally came from), Namakkal lay at the foot of a great white rock, two hundred feet straight up and half a mile around. Throughout the rock were tiny sacred grottos, *jonais*, formed over the eons where rock-rooted greenery and rainfall had conspired to wear away fissures. At the top, reached by narrow steps hewn into its southwest slope, stood an old fort protected by brick battlements three feet thick. To someone, the vertical gash that the great rock made in the sky suggested the vertical white streaks of the Vaishnavite caste mark, known as the *namam*. Hence, Namakkal.

It was for here that in late December of 1913, Ramanujan, his mother, Narayana Iyer, and Narayana Iyer's son set out; Janaki had asked to go, too, but Ramanujan told her she was too young. They got off the train in Salem, a city of seventy thousand set in a picturesque valley rimmed by mountains. There they stayed at the house of Ramaswami Iyer, who was deputy collector as well as founder of the Indian Mathematical Society. From there, Ramanujan wrote back home to Kumbakonam, then set out alone with Narayana Iyer, probably in a bullock cart, for Namakkal, about thirty miles due south.

In Namakkal, they took a little road, flanked with houses built up to the edge of it, that gently climbed and curved up from the middle of town. Soon, near the base of the great rock, they reached the stone-columned facade and giant wooden doors that guarded the temple of Lord Narasimha, the lion-faced, fourth incarnation of Vishnu—and, in a separate, smaller, pillared shrine off to the left, his consort, the goddess Namagiri.

Ramanujan's family was not the only one upon whom the goddess exerted a powerful hold. On certain days of the week, zealous, frenzied women would wend their way to the shrine to be exorcised of devils. "The hall in front of the goddess," according to one account, would be "filled with their shrieks and convulsions, until a sprinkling of sacred water over their heads by the pujaris [priests] silences them." Within the cramped passage where the sweating priests performed their devotions, the air was full of smoke, the stone walls black with incense.

For three nights, Narayana Iyer and Ramanujan slept on the temple grounds. Sitting there on the stone slab floor, they could look up at the sheer rock face that formed the back wall of the temple, see the scalloped battlements of the old fort silhouetted against the sky. The first two nights, nothing happened. But on the third, Ramanujan rose from a dream and woke Narayana Iyer with word that, in a flash of brilliant light or some such similar revelation, he had received the *adesh*, or command, to bypass the injunction against foreign travel.

Narayana Iyer's family today believes that, with shrewd insight into Ramanujan's psyche and mindful of his devotion to the goddess Namagiri, he conceived the trip to Namakkal; that so strong was Ramanujan's wish to go to England and so strong his devotion to Namagiri, that something like what happened *had* to happen; and that when it did,

Narayana Iyer calculated, he would be there to "correctly" interpret any revelation Ramanujan might have.

Did a vision of Namagiri actually come to Ramanujan, in all earnestness, while he slept on the columned temple grounds at Namakkal? Or, as Neville heard it, did his mother have the dream, her importunings to obey its message firmly settling matters? Or did both mother and son dream of Namagiri? Or did Ramanujan now more ardently than ever wish to go to England and, having deferred to his family for most of a year, seek a socially acceptable way to do it?

Certainly Ramanujan always *attributed* his decision to divine inspiration. (So did Joan of Arc attribute hers: "I locked myself up in the attic for a day and a night and God told me that I was to become a General of Heaven and lead the armies of France and expel the English.") Known, too, is that in February 1913, so far as the India Office was concerned, Ramanujan refused to travel to England; that in December, he visited Namakkal; that a week or two later he met Neville at the Senate House and surprised him with his willingness to embark for England; that on January 22, he wrote Hardy, asking him and Littlewood to "be good enough to take the trouble of getting me there—within," he scribbled between the lines, "a very few months."

In his letter, Ramanujan sought to distance himself from what, in Cambridge's view, had been a year's obdurate refusal to go. "Now I learn from your letter and Mr. Neville that you are anxious to get me to Cambridge," Ramanujan wrote, as if hearing of it for the first time. "If you had written to me previously I would have expressed my thoughts plainly to you." Meeting with Davies back in February, he suggested, he'd been a helpless pawn under the sway of his "superior officer." Why, he had not even supplied the language of his earlier letters; that, too, had been the work of his superior. As for his own religious scruples, or that of his family, or the trip to Namakkal, Ramanujan said nothing.

The whole scheme may have been the work of Narayana Iyer, or at least carried out with his willing complicity. In any case, it would not be the only time Ramanujan bent the truth to avert embarrassment. Visiting Kumbakonam early in 1914, just before leaving for England, he told the family of his friends Anantharaman and Subramanian that he had come to say good-bye. He was leaving, he announced—for Calcutta. Under the circumstances, the lie may have been well-advised. When their father learned the truth, and fearing that Ramanujan's example might lure his sons to England, he went to Madras to try to dissuade Ramanujan from going.

The key obstacle removed, Neville set about addressing Ramanujan's other doubts. Money to get to England and live there? Don't worry, Neville assured him, that would be taken care of. His English was not very good? It was, Neville said, good enough. His vegetarianism? That would be respected. And examinations? Having flunked virtually every one he'd ever taken in college, he was pained at the prospect of taking any more—yet knew that doing so was the inevitable lot of Indian students in Europe. No, Neville reassured him, *he* would not have to take any.

In opting for England, Ramanujan still went against a large body of opinion. His father-in-law, for example, wondered why he could not pursue mathematics in India. His mother worried that his finicky health might suffer in the English cold; that remaining vegetarian, without good Indian food available, would be difficult; that he might encounter outright prejudice from the locals—or, on the other hand, that he'd be beset by English girls. (Before he left, she was upset when some English women, come to meet the dark genius bound for Cambridge, actually shook his hand.)

Some of Ramanujan's friends, meanwhile, saw the trip, in Neville's words, as "a mean attempt to transfer to the English university the glory that belonged to Madras." Neville, taking no chances, set about wooing them, too. "Lest he should be harassed by attempts to dissuade him, I addressed myself to the task of convincing his Indian friends that the proposal was in Ramanujan's own interest, and in fact embodied the only chance of placing him on the pinnacle where they longed to see him."

Next, Neville wrote Hardy to say that it was now time to address the financial obstacles to Ramanujan's visit to England. He, Neville, would try to find money in Madras. But should he fail, as he later paraphrased his letter, "the money must somehow be found in England. . . . Financial difficulties must not be allowed to interfere."

Hardy apparently forwarded Neville's pronouncement to the India Office, because he soon received from C. Mallet, secretary for Indian students (who a year before had relayed word of Ramanujan's refusal to go to England), a worried reply. "Mr. Neville's letter rather alarmed me, because it seemed to me that he was encouraging Ramanujan to come to England without any real prospect of providing for him when he got here." Too often, he had found, Indian students arrived without enough money, only to meet with "disappointment and misery."

In the bluntest terms, Mallet advised Hardy that "no money for this purpose can be got from the India Office." Furthermore, he doubted whether Trinity or Cambridge would come up with any, and he didn't think Madras could either. He was not sanguine. And he infected Hardy with his pessimism. "I'm writing in a hurry to catch the mail," Hardy wrote Neville, a bit frantically, "and warn you to be a little careful"; the money *had* to be there, else Ramanujan couldn't come. He and Littlewood might together contribute fifty pounds a year for the contemplated two years of Ramanujan's visit—"Don't tell [Ramanujan] so"—but that came to only about a fifth of Ramanujan's needs.

Perhaps influenced by correspondence with Hardy or the India Office that does not survive, Neville would later cast Hardy's concern as a case of intellectual, not financial, cold feet, as jitters about Ramanujan's real abilities. "We have heard of these unknown geniuses before," he paraphrased (or misconstrued, or misrepresented) the India Office letter Hardy had forwarded. "They dazzle their friends in India, and when we bring them to England we see them for the precocious schoolboys they are; in a few weeks they fizzle out, and more harm than good has been done by our benevolence." Neville would laugh at the timidity he imputed to Hardy for, presumably, endorsing such doubts. Of course, he added, "I had seen the note-books and talked with Ramanujan, and Hardy had not."

In any case, by the time he'd heard from Hardy, Neville already had money matters well in hand. Littlehailes had introduced him to people influential in the university or government, and everywhere he talked up Ramanujan. "The discovery of the genius of S. Ramanujan of Madras," he'd written Francis Dewsbury, registrar of the university, on January 28, "promises to be the most interesting event of our time in the mathematical world." It was a thoughtful, rather grandly stated letter, all aimed at precisely one end—funding Ramanujan's stay in England. "I see no reason to doubt," it concluded,

> that Ramanujan himself will respond fully to the stimulus which contact with Western mathematicians of the highest class will afford him. In that case, his name will become one of the greatest in the history of mathematics, and the University and City of Madras will be proud to have assisted in his passage from obscurity to fame.

Next day, Littlehailes himself took up the attack, formally asking Dewsbury for a 250-pound-per-year scholarship, coupled with a 100-pound grant to equip Ramanujan with Western clothes and book passage to England. "Ramanujan," he wrote, "is a man of most remarkable mathematical ability, amounting I might say to genius, whose light is metaphorically hidden under a bushel in Madras."

The following week Lord Pentland himself, governor of Madras, became the target of this bombardment of Madras officialdom on Ramanujan's behalf. Sir Francis wrote Pentland's private secretary, C. B. Cotterell:

> I am anxious to interest him in a matter which I presume will come before him within the next few days—a matter which under the circumstances is, I believe, very urgent. It relates to the affairs of a clerk of my office named S. Ramanujan, who, as I think His Excellency has already heard from me, is pronounced by very high mathematical authorities to be a Mathematician of a new and high, if not transcendental, order of genius.

Spring had just learned that the university was prepared to set aside ten thousand rupees, equivalent to more than six hundred pounds, or enough for two years in England. But the decision hinged on higher approval. And here, he said, "His Excellency may perhaps be able to interfere with advantage."

"The best gentleman and by no means the worst brain we ever sent to India," it was once said of Lord Pentland. Born John Sinclair, he was a slim, slight man with a full mustache who had graduated from Sandhurst, the British West Point, and served eight years in the Royal Irish Lancers. Then he'd turned to politics, most recently serving as secretary for Scotland under Prime Minister David Lloyd George. Only the previous October he'd come to Madras as governor of Fort St. George, in which capacity he ruled over forty million people. He believed that the state's function, according to one who knew him, should be "to secure for all its members the best procurable conditions for the full development of personality." Now, in the case of Ramanujan, he had the chance to act on it.

He had already gone to bat for Ramanujan once when, the year before, he had consented to his special research scholarship. Now he was ready to "interfere with advantage" again. "His Excellency cordially sympathizes with your desire that the University should provide Ramanujan with the means of continuing his researches at Cambridge," his secretary wrote back to Spring, "and will be glad to do what he can to assist."

The scholarship was approved. The last roadblock was gone. Ramanujan was going to England.

5. AT THE DOCK

On February 26, Binny & Co. sent Ramanujan his second-class ticket.

On March 11, Sir Francis wrote the steamer agents to make sure he got vegetarian food en route.

On March 14, Ramanujan accompanied his wife and mother to Madras's Egmore Station. There, in its compact waiting room, flanked by two rows of columned arches, they awaited the train. Ten years before, he had arrived here to begin his studies at Pachaiyappa's College. Now, he wept: he was dispatching his family to Kumbakonam, so they would not have to witness his painful transformation into a European gentleman.

Janaki, one day while her mother-in-law was at the temple, had asked him to take her with him to England. But influenced by Ramachandra Rao, among others, Ramanujan said no, explaining to her that if he had to tend to her in England he would not be able to concentrate on mathematics; that, besides, she was so young and pretty he'd have only to turn his back and the Englishmen would be upon her. . . .

Nobody who saw Ramanujan during those last busy days recalled seeing in him any exhilaration, anticipation, or joy. "He was not very jubilant over his future journey," Ramachandra Rao would recall. "He seemed to [move] as if . . . obeying a call."

His friends coached him in Western ways. Still taking a proprietary interest in his young protégé, Ramachandra Rao decreed that his *kutumi*, the long bunched-up knot of hair at the back of his head, had to go. And it was done. Further, Ramanujan must wear Western clothes. Soon Richard Littlehailes was driving him around town on his motorcycle, Ramanujan in the sidecar, shopping for collars and ties and stockings and shoes and shirts.

For a few days Ramanujan stayed in the country, at the house of a friend of Ramachandra Rao's who lived European-style, learning how to use knife and fork—though "under the strict stipulation," as his patron observed, "that nothing but vegetable food should be served." But even this tentative step into alien ways left Ramanujan unhappy. "He did not relish food being served by strange servants."

Ramanujan worried about how he would stay vegetarian in England.

He hated his Western haircut. He was miserable about the clothes he had to wear. The day before he was to leave, he walked into the faculty common room at Presidency College with a big suitcase, opened it up, and laid out on the table the Western clothes purchased for him. How, he pleaded, was he to wear them? Making the knot on his tie confounded him, to everyone's amusement. Ramanujan tried to joke about it, but his old Kumbakonam friend, Raghunathan, now on the staff of the college, thought he looked distinctly uncomfortable. Later, when his mother got a photo of him from England, all squeezed into collar, tie, and jacket, she wouldn't recognize him.

That night, K. Narasimha Iyengar and his cousin had Ramanujan over to their place in Triplicane. Outside, the streets were filled with carts drawn by bullocks with bells jingling from painted horns, with bare-chested men in dhotis, with women in saris, their nose-rings and bangles gleaming against their dark skin. The pungent smell of burning cow dung filled the air. To Ramanujan, it was all he'd ever known. But England? What was to come? All night his friends stayed up with him, tried to calm his jitters, prepare him for the great adventure.

On March 15, the British India Lines ship S. S. *Nevasa* had arrived in Madras through the new northern entrance of the harbor, built during Sir Francis's tenure. Most ships still tied up within the protected breakwaters of the harbor, then had their cargoes transported by lighter to the docks. But a special wharf to bring in passengers, troops, and horses had been built at the south end of the harbor, with a timber-decked dock, like a little boardwalk, projecting into the harbor from the breakwater. It was here that the *Nevasa* tied up.

The *Nevasa* was brand-new, designed expressly for the India run. Her hull was painted black, except for red accents and a thin ribbon of white running fore and aft. She was a smart-looking vessel, graceful in her way, her single funnel, about midships, set with a modest rake.

The morning of its departure, an official send-off was held in Ramanujan's honor, organized by Srinivasa Iyengar, the advocate general. On hand were Professor Middlemast and Sir Francis Spring, prominent judges, and Kasturirangar Iyengar, publisher of the *Hindu*. So was Narayana Iyer who had worked so closely with Ramanujan, the incessant clicking and scraping on their slates keeping people in his house up all night. "My father made a strange request to him," his son N. Subbanarayanan would record many years later. "As a memento my father wanted to exchange his slate with Ramanujan's slate, [a request that] was granted. Perhaps my father thought that he may get an inspiration from the slate during [Ramanujan's] absence."

Ramanujan was introduced around. Madras's director of public instruction, J. H. Stone, wished him success, told him he had written friends in England who would take care of him. Among the passengers, Ramanujan met a man in the Salvation Army bound for Southampton, and a Dr. Muthu, a tuberculosis specialist. He met the captain, too, who joked that they'd get along fine so long as Ramanujan didn't bother him with mathematics.

For most everyone, it was a time of good cheer and light banter. But not for Ramanujan who, recalled a friend, "was in tears."

Finally, there was nothing left to do. Ramanujan was on board, his well-wishers left behind. At about ten o'clock on the morning of March 17, 1914, the *Nevasa* slipped slowly away from the dock.

CHAPTER SIX

Ramanujan's Spring

[1914 to 1916]

1. OUT OF INDIA

The *Nevasa*'s paint still gleamed. Barely a year had elapsed since she'd been delivered to her owners by a Glasgow shipyard and set out on her maiden voyage to the East. At nine thousand tons, she was by far the largest ship in the British India Lines fleet. Her four wide decks were airy and comfortable. She had been designed expressly for service in the tropics. Yet Ramanujan could find no comfort aboard her, no relief from the pitching seas that left him, on his first ocean voyage, seasick and unable to eat.

Temporary relief came with the ship's first stop in Colombo, capital of Ceylon (today's Sri Lanka), the large island just off India's southeast coast. Long before the *Nevasa* docked, passengers could smell Colombo's cinnamon gardens. Blue hills rose above the harbor and the houses of the city, with their red roofs and walls of shimmering white.

On March 19, the *Nevasa* steamed out of port, skirted south of Cape Comorin, at the tip of the subcontinent, and made direct for Aden, a week's passage across the Arabian Sea. For Ramanujan, now past his seasickness, the voyage grew pleasant. Now he could enjoy the ship's roominess. He had his vegetarian food. He'd met several among his two hundred or so fellow passengers. Sometimes, the story goes, he withdrew to his second-class cabin and played with numbers suggested by the dimensions of the cabin or the number of passengers.

Ramanujan was not an introspective man—was not, as Hardy would put it, "particularly interested in his own history or psychology." But now, in the coat and collar that tortured him, midst a limitless expanse of sea, as day by day the *Nevasa* steamed west at fourteen knots and fresh sea breezes swept across the deck, it would have been hard for half-formed thoughts not to spill over into consciousness.

Five years ago, he was alone on the *pial* of his house in Kumbakonam—unknown, unmarried, boy more than man; now, the Madrasi elite had turned out to see him off to England.

Then, he was a dropout from Government College, Kumbakonam; now, he was bound for Trinity College, Cambridge.

Back then, getting to Madras by train, a matter of about three rupees, was no trifling expense; now, to the four-hundred-rupee fare to England aboard this great steamship he need give no thought.

Then, the thought of violating the bar on overseas travel would scarcely have crossed his mind; now, influential men, orthodox Brahmins among them, had *urged* him to break it.

What had wrought such changes? Plainly, he had not overnight become a better mathematician. What *had* changed was that he had thrust himself onto the world. His mother ordaining that he marry, he had set out from door to mathematician's door in search of livelihood. Then, with much persistence and a little luck, one thing had led to another. . . .

And the future—is it conceivable he never thought of it? Probably, he felt a vague fear of the unknown. Possibly, his disquiet took more concrete form. Until now, even in that impetuous flight to Vizagapatnam, he had never left South India, where fair-skinned Englishmen were the one-in-a-thousand exception; soon, in England, *he* would be the conspicuous one, *his* face would stand out, *his* accent would seem alien. The future meant Neville, and Hardy, and Littlewood, and Cambridge, and Trinity. But only Neville was a face, a person; Hardy was just a few letters. The rest were names, disembodied abstractions, mysteries.

After a week of steady steaming, more than two thousand miles out of Colombo, the *Nevasa* put in at Aden, at the southernmost end of the Red Sea. Now began the part of the journey that many among the English passengers, at least, had learned to dread—the fourteen-hundred-mile passage up the Red Sea, where day after day the ship baked in a desert heat that never deviated much from a hundred degrees. The passage gave the language a new word: if, outwardbound to India, you got stuck in a starboard cabin, you absorbed the full heat of the sun all afternoon and by bedtime your cabin smoldered; portside cabins, meanwhile, had the whole afternoon to cool down. On the homeward passage, it was all reversed. Not surprisingly, VIPs managed to land the cooler cabins, which were designated "Port Outward—Starboard Homeward" and granted the acronym POSH.

Whether bored by the long voyage or inspired by the prospect of soon breaking into the Mediterranean, Ramanujan grew expansive. From Suez, at the entrance to the canal, and Port Said at its other end, exactly two weeks out of Madras, he posted at least four letters back to India. One, to Viswanatha Sastri, bore stamps showing the pyramids. One went to R. Krishna Rao, nephew of Ramachandra Rao. "I do not know whether I have to go to Cambridge directly or stay at London and then go," he wrote after telling of the voyage thus far. "I shall write to you after I reach England and everything is settled. My best compliments to your brother and respects and warmest thanks to your uncle."

The next day, the *Nevasa* sailed from Port Said and steamed into the Mediterranean. On April 7, after a stop in Genoa, from which Ramanujan posted another letter home, she left Marseilles.

Then it was through the Strait of Gibraltar and up along the Spanish coast through the Bay of Biscay to England. The *Nevasa* docked first at Plymouth, then steamed up the English Channel and arrived at the mouth of the Thames on April 14.

It was a bright, lovely day, a little warmer than usual, without so much as a trace of overcast—more of the run of fine weather that, on Easter Sunday, two days before, had brought Londoners out to the parks and streets of the city by the hundreds of thousands. Now, waiting for Ramanujan at the dock was Neville and his older brother, who had arrived there by car. They drove to 21 Cromwell Road, in the South Kensington district of London, a reception center for Indian students just arrived in England.

London was a city of five million people, spilling over its ancient borders into the villages and hamlets of Surrey and Middlesex. In population, it was ten times larger than Madras. Madras was the capital of South India? Then London was the capital of the world, the nerve center from which the empire was directed. The clopping of horses' hooves, the jingle of harnesses, and the clatter of hansom cabs over cobblestoned streets could still be heard in London, but these had begun to give way to the roar and smoke of Studebaker Cabriolets, of Wolsley "Torpedo Phaetons," of double-decker buses bearing signs advertising Nestlé's chocolate. London was *fast*. Even Hardy's friend Leonard Woolf, who had returned to London in 1911 after seven years in Ceylon, found London aquiver, marching to "a tempo clearly faster and noisier than what I was accustomed to."

Back in Madras, the Englishmen Ramanujan had known were mostly educated and upper-class, with accents to match, never seen to stoop to manual labor. Now, on the streets of London, he heard the nasal Cockney twang of rag merchants. He saw lamplighters who patrolled the streets at dusk with their long poles. He saw knife grinders manning little two-wheeled carts, men selling muffins who heralded their wares by ringing a bell. Here, there was every sort of Englishman—men in bowlers and flat workingmen's caps, women in finery and in rags.

Just off the boat and England was already a strange new world for any Indian. Cromwell Road, to which Neville now took Ramanujan, was supposed to ease the transition. Of course it didn't. The National India Association had offices there. Several rooms in the stately Georgian-styled corner building were available to students passing through. And across the street, the imposing edifice of the Natural History Museum, fairly glowing in its two-toned stone and adorned with sculpted griffins, lent an appropriately imperial luster to the immigrant's first days in England. But Cromwell Road typically failed in its mission; a study a few years later would chide the reception center for invariably making things worse for newcomers, not better.

But unlike most Indians in England, Ramanujan had by his side, in Neville, a Cambridge don to help smooth the way. He also met A. S. Ramalingam, a twenty-three-year-old engineer from Cuddalore, south of Madras, who had been in England for four years and who also tried to help him feel at home. In any case, Ramanujan survived whatever rigors Cromwell Road could inflict and on April 18 went with Neville to Cambridge. Soon he was settled in Neville's house on Chesterton Road, in a little suburb of Cambridge just across the River Cam from the town itself.

Chesterton Road was a street of fine townhouses, some turreted and cupolaed, their front yards typically set off from the street by wrought-iron fences. The bay-windowed Neville house sat one in from the end of the block on a gentle arc of street beside the river, occupying a peculiarly shaped plot just sixteen feet wide on the street side but fanning out to more than fifty in the back. Neville and his new wife, Alice, had moved in the year before and now, for two months in the spring of 1914, it was Ramanujan's introduction to the English home.

It had been a modestly scaled two-story affair when first built around the middle of the previous century. Early on, though, it had been enlarged. A long, fingerlike projection, built of the same tan brick, jutted out the back of the house and followed the oddly angled property line. An added third story, with three more rooms, gave the house a height that made it like an exclamation point to the little group of houses at the end of the block.

In any case, the Neville house was by now quite spacious, and wherever they settled him in the sprawling place, Ramanujan was bound to have a measure of privacy he had never enjoyed before. The back of the house faced a large garden, which had once been a pear orchard. From the second-floor sitting room, Ramanujan could look out over the River Cam and Victoria Bridge and at the broad expanse of Jesus Green with its crisscross of pedestrian paths leading to the old stone courts and cloisters of the colleges.

There was, of course, business to attend to—fees to pay, paperwork to work through. Hardy and Neville took care of most of it. A printed list of first-year students, prepared after Michaelmas Term (pronounced *Mih*-kel-miss, and starting in mid-October) of the previous year, listed each student alphabetically. Now, squeezing it in between Pugh, F. H. and Rawlins, J. D., someone dipped his pen in black ink and wrote in Ramanujan's name by hand.

Those early weeks were rich with new promise, graced by a wondrous spring. Day after day the weather was lovely and warm. May flowers bloomed in April. Tracts of open countryside were transformed into great seas of bluebells. At the end of the month, King George visited Cambridge, where he was greeted by thousands of schoolchildren, waving tiny Union Jacks, trying to gain a glimpse of the royal Daimler.

Meanwhile, Ramanujan had already set to work with Hardy and Littlewood; Littlewood, for one, saw him about once a week, Hardy much more often. Ramanujan was productive, working hard, happy. "Mr. Hardy, Mr. Neville and others here are very unassuming, kind and obliging," he wrote home in June.

Ramanujan had not come to Cambridge to go to school. But arriving in time for the Easter term, which began in late April, he did attend a few lectures. Some were Hardy's. Others, on elliptic integrals, were given by Arthur Berry, a King's College mathematician in his early fifties. One morning early in the term, Berry stood at the blackboard working out some formulas and at one point looked over to Ramanujan, whose face fairly glowed with excitement. Was he, Berry inquired, following the lecture? Ramanujan nodded. Did he care to add anything? At that, Ramanujan stood, went to the blackboard, took the chalk, and wrote down results Berry had not yet proved and which, Berry concluded later, he could not have known before.

Soon the word was getting out about Ramanujan. W. N. Bailey, then an undergraduate, heard "strange rumors that he had been unable to pass examinations, and that he had run away from such terrors. But apart from these rumors we only knew that his name was Ramanujan, and even this was pronounced wrongly," probably Rah-ma-*noo*-jn. People didn't often see him; he was usually busy in his rooms. But when they did, they noticed him—remembered his squat, solitary figure as, in the words of one, he "waddled" across Trinity's Great Court, his feet in slippers, unable yet to wear Western shoes.

It was Hardy's rooms in New Court to which Ramanujan was apt to be bound. This smaller quadrangle was "new" only by Cambridge standards, of course, having been built in 1823, two centuries or so after most of the rest of the college. Hardy lived on the second floor of Staircase A, just over the portal leading out to the Avenue, a double row of two-hundred-year-old lime trees parading across the Backs. It was a long haul from the Nevilles' to Hardy's rooms—across the bridge at the far end of town to Jesus Green, along one of several footpaths crossing it, and onto Park Parade; then by one or another old cobblestoned street to the Great Gate, and only then into the college itself. All in all, perhaps a twenty-minute hike to New Court at the far southwestern edge of the college.

That, apparently, was too far. In early June, after about six weeks on Chesterton Road, Ramanujan moved into rooms on Staircase P in Whewell's Court. It would be "inconvenient for the professors and myself if I stay outside the college," he wrote to a friend.

Probably, he was sad to leave the Nevilles. Neville was the first Englishman to win Ramanujan's confidence and, from the moment Ramanujan disembarked from the *Nevasa*, had done much to ease his adjustment to English life. Then, too, if later accounts are any guide, he and his wife Alice were the consummate hosts, their hospitality a legend. They were young, liberal-minded, and by now had an emotional stake in Ramanujan. In all likelihood, they doted on him.

In Whewell's Court, only about five minutes from Hardy's rooms, Ramanujan could look out his window across to where Hardy had lived as an undergraduate twenty years before. But Ramanujan had more than twenty years' worth of mathematical catching up to do. His education had ended, in a sense, when George Shoobridge Carr put the finishing touches on his *Synopsis* in 1886. And Carr's mathematics was old when it was new, mostly barren of anything developed past about 1850.

Ramanujan, then, had much to learn. But, then again, so did Hardy.

2. TOGETHER

Together now in Cambridge, there was no longer the need for those long, awkward letters, across a gulf of culture and geography, with all their chance for misunderstanding. Now, as he would for the next few years, Ramanujan saw Hardy nearly every day and could show him the methods he had developed in India that he'd been loath to describe by international post. Meanwhile, Hardy had the notebooks themselves before him and, with their author by his side, could study them as much as he wished.

Many of the 120 theorems Ramanujan had sent him in those first two letters, Hardy could see now, had been plucked intact from the notebooks. Here, in chapter 5, section 30 of the second notebook, was what Ramanujan had written in the first letter about a class of numbers built up from "an odd number of dissimilar prime divisors." From chapter 5 also came much of the work going into Ramanujan's first published paper on Bernoulli numbers. In chapter 6 was that bizarre stuff from the first letter about divergent series that, Ramanujan had feared, might persuade Hardy he was destined for the lunatic asylum—the one where $1+2+3+4+\ldots$ unaccountably added up to $-1/12$. On its face, that was ridiculous; yet it sought to give meaning to divergent series—which at first glance added up to nothing more revealing or precise than infinity. But now Hardy found something like Ramanujan's reasoning behind it, which involved a "constant" that, as Ramanujan wrote, "is like the center of gravity of a body"—a concept borrowed from, of all places, elementary physics.

A few of Ramanujan's results were, Hardy could see, wrong. Some were not as profound as Ramanujan liked to think. Some were independent rediscoveries of what Western mathematicians had found fifty years before, or a hundred, or two. But many—perhaps a third, Hardy would reckon, perhaps two-thirds, later mathematicians would estimate—were breathtakingly new. Ramanujan's fat, mathematics-rich letters, Hardy now saw, represented but the thinnest sampling, the barest tip of the iceberg, of what had accumulated over the past decade in his notebooks. There were *thousands* of theorems, corollaries, and examples. Maybe three thousand, maybe four. For page after page, they stretched on, rarely watered down by proof or explanation, almost aphoristic in their compression, all their mathematical truths boiled down to a line or two.

The notebooks would frustrate whole generations of mathematicians, who were forever underestimating the sheer density of mathematical riches they contained. In 1921, after having for seven years been exposed to them, Hardy would note that "a mass of unpublished material" still awaited analysis. Two years later, having devoted a paper to Ramanujan's work in chapters 12 and 13 of his first notebook, on hypergeometric series, he had to report that those were, in fact, "the only two chapters which, up to the present, I have been able to subject to a really searching analysis."

Around that time, Hardy was visited by the Hungarian mathematician George Polya, who borrowed from him his copy of Ramanujan's notebooks, not yet then published. A few days later, Polya, in something like a panic, fairly threw them back at Hardy. No, he didn't want them. Because, he said, once caught in the web of Ramanujan's bewitching theorems, he would spend the rest of his life trying to prove them and never discover anything of his own.

In 1929, G. N. Watson, professor of pure mathematics at the University of Birmingham and a former Trinity fellow, and B. M. Wilson, who had known Ramanujan in Cambridge and was then at Liverpool University, set out on a mathematical odyssey through Ramanujan's notebooks. Two years later, recounting their progress, Watson admitted the task was "not a light one." A single pair of modular equations, for example, had taken him a month to prove. Yet Ramanujan was so rewarding, he wrote, that he and Wilson thought it "worth while to spend a fairly substantial fraction of our lives in editing the Note Books and making Ramanujan's earlier discoveries accessible." He estimated the job might take another five years. In fact, before his energy waned in the late 1930s, Watson had devoted most of a decade to the job, producing more than two dozen papers and masses of notes never published. (Wilson could give the project only fours years more; he died, after routine surgery, in 1935.)

In 1977, the American mathematician Bruce Berndt took up where Watson and Wilson left off. After thirteen years of work, having published three volumes devoted solely to the notebooks, he is still at it today, the task unfinished.

Plainly, then, in the months after Ramanujan arrived in England, Hardy and Littlewood could hardly have more than skimmed the surface of the notebooks, dipping into them at points, lingering over particularly intriguing results, trying to prove this one or simply understand that one. But this first glance was enough to reinforce the impression left by the letters. After the second letter, Littlewood had written Hardy, "I can believe that he's at least a Jacobi." Hardy was to weigh in with a tribute more lavish yet. "It was his insight into algebraical formulae, transformation of infinite series, and so forth, that was most amazing," he would write. In these areas, "I have never met his equal, and I can compare him only with Euler or Jacobi."

Euler and Jacobi were both towering figures on the stage of mathematical history. Leonhard Euler has been called "the most productive mathematician of the eighteenth century," the author of almost eight hundred books and papers, many of them after he was blind, in every field of mathematics known in his day. It was Euler who, in his 1748 book, *Introductio in analysin infinitorum*, gave the trigonometric functions the form they have today. Today's mathematics texts fairly spill over with Euler's constant, and Euler's criterion, and the Euler-Maclaurin formula, and Eulerian integrals, and Euler numbers. Born twenty years after Euler's death, Karl Gustav Jacob Jacobi was nearly his equal in genius. The son of a Berlin banker, he pioneered elliptic functions and applied them to number theory. His name, too, is enshrined in mathematics, in Jacobi's theorem and Jacobi's polynomials.

But Euler and Jacobi were not just generic "great mathematicians"; it was not capriciously that Hardy and Littlewood had compared Ramanujan to them. Rather, these two men represented a particular mathematical tradition of which Ramanujan, too, was part—that of "formalism." *Formal*, here, carries no suggestion of "stiff" or "stodgy." Euler, Jacobi, and Ramanujan had (along with deep insight) a knack for manipulating formulas, a delight in mathematical *form* for its own sake. A "formal result" suggests one fairly bubbling up from the formulas themselves, almost irrespective of what those formulas mean. Computers today manipulate three-dimensional contours regardless of whether they represent economic forecasts or car bumpers. Some painters care as much for form, line, or texture as they do subject matter. The mind of the mathematical formalist works along similar lines.

All mathematicians, of course, manipulate formulas. But formalists were almost magicians at it, uncannily selecting just the tricks and techniques needed to obtain intriguing new results. They would replace one variable in an equation by another, thus reducing it to simpler form. They would know when to integrate a function, when to differentiate it, when to construct a new function, when to worry about rigor, when to ignore it.

But already by 1914, a faint odor of derision clung to them. For one thing, mathematicians of more finicky tastes clucked at how formalists sometimes steamrolled over certain mathematical niceties. By this light, they were holdovers from a prerigorous past whose ingenious formulas sometimes failed to stand the test of close reasoning.

But more, formalists were seen to inhabit a mathematical backwater. Useful formulas tended to be found early in the development of a new mathematical field, pointing the way to future work. But as a field matured and these early formulas were applied and extended, they often grew too complicated to be useful. By Ramanujan's time, something like this had happened in branch after branch of mathematics.

And so the smart money, as it were, had largely abandoned the search for formulas. The formalist came to be seen as one who stopped short of deep thinking and churned out narrow results, through mere conjurer's tricks, that failed to turn over important new ground. His was a style of mathematics not so much profound as clever; that smacked not of High Science but Low Art—or Black Magic.

Ramanujan's mathematics, if it fit any category, fit this one. And yet, Hardy could see that if Ramanujan possessed conjurer's tricks, they were ones of almost Mephistophelean potency. Ramanujan was a formalist who undermined the stereotypes. "It is possible that the great days of formulae are finished and that Ramanujan ought to have been born 100 years ago," Hardy would write. But, he acknowledged, "He was by far the greatest formalist of his time," one whose mathematical sleight of hand no one could match, and whose theorems, however he got them, later generations of mathematicians would esteem as elegant, unexpected, and deep.

It was with some sense, then, of mingled mystery and awe that Hardy and Littlewood came away from their first long look at Ramanujan's notebooks. "The beauty and singularity of his results is entirely uncanny," Littlewood would comment in a review of Ramanujan's papers published later. "Are they odder than one would expect things selected for oddity to be? The moral seems to be that we never expect enough; the reader at any rate experiences shocks of delighted surprise."

For Hardy's part, confronting the mystery of Ramanujan's mind would constitute, as his friend Snow had it, "the most singular experience of his life: what did modern mathematics look like to someone who had the deepest insight, but who had literally never heard of most of it?"

Ramanujan "combined a power of generalization, a feeling for form, and a capacity for rapid modification of his hypotheses, that were often really startling, and made him, in his own peculiar field, without a rival in his day," Hardy would conclude. As for his ultimate influence, Hardy couldn't, at the time he wrote, say; in a sense, its very peculiarity undercut it. "It would be greater," he suggested, "if it were less strange."

But, he added, "One gift it has which no one can deny—profound and invincible originality."

Having gone so out on a limb to bring Ramanujan to Cambridge, Hardy, after familiarizing himself with the notebooks, probably felt a little relieved, too. And proud. "Ramanujan was," he would write, "my discovery. I did not invent him—like other great men, he invented himself—but I was the first really competent person who had the chance to see some of his work, and I can still remember with satisfaction that I could recognize at once what a treasure I had found."

It didn't take long to see that much in Ramanujan's notebooks was well worth publishing. Editorial work was needed, of course; his results needed to be shaped, cast into lucid English, their notation made more familiar. Hardy, ever the mathematical journalist, now proceeded to do just that. "All of Ramanujan's manuscripts passed through my hands," he wrote, "and I edited them very carefully for publication. The earlier ones I rewrote completely." (But, he added, he contributed nothing to the mathematics itself—an assertion made entirely credible by his readiness to take credit, in their jointly bylined papers, whenever he *did* contribute. "Ramanujan," he wrote, "was almost absurdly scrupulous in his desire to acknowledge the slightest help.") Now, in any event, Ramanujan's bare notebook entries began to take new form, as mathematical papers fit to be seen and read by the world. By June, he and Hardy had the beginnings of two papers, one of which was ready enough to show around.

The second Thursday of each month normally saw Hardy board the

2:15 P.M. train out of Cambridge to attend that evening's London Mathematical Society meeting. When on one particular second Thursday—June 11, 1914—he greeted his friends at the society's meeting room near Piccadilly, he bore a manuscript with theorems by Ramanujan. Among those there to hear them presented was Hobson, one of those to whom Ramanujan had written a year and a half before. Bromwich was there, too; his book on infinite series was the one Ramanujan had been so sternly advised to read. So was Professor Love, Hardy's advisor from his undergraduate days. So was Littlewood.

Ramanujan himself was not.

The following year, 1915, would see a flood of papers published by Ramanujan, including the one Hardy presented that June evening to the London Mathematical Society. But 1914, the year of his arrival in England, saw only one. Comprising mostly Indian work, it appeared in the *Quarterly Journal of Mathematics* under the title "Modular Equations and Approximations to Pi."

Every schoolchild knows that pi gives the ratio of a circle's circumference to its diameter, about 3.14. Why waste time pursuing new ways to approximate it? Surely not for the sake of fixing it more precisely; even by the midnineteenth century, mathematicians had determined pi to five hundred decimal places, far in excess of any practical need. (Two Canadian mathematicians, the brothers Jonathan M. Borwein and Peter B. Borwein, have noted that thirty-nine decimal places will fix the circumference of a circle around the known universe to within the radius of a hydrogen atom.)

But pi is not merely the ubiquitous actor in high school geometry problems; it is stitched across the whole tapestry of mathematics, not just geometry's little corner of it. Since mathematicians find it more convenient to express angles in pi-based "radians" than in everyday degrees, pi occupies a key place in trigonometry, too. It is intimately related to that other transcendental number, e, and to "imaginary" numbers through Euler's elegant relationship,

$$e^{i\pi} = -1$$

which in a single, strange, beautiful statement of mathematical truth ties trigonometry and geometry to natural logarithms and thence to the whole world of "imaginary" numbers. Pi even shows up in the mathematics of probability. Drop a needle onto a table finely scored by parallel lines each separated by the length of the needle and the chance of its intersecting a line is 2/pi. Again and again pi pops up. So finding new ways to express it can reveal hidden links between seemingly disparate mathematical realms.

Ancient societies were usually content to figure pi at, simply, 3. The seventh-century Indian mathematician Brahmagupta put it as the square root of 10, which is about 3.16. In the West, early efforts to define pi were pursued geometrically; you circumscribe circles, drop perpendiculars, bisect angles, draw parallels and wind up with pi as the length of some line segment. "Squaring the circle," as this classic problem is called, turned out to be impossible. But Archimedes took another geometric approach and came up with a value of pi equal to between $3\frac{10}{70}$ and $3\frac{10}{71}$.

In the midseventeenth century, the powerful tools of the calculus were brought to bear, leading to a variety of infinite series that converged to pi. Newton himself came up with one that gave pi to fifteen decimal places. "I am ashamed to tell you," he confessed to a colleague, "to how many figures I carried these calculations, having no other business at the time."

Series yielding pi, or approximations to it, can be of surpassing grace, like this one, attributed variously to Leibniz, or the Scottish mathematician James Gregory, or to mathematicians in Kerala:

$$\frac{\pi}{4} = 1 - \frac{1}{3} + \frac{1}{5} - \frac{1}{7} + \frac{1}{9} + \ldots$$

John Wallis came up with this infinite product at about the same time:

$$\frac{\pi}{2} = \frac{2}{1} \times \frac{2}{3} \times \frac{4}{3} \times \frac{4}{5} \times \frac{6}{5} \times \frac{6}{7} \times \ldots$$

Another pretty one is:

$$\frac{\pi - 3}{4} = \frac{1}{2\times3\times4} - \frac{1}{4\times5\times6} + \frac{1}{6\times7\times8} - \ldots$$

Thus pi, about as unruly a number as you can imagine (no pattern in its digits, even out to millions of decimal places, has ever been found), can be represented by series of the most appealing simplicity.

Ramanujan's early letters to Hardy included several such series approximations. Now, his twenty-three-page paper was filled with other routes to pi. Many rested on modular equations, a subject, going back to the work of the French mathematician Legendre in 1825, that he had exhaustively surveyed in his notebooks. Roughly, a modular equation relates a function of a variable, x, with the same function expressed in terms of x raised to an integral power (x^3 or x^4, for instance, but not $x^{3.2}$). The trick, of course, is to find a function, $f(x)$, to satisfy it. Not surprisingly, they are rare. But when they do show up, it turns out, they often display special properties mathematicians can exploit. Ramanujan found that certain such functions, satisfying certain modular equations, gave solutions that, under certain circumstances, could be used to closely approximate pi.

Some of his results, it turned out, had been anticipated by European mathematicians, like Kronecker, Hermite, and Weber. Still, Hardy would write later, Ramanujan's paper was "of the greatest interest and contains a large number of new results." If nothing else, it was astounding how *rapidly* some of his series converged to pi. Leibniz's series, on page 209, is lovely—but almost worthless for getting pi; three-decimal-point accuracy demands no fewer than five hundred terms. Some of Ramanujan's series, on the other hand, converged with astonishing rapidity. In one, the very first term gave pi to eight decimal places. Many years later, Ramanujan's work would provide the basis for the fastest-known algorithm, or step-by-step method, for determining pi by computer.

During most of the ten years since he'd encountered Carr's *Synopsis,* Ramanujan had inhabited an intellectual wilderness. In India, he'd been surrounded by family, friends, familiar faces. He was a South Indian among other South Indians, a Tamil-speaker among other Tamil-speakers, a Brahmin among other Brahmins. And yet, he was also a mathematical genius of perhaps once-in-a-century standing cut off from the mathematics of his time. He roused wonder and admiration among those, like Narayana Iyer and Seshu Iyer, who could glimpse into his theorems. Yet no one had been able to truly appreciate his work. He had been alone. He had had no peers.

And now, in Littlewood and Hardy, he did have them. In Cambridge he had at last found an intellectual home, a community of mathematicians who saw in his work all that he saw in it. And, at least in the beginning, that more than made up for being a stranger in an alien land.

The Cambridge into which he stepped that day in April 1914 was a Cambridge where the Playhouse, on Mill Road, played *The Fatal Legacy,* billed as "a grand and absorbing drama, with an exciting foxhunt." A barbed-wire fence was going up on the Old Chestertown Recreation Ground to protect its green for the playing of bowls. And the big story in the paper was that Caesar, the late King Edward's favorite dog, the Irish terrier who had accompanied his Royal Master everywhere, had died the previous Saturday.

To a South Indian, the knives and forks the English used and in which Ramachandra Rao had tried to school Ramanujan back in India seemed like an invasion—hard metallic *things* penetrating the mouth. Feet long unconstrained by anything more than sandals felt pinched by shoes; it took months to get used to them. English names all ran together in a blur. And whether oval or square, topped with brown hair or blond, their faces seemed so alike in their essential *whiteness;* you could talk for hours with an Englishman yet fail to recognize him later on the street. Then, of course, a South Indian found that when he gave that little undulating jiggle of the head that back home meant something between a simple acknowledgment and a *yes,* the English were apt to take it as a *no.*

Still, in the glow of Ramanujan's arrival and his first few months, any feelings of homesickness, loneliness, or frustration must have been fairly swept under the emotional rug. Yes, wrote Neville, who witnessed his adjustment up close, "He felt the petty miseries of life in a strange civilization, the vegetables that were unpalatable because they were unfamiliar, the shoes that tormented feet that had been unconfined for twenty-six years. But he was a happy man, reveling in the mathematical society which he was entering and idolized by the Indian students." He had all the money his tastes required. He had perfect leisure to pursue his work. In this old town of cobbled walks, grassy courts, and medieval chapels, whole universes away from Madras, Ramanujan had found a kind of intellectual nirvana.

But then, the first cannon sounded.

3. THE FLAMES OF LOUVAIN

The Great War was both expected and unexpected.

Everyone knew it was coming. All during 1913 and 1914, Europe seethed. The assassination of the Austrian crown prince Franz Ferdinand on June 28, 1914 set in motion a chain of events which ultimately proved irresistible. Germany and France, locked into alliances, declared war. When German armies bound for France swept through Belgium, violating its neutrality, England declared war, too. That came at 11 A.M. on August 4, 1914. All Europe was soon engulfed, Germany and the Central Powers marshaled against France, Britain, and the other Allied Powers.

To many, the war was something hotly sought, a chance to burn off tensions built up over forty years of peace. Europe marched into war, flags flying, to the sound of martial music. The uniforms were fresh, the ranks and files still intact. The enemy would be taught its lesson and the war itself would be over in a month or two. The troops, somebody said, would be "home before the leaves fall."

That was the unexpected part, the war's terrible surprise—that it was no brief but glorious orgasm of arms but rather ground along, month after month, year after cruel year. During August, German armies roared through Belgium, in obedience to the Schlieffen Plan with which they hoped to humble France. The French, with their Plan Seventeen, took a similarly offensive stance, aiming straight for Berlin. But hopes for rapid victory on both sides were dashed within the war's first six weeks. The great armies met, fought, bled. Plans went wrong. And after the Battle of the Marne in September, trench warfare replaced great sweeping battlefield maneuvers. "Running from Switzerland to the Channel like a gangrenous wound across French and Belgian territory," Barbara Tuchman wrote in *The Guns of August*, "the trenches determined the war of position and attrition, the brutal, mud-filled, murderous insanity known as the Western Front that was to last for four more years."

On September 11, while Germany and France grappled at the Marne, Ramanujan wrote his mother, reassuring her that "there is no war in this country. War is going on only in the neighboring country. That is to say, war is waged in a country that is as far as Rangoon is away from the city [Madras]."

That wasn't true; the war was much closer. In fact, Cambridge had already felt the war's impact. In its first week, back in early August, the Sixth Division, from Ireland, converged on Cambridge and set up tents on Midsummer Common, just across from Neville's house in Chestertown. Corpus Christi College became temporary headquarters for the Officers Training Corps; professors, undergraduates, and fellows all volunteered to help. At Trinity, the columned area under Wren's great library was made into an open-air hospital; wooden boards laid on the uneven stone floor, to keep the beds level, now muffled the echoes that had been a fixture of those stone precincts for centuries. In the court's northwest corner, near the winding staircase that climbed up to the library, bathrooms were installed. The south cloister, near where Littlewood lived, had lights strung from the ceiling and became an operating theater. Meanwhile, in Hardy's New Court, rooms were made into offices.

On August 14, an ambulance with a great red cross on its side bore the first wounded patient to what was, officially, the First Eastern General Hospital—the open-air hospital at Trinity. Later in the month, the Germans burned Louvain, and two British divisions retreated from positions they had defended along the Mons Canal in Belgium. In the nine hours of battle before the retreat began, the fighting cost sixteen hundred British casualties. Of the wounded, many were soon in Nevile's Court, in double rows of beds under the library, where Henry Jackson, the college's vice-master, a venerable classicist, would see them on his way to Hall.

"We have a new Cambridge," wrote Jackson, "with 1700 men *in statu pupillari* instead of 3600. . . . Medical Colonels and Majors and Captains dine in hall in khaki."

In early September, the Sixth Division left for France. But Cambridge was still crowded with men under arms and would remain the final training station for a whole succession of army divisions bound for the front. When it rained, horse-drawn gun carriages and other army vehicles churned up inches-deep mud on unasphalted roads.

On September 20, the Chapel Choir met on the lawn in the middle of Nevile's Court to sing hymns to the wounded soldiers in the surrounding cloisters.

In the early days of the war, jingoism had not yet been buried by cynicism. "The depravation of Germany—its gospel of iniquity and selfishness—is appalling," wrote Jackson. "For though I never thought the Prussians gentlemen, I had a profound respect for their industry and efficiency, and I attributed to them domestic virtues. As it is, their good qualities subserve what is evil." Henry Butler, master of Trinity, had no trouble believing "that the German infantry can neither shoot nor stand the bayonet. As to the last, they turn and run and get stuck in the back. Many of [the Nevile's Court wounded] assure me that they have seen women and children driven in front of the enemy when they charge."

Feeling against Germany swelled. Even Ramanujan was caught up in it, writing his mother about the German advance across Belgium. "Germans set fire to many a city, slaughter and throw away all the people, the children, the women and the old."

* * *

The popular English magazine *Strand* had long carried a page, entitled "Perplexities," devoted to intriguing puzzles, numbered and charmingly titled, like "The Fly and the Honey," or "The Tessellated Tiles," the answers being furnished the following month. Each Christmas, though, "Perplexities" expanded, the author fitting his puzzles into a short story. Now, in December 1914, "Puzzles at a Village Inn" took readers to the imaginary town of Little Wurzelfold, where the main topic of interest was what had just happened in Louvain.

In late August, pursuing an explicit policy of brutalization against civilian populations, German troops began burning the medieval Belgian city of Louvain, on the road between Liège and Brussels. House by house and street by street they set Louvain to the torch, destroying its great library, with its quarter million books and medieval manuscripts, and killing many civilians. The burning of Louvain horrified the world, galvanized public opinion against Germany, and united France, Russia, and England more irrevocably yet. "The March of the Hun," English newspapers declared. "Treason to Civilization." It was an early turning point of the war, doing much to set its tone. *Louvain* came to symbolize the breakdown of civilization. And now it reached even the "Perplexities" page of *Strand*.

One Sunday morning soon after the December issue appeared, P. C. Mahalanobis sat with it at a table in Ramanujan's rooms in Whewell's Court. Mahalanobis was the King's College student, just then preparing for the natural sciences Tripos, who had found Ramanujan shivering by the fireplace and schooled him in the nuances of the English blanket. Now, with Ramanujan in the little back room stirring vegetables over the gas fire, Mahalanobis grew intrigued by the problem and figured he'd try it out on his friend.

"Now here's a problem for you," he yelled into the next room

"What problem? Tell me," said Ramanujan, still stirring. And Mahalanobis read it to him.

> "I was talking the other day," said William Rogers to the other villagers gathered around the inn fire, "to a gentleman about the place called Louvain, what the Germans have burnt down. He said he knowed it well—used to visit a Belgian friend there. He said the house of his friend was in a long street, numbered on this side one, two, three, and so on, and that all the numbers on one side of him added up exactly the same as all the numbers on the other side of him. Funny thing that! He said he knew there

was more than fifty houses on that side of the street, but not so many as five hundred. I made mention of the matter to our parson, and he took a pencil and worked out the number of the house where the Belgian lived. I don't know how he done it."

Perhaps the reader may like to discover the number of that house.

Through trial and error, Mahalanobis (who would go on to found the Indian Statistical Institute and become a Fellow of the Royal Society) had figured it out in a few minutes. Ramanujan figured it out, too, but with a twist. "Please take down the solution," he said—and proceeded to dictate a continued fraction, a fraction whose denominator consists of a number plus a fraction, *that* fraction's denominator consisting of a number plus a fraction, ad infinitum. This wasn't just the solution to the problem, it was the solution to the whole class of problems implicit in the puzzle. As stated, the problem had but one solution—house no. 204 in a street of 288 houses; 1+2+ . . . 203 = 205+206+ . . . 288. But without the 50-to-500 house constraint, there were other solutions. For example, on an eight-house street, no. 6 would be the answer: 1+2+3+4+5 on its left equaled 7+8 on its right. Ramanujan's continued fraction comprised within a single expression *all* the correct answers.

Mahalonobis was astounded. How, he asked Ramanujan, had he done it?

"Immediately I heard the problem it was clear that the solution should obviously be a continued fraction; I then thought, Which continued fraction? And the answer came to my mind."

4. THE ZEROES OF THE ZETA FUNCTION

The answer came to my mind. That was the glory of Ramanujan—that so much came to him so readily, whether through the divine offices of the goddess Namagiri, as he sometimes said, or through what Westerners might ascribe, with equal imprecision, to "intuition." And yet, it was the very power of his intuition that, in one sense, undermined his mathematical development. For it blinded him to intuition's limits, gave him less reason to learn modern mathematical tools, shielded him from his own ignorance.

"The limitations of his knowledge were as startling as its profundity," Hardy would write.

> Here was a man who could work out modular equations and theorems of complex multiplication, to orders unheard of, whose mastery of continued fractions was, on the formal side at any rate, beyond that of any mathematician in the world, who had found for himself the functional equation of the Zeta-function, and the dominant terms of many of the most famous problems in the analytic theory of numbers; and he had never heard of a doubly periodic function or of Cauchy's theorem, and had indeed but the vaguest idea of what a function of a complex variable was. His ideas as to what constituted a mathematical proof were of the most shadowy description. All his results, new or old, right or wrong, had been arrived at by a process of mingled argument, intuition, and induction, of which he was entirely unable to give any coherent account.

That mysterious "process" sometimes led him seriously astray. And Hardy's reference to the zeta function embodied one notorious example of it.

In Ramanujan's first letter, he'd referred to a statement in Hardy's *Orders of Infinity* concerning prime numbers. He had, he wrote, "found a function which exactly represents the number of prime numbers less than x," in the form of an infinite series. Very interesting, Hardy had written back, let's see some proof. In Ramanujan's next letter, he elaborated. "The stuff about primes is wrong," Littlewood shot back to Hardy when he saw it. And now, with Ramanujan in England, Hardy came to see, up close, just how he had gone wrong.

Ramanujan was not the first mathematician to be bewitched by prime numbers. The primes—numbers like 2, 3, 5, 7, and 11, but not 6 or 9 (respectively, 2×3 and 3×3, and so "composite")—were the building blocks of the number system. Start counting and it was hard to help wondering, when would you get to a prime? Was there any sort of pattern? If you laid out all the numbers on a giant grid, colored in the primes and left the others blank, then stood back and took a picture of the resulting pattern, would there be anything to see, any pattern you could call a pattern?

At first glance there wasn't, certainly nothing obvious. And yet, then again, there was *something* there, right in front of your eyes: as you kept on counting, you found more primes. There was *no last prime*. Euclid had proved as much twenty-three hundred years ago. "The primes are the raw material out of which we have to build arithmetic," Hardy would write, "and Euclid's theorem assures us that we have plenty of material for the task"; just as you never run out of numbers, you never run out of primes.

Well, then, that was something. But could you say anything more? Yes: as you kept counting it looked, on average, as if you encountered a lower density of primes. There were always more of them, but the *rate* at which you encountered them dropped off. In the first one hundred numbers, for example, there were twenty-five primes; in the second hundred, twenty-one; in the ninth hundred, fifteen. Occasionally, you ran into a span of higher density; there are sixteen primes between 1100 and 1200, for example. But on average the falloff held.

Anything else? Yes, again. While the density of primes fell off, it fell off slowly. Deliberately, inexorably, the slowing effect could be felt; but it took hold oh-so-gradually. This was not like a campfire roaring away at dusk that is nothing but embers by midnight, and cold by morning; but rather one that by sunrise still burned brightly, only not *quite* so brightly. With primes, by the time you were in the billions, you still got five of them in every 100 numbers, compared to fifteen per hundred in the first thousand. So the braking effect was a gentle one.

Could you be more precise than that? Could you say just how slow the slowing was? Could you, to put it another way, make a mathematical statement that gave the number of primes you encountered in counting to any given number?

In time, mathematicians guessed that the restraining influence, the mathematical "force" slowing the increase in the number of primes, was logarithmic; that the mathematical function known as the logarithm was somehow at work.

To say that something "drops off logarithmically" is the reverse of saying that it "increases exponentially." Exponential growth implies a kind of takeoff, rising and re-rising, ever more rapidly, on the strength of its own growth; compound interest, the kind that yields those thick retirement larders thirty years down the road, is an exponential process, rising slowly at first, then accelerating. But with logarithmic growth, quite the opposite occurs: you get less and less bang for the buck.

The logarithmic response is ubiquitous in nature, as, for example, in the realm of the human senses. Double the amount of light in a room and you scarcely notice; the eye can respond to both the glare of the noonday sun and to the flicker of a match a mile away because its response is logarithmic. If light intensity rises by a factor of a thousand, which can be written as 10^3, the response, in principle, may be only about three times as much instead of a thousand.

Well, it was just such a slow logarithmic growth in this braking effect that mathematicians long thought they saw with prime numbers. Or, mathematically

$$\pi(x) = \frac{x}{\ln x}$$

Here $\pi(x)$, read "pi of x," means the number of primes encountered in the first x numbers. And this, the equation says, is equal to simply x divided by the ("natural") logarithm of x. In fact, this is not quite right. Rather,

$$\pi(x) \sim \frac{x}{\ln x}$$

This expresses the same idea except as an approximation, thought to get better and better as x grows larger. (Gauss actually wrote it as a logarithmic integral, which is similar in principle.)

In 1896, the French mathematician Hadamard and the Belgian de la Vallée-Poussin proved the long-conjectured prime number theorem, as this relationship was known. And that's where things stood in 1914: As x grew large, the theorem said, a key ratio—$\pi(x)\ \ln x/x$—approached one. That was pretty good, certainly putting things on a firm mathematical basis. And yet, it was still an approximation, one that gave no clue to its accuracy; it failed to specify how "fast" the ratio approached one, much less give you a formula that said, give me the number to which you counted, and I'll tell you how many primes you've passed along the way. It couldn't say, for example, that among the first million integers there are, exactly, 78,498 primes. And it was this that Ramanujan in his letters to Hardy confidently declared he could do: *I have found a function which exactly represents the number of prime numbers less than x.*

Well, he hadn't.

Ramanujan's formula, three versions of which he gave in his second letter, was an infinite series which, for values of x up to 1000 for example, gave virtually exact agreement. Even larger values, Hardy found later, were strikingly close. Of the first nine million numbers, it was known that 602,489 were prime. Ramanujan's formula gave a figure off by just 53—closer than the canonical version of the prime number theorem. What was more—and this was the crucial point—any error accompanying its use, he claimed, was bounded, stayed within given limits. *This* was why, in his first letter, he'd objected to Hardy's assertion that the "precise order" of the error term "has not yet been determined"; his formulas, he insisted, *did* determine it.

But they didn't. Back in 1913, shortly after Hardy's first encouraging letter had arrived, Ramanujan's friend Narayana Iyer had slipped into the *Journal of the Indian Mathematical Society* some of Ramanujan's results on primes. "Proofs," he added, "will be supplied later." But they never were. Because they didn't exist—*couldn't* exist. Because Ramanujan's conclusions were wrong.

Once Ramanujan was in England, Hardy studied Ramanujan's notebooks, listened as he worked through his arguments—and came to see where he had stumbled. Ramanujan had been misled by undue reliance on the low values of x for which he'd tried his formula; the error, for higher values of x, was much larger than he thought. These, had he tried them, might have alerted him to the more basic flaws in his approach, which Hardy would find so illuminating that one day he'd build a whole lecture around them. Ramanujan's theory, Hardy would write, "was (so to say) what the theory might be if the zeta function had no complex zeroes."

The Riemann zeta function was a simple enough looking infinite series expressed in terms of a complex variable. Here, "complex" means not difficult or complicated, but refers to a variable of two distinct components, "real" and "imaginary," which together could be thought to range over a two-dimensional plane. In 1860, Georg Friedrich Bernhard Riemann made six conjectures concerning the zeta function. By Ramanujan's time, five had been proven. One, enshrined today as the Riemann hypothesis, had not.

If you set the zeta function equal to zero, Riemann conjectured, then certain solutions to the resulting equation—its "complex zeroes"—would, when graphed, all lie along a particular line, one parallel to the "imaginary" axis and half a unit to its right. And from this hypothesis, if valid, certain important conclusions about the distribution of primes, going beyond the prime number theorem itself, would automatically flow.

But *was* the Riemann hypothesis true? To this day, no one knows, and it remains one of the great unproved conjectures in mathematics. Around the time Ramanujan arrived in England, Hardy proved something just short of it—that an infinite number of solutions lay on the crucial line.

But this was not the same as saying they *all* did. The renowned German mathematician David Hilbert once said that, were he awakened after having slept for a thousand years, his first question would be, *Has the Riemann hypothesis been proved?*

Ramanujan had come up with something like Riemann's zeta function on his own but had misunderstood what he'd found. He had ignored the crucial complex zeroes, acted as if they didn't exist—leading to a version of the prime number theorem that was, simply, wrong. "There are regions of mathematics in which the precepts of modern rigour may be disregarded with comparative safety," Hardy would write, "but the Analytic Theory of Numbers is not one of them."

As Littlewood would write of Ramanujan's effort on primes, "These problems tax the last resources of analysis, took over a hundred years to solve, and were not solved at all before 1890 [sic; he probably meant 1896, when the prime number theorem was proved]; Ramanujan could not possibly have achieved complete success. What he did was to perceive that an attack on the problems could at least be begun on the formal side, and to reach a point at which the main results become plausible. The formulae do not in the least lie on the surface, and his achievement, taken as a whole, is most extraordinary."

Hardy would be similarly indulgent of Ramanujan's errors. "I am not sure," he would write, "that, in some ways, his failure was not more wonderful than any of his triumphs." He later regretted saying that, dismissing it as unduly sentimental. But he was getting at something—that in reaching his faulty conclusions, Ramanujan had rediscovered the prime number theorem; and that, in trying to reach beyond it, he had pursued arguments that, though technically flawed, were, in their own way, brilliant.

Still, Ramanujan *was* wrong, which now, in England, under Hardy's tutelage, he came to understand. "His instincts misled him," wrote Hardy. And *that* was the point. Ramanujan's "instincts," sure as they were, in some ways better than those of any other mathematician of his day, were not good enough.

A car mechanic reliant on mechanical instinct may "know" how an engine works yet be unable to set down the physical and chemical principles governing it. For a writer, it may be enough to "know" that one scene should precede another and not follow it, without being able to explain why. But mathematicians are not normally content to guess, or assume, or assert that something is true; they must prove it, or feel they have—or as Hardy would put it, "exhibit the conclusion as the climax of a conventional pattern of propositions, a sequence of propositions whose truth is admitted and which are arranged in accordance with rules."

Proof is no mere icing on the cake. Take the sequence of integers 31, 331, 3331, 33331, 333331, 3333331. Each is a prime number. So is the next in the sequence. Have we hit upon some hidden pattern? No, the pattern self-destructs with the next in line, the product of 17 and 19,607,843. Or what about numbers of the form $2^{2^n} + 1$. For $n = 0, 1, 2, 3$, and 4, the resulting numbers are all prime. Are they for all n? Pierre de Fermat conjectured as much. But he conjectured wrong. Because, as Euler found, even the next number is not prime, but the product of 641 and 6,700,417.

Many other examples like this, where seemingly "obvious" patterns prove not to be patterns at all, appear all through number theory and elsewhere in mathematics. One Hardy liked to cite also emerged from the theory of primes. Over the years, in comparing the approximations of the prime number theorem with the actual number of primes calculation revealed, the approximation always proved higher; the error was always in the same direction. You could try a thousand, you could try a million, you could try a billion, you could try a trillion, you could try a billion trillion, and it always came out the same, making all the forces of intuition argue that it was always so—and that a theorem embodying it would be a great one to run off and prove. But no such theorem could ever be proved. Because, intuitively obvious though it might seem, it simply wasn't true.

A year before he heard from Ramanujan, Littlewood had proved that, if you went far enough, the prime number theorem was destined to sometimes predict less, not more, than the actual number of primes. Later, someone found the number below which this reversal was guaranteed to take place. And it was a number so big you had to laugh—a number more than the number of particles in the universe, more than the possible games of chess. It was, Hardy would say, "the largest number which has ever served any definite purpose in mathematics." And it made for the ultimate illustration of how intuition could serve you badly, and so must always be subject to proof.

To "prove" something, then, is a kind of guarantee—that, for mathematical entities *A, B,* and *C,* and subject to constraints *D, E,* and *F,* the theorem holds. A theorem must stand up to hard use, will often be applied to unanticipated new situations. Fail to precisely fix the conditions under which it applies, and you're apt to go wrong. A mathematician with an insufficiently ironclad proof is a little like the brash young police lieutenant in the movies convinced of the butler's guilt but brought up short by his boss's caution, *Yeah, but that won't convince a jury.*

Throughout his notebooks, all during his time in India, and all through his early letters, Ramanujan had proclaimed a thousand versions of *the butler did it.* Most of the time he was right, and the butler *had* done it: his results were true. And yet before coming to England, he would have been unable to secure a conviction: he had—to extend the metaphor—scant knowledge of the rules of evidence, or the relevant criminal law, or even the standards of sound legal argument. Ramanujan's "proof" of his new version of the prime number theorem was no proof at all.

How can a "proof" be wrong? How can you dutifully march through its reasoning, convince yourself that what you've said is right, only to have some other mathematician show it is not? What, in other words, can go wrong with a mathematical proof seemingly laid out in obeisance to a relentlessly "logical" sequence of clear-cut steps?

In a word, everything.

Perhaps the most familiar, if trivial, example of such a failure is the "proof" that purports to show that $2 = 1$. For starters, let $a = b$. Now, multiply both sides of the equation by a, leaving

$$a^2 = ab$$

Add $a^2 - 2ab$ to both sides of the equation:

$$a^2 + (a^2 - 2ab) = ab + (a^2 - 2ab)$$

Which reduces to:

$$2(a^2 - ab) = a^2 - ab$$

Now divide both sides by $a^2 - ab$, leaving

$$2 = 1.$$

Voilà. Treating both sides of the equation with scrupulous equality we reach a result defying common sense.

What's gone wrong? Nothing but that scourge of many an elementary proof: we have divided by zero, a mathematically impermissible operation that gives an "answer" devoid of meaning.

But nowhere, a voice interrupts, does the proof say we should divide by zero.

On the contrary: we started off with $a = b$, then toward the end divided by $a^2 - ab$, which *is* zero.

Even at his most innocent, Ramanujan would never commit such a gaffe. And yet it suggests how, busy manipulating symbols and without appreciating every nuance of what he was doing, he could go wrong.

As a mathematician, you can slide into trouble in numerous ways. You can differentiate a function without realizing the function cannot be differentiated. Or you can write off later terms of a series on the assumption that they are of a lower "order" than earlier terms, when, in fact, they contribute substantially to the series sum. Or you can assume that an operation correct for a finite number of terms is correct for an infinite number. Or you can integrate a function between two points, yet fail to note where the integral may be undefined, and so carry through your proof such meaningless quantities as "infinity minus infinity."

In his work with primes, Hardy wrote, Ramanujan's proofs

> depended upon a wholesale use of divergent series. He disregarded entirely all the difficulties which are involved in the interchange of double limit operation; he did not distinguish, for example, between the sum of a series $\Sigma\, a_n$ and the value of the Abelian limit
>
> $$\lim_{x \to 1} \sum a_n x^n$$
>
> or that of any other limit which might be used for similar purposes by a modern analyst.

These were all quite technical, of course—like legal loopholes of which the police lieutenant is unaware. Ramanujan's intuition steered him clear of many obstacles of which his truncated education had failed to warn him. But not all. The problem was not only that he was sometimes wrong; it was that he lacked mathematical knowledge enough to tell when he was right and when he was wrong. He stated correct and incorrect theorems with the same aplomb, the same sweet, naive confidence. And when he did offer proofs, they scarcely warranted the name.

For Ramanujan, Littlewood wrote later, "the clear-cut idea of what is meant by a proof, nowadays so familiar as to be taken for granted, he perhaps did not possess at all; if a significant piece of reasoning occurred somewhere, and the total mixture of evidence and intuition gave him certainty, he looked no further."

It was just Ramanujan's luck, then, to be thrown in with Hardy, whose insistence on rigor had sent him off almost single-handedly to reform English mathematics and to write his classic text on pure mathematics; who had told Bertrand Russell two years before that he would be happy to prove, *really* prove, anything: "If I could prove by logic that you would die in five minutes, I should be sorry you were going to die, but my sorrow would be very much mitigated by pleasure in the proof." Ramanujan, Intuition Incarnate, had run smack into Hardy, the Apostle of Proof.

And now, as he would over the succeeding months and years, Hardy set to work on him, trying to overcome the deficit that was the price Ramanujan had paid for his intellectual isolation. At twenty-six, Ramanujan had long-established ways of doing things. Nonetheless, he responded. As Hardy wrote later,

> His mind had hardened to some extent, and he never became at all an "orthodox" mathematician, but he could still learn to do things, and do them extremely well. It was impossible to teach him systematically, but he gradually absorbed new points of view. In particular he learnt what was meant by proof, and his later papers, while in some ways as odd and individual as ever, read like the works of a well-informed mathematician.

By early 1915, about the time Mahalanobis read him the Louvain street problem in *Strand* magazine, Ramanujan had already begun to shift gears, to redirect his work in ways Hardy was urging. "I have changed my plan of publishing my results," he wrote Krishna Rao in November 1914. "I am not going to publish any of the old results in my note book till the war is over. After coming here I have learned some of their methods. I am trying to get new results by their methods." By early the following year, he had literally set aside much of his old work, was drinking in the new: "My notebook is sleeping in a corner for these four or five months," he wrote his childhood friend S. M. Subramanian, with whom he had briefly roomed in Summer House two years before. "I am publishing only my present researches as I have not yet proved the results in my notebook rigorously."

Proof and *rigor:* he was absorbing the Gospel according to Hardy.

But while taking in the new, as Hardy would write, Ramanujan's "flow of original ideas showed no symptom of abatement." In his letter to Subramanian, for example, Ramanujan at one point abruptly broke into a paragraph with: "I shall now tell you [about] a very curious function," then wrote this pattern of fractions across the page:

1, 1/2, 2/1, 1/3, 3/1, 2/3, 3/2, 1/4, 4/1, 2/4, 4/4, 3/4, 4/3

Guided by it, he ingeniously constructed a function which, bizarrely, was mathematically undefined for the fractions in his series, and existed at all only for "irrational" values, those not representable as a fraction. "A very curious function," he had called it. Then, again: "There is another curiosity here . . ." And "Just imagine [how] this function [behaves] . . ." This was vintage Ramanujan, his delight in the function's peculiar behavior fairly spilling across the pages of the letter.

In what, in some ways, was his greatest achievement, then, Hardy brought Ramanujan mathematically up to speed without muzzling his creativity or damping the fires of his enthusiasm. It would have been easy to sniff at his shortcomings and dutifully correct them, like a bad editor who crudely blue-pencils his way through a delicate manuscript. But he knew that Ramanujan's mathematical insight was rarer by far than even the most formidable technical mastery. It was fine to know all the mathematical tools needed to prove a theorem—but you had to have a theorem to prove in the first place.

That was easy to forget as you flipped through the *Proceedings of the London Mathematical Society*. There, as in any mathematics journal, the proof was made to seem the culmination of a hundred closely reasoned steps ranging over a dozen pages. There, mathematics could seem no more than a neat lockstep march to certainty, *B* following directly from *A*, *C* from *B*, . . . *Z* from *Y*. But no mathematician actually worked that way; logic like that reflected the demands of formal proof but hinted little at the insights leading to Z. Rather, as Hardy himself would write, "a mathematician usually discovers a theorem by an effort of intuition; the conclusion strikes him as plausible, and he sets to work to manufacture a proof."

The theorem itself was apt to emerge just as other creative products do—in a flash of insight, or through a succession of small insights, preceded by countless hours of slogging through the problem. You might, early on, try a few special cases to informally "prove" the result to your own satisfaction. Then later, you might go back and, with a full arsenal of mathematical weapons, supply the kind of fine-textured proof Hardy championed. But all that came later—after you had something to prove. Besides, it was mostly technical, like the laws of evidence; you could learn it. Rigor, Littlewood would observe, "is not of first-rate importance in analysis beyond the undergraduate state, and can be supplied, given a real idea, by any competent professional."

Given a real idea—that was the rare commodity.

"Mathematics has been advanced most by those who are distinguished more for intuition than for rigorous methods of proof," the German mathematician Felix Klein once noted. (Added Louis J. Mordell, an American mathematician who would ultimately succeed Hardy in his chair at Cambridge: "To very few other mathematicians are Klein's remarks . . . so appropriate as to Ramanujan.") A "real idea" wasn't dished up, like a Tripos problem, by some anonymous mathematical Intelligence. It had to *come from somewhere,* had to be seen before it could be proved. But *where* did it come from? That was the mystery, the source of all the circular, empty, ultimately unsatisfying explanations that have always beset students of the creative process. Here, "talent" came in, and "genius," and "art." Certainly it couldn't be taught. And certainly, when in hand, it had to be nurtured and protected.

Plenty of mathematical technicians, Hardy knew, could follow a step-by-step discursus unflaggingly—yet counted for nothing beside Ramanujan. Years later, he would contrive an informal scale of natural mathematical ability on which he assigned himself a 25 and Littlewood a 30. To David Hilbert, the most eminent mathematician of his day, he assigned an 80.

To Ramanujan he gave 100.

"It was impossible to ask such a man to submit to systematic instruction, to try to learn mathematics from the beginning once more," Hardy would write. "I was afraid too that, if I insisted unduly on matters which Ramanujan found irksome, I might destroy his confidence or break the spell of his inspiration." And so he sought an elusive middle ground where, without crimping Ramanujan's creativity, he could teach him, as he wrote, the "things of which it was impossible that he should remain in ignorance. . . . It was impossible to allow him to go through life supposing that all the zeroes of the zeta function were real. So I had to try to teach him, and in a measure I succeeded, though obviously I learnt from him much more than he learnt from me." Teaching Ramanujan, mathematician Laurence Young has written, "was like writing on a blackboard covered with excerpts from a more interesting lecture."

Hardy held in his hands a rare and delicate flower. And the responsibility he bore for nurturing it was only redoubled by the war.

5. S. RAMANUJAN, B.A.

"At Cambridge we are in darkness," wrote Trinity vice-master Jackson in January 1915. "No gas in the streets or courts; few electric lights and those shaded; candles on the high tables. The roads into Cambridge are blocked to prevent the approach of motors such as those which guided the East Coast Zeppelins. Rumor says that an attack on Windsor Castle was expected last week."

In April, about a year after Ramanujan's arrival, Jackson wrote: "In France and Flanders we make no progress. In the Dardanelles we are at a standstill. The army does not grow as it ought. We have not got ammunition for the existing army. The Germans have been preparing villainies for years."

Wounded soldiers flooded Cambridge, almost twelve thousand being admitted to the First Eastern General Hospital by July 1915. They'd arrive on trains late in the evening at the station, where they were met by white-hooded nurses and motorized ambulances. "When they succeeded each other at frequent intervals for some time," a Cambridge schoolgirl recalled later, "I knew that there must have been heavy fighting in Flanders or France."

The maimed, hurt, and sick streamed into Cambridge, the healthy and strong out. Of undergraduates, Cambridge was largely deserted; normally some thirty-five hundred, they now numbered five or six hundred. College fellows served in the Foreign Office, in the War Office, in the Treasury, as well as in the army itself. One Trinity man wrote, in Jackson's words, that "the front was like a first-rate club, as you met all your friends there." The university's medical laboratories, meanwhile, were put at the disposal of the First Eastern General Hospital. The chemical laboratory did research in gas warfare. The engineering laboratories began making shells and gauges for the Ministry of Munitions.

Among the many Trinity fellows to leave was Littlewood, now a second lieutenant in the Royal Garrison Artillery. He had little use for the war, but adapted, just as he had to the Tripos system, affecting a "cheerful indifference." By late 1915, he'd been put to work on a fresh mathematical approach to fashioning antiaircraft range tables. ("Even Littlewood could not make ballistics respectable," Hardy would write, in a gibe at applied mathematics, "and if he could not, who can?") Almost alone among Trinity fellows he was never promoted, and ultimately was relieved of routine chores and allowed to live with friends in London. "Ballistics," one biographer would gently put it, "did not fill all of Littlewood's working hours during the war years"; between 1915 and 1919, he managed to collaborate with Hardy on ten joint papers, in areas of mathematics as distant as could be imagined from the war.

Still, he was away from Cambridge—and away from Ramanujan. And he, and his formidable mathematical powers, had been a prime reason to bring Ramanujan to Cambridge in the first place. So that almost before he had caught his breath in England, after barely four months, Ramanujan had been thrown, more dependently than ever, into the arms of Hardy.

Hardy would later register for military service under the "Derby Scheme," Lord Derby's politically shrewd move to forestall unpopular conscription through a voluntary, but socially pressured, "attestation" of readiness to serve. But Hardy, thirty-eight when the Derby Scheme was launched in October 1915, was, according to Littlewood, deemed "unfit" to serve, and spent most of the war in Cambridge.

In the First World War, unlike the Second, antiwar feeling ran high. Hardy's activities would lead at least one obituary to assert he'd been a conscientious objector. He was not; indeed, late in the war he would write, "I don't like conscientious objectors as a class." Still, as Littlewood tells it, Hardy "wrote passionately about the notorious ill-treatment [to which] objectors were subject." He belonged to an earnest, high-minded group called the Union of Democratic Control, which focused on the peace to follow what was still assumed would be a brief, if bloody conflict. Hardy was secretary of its Cambridge chapter, his old friend from the Apostles, G. Lowes Dickinson, its president. Its first public meeting was held on March 4, 1915. Later, in November, when it was announced that a meeting would take place in the Trinity rooms of Littlewood—who was a U.D.C. member in absentia and had supplied written permission—the authorities moved to block it.

Just outside the gates of Ramanujan's castle the war raged, coming closer day by day. Comely Cambridge had been transformed into a training camp and hospital. In May 1915, the *Lusitania* had been sunk, hardening sentiment against Germany. By June, food prices had risen 32 percent over the previous year's. Intellectual commerce with German mathematicians was cut off. Littlewood and other mathematicians were gone. Hardy was distracted by extra-mathematical concerns.

And yet, so far, the war had not yet reached Ramanujan. Whatever his private revulsion toward it, he yet basked in a kind of warm intellectual spring.

Since his arrival in England, he'd been writing home regularly—at first, three or four times a month, and even now, during 1915, twice a month or so, regularly assuring his family that he maintained his vegetarianism and his religious practices. His letters to friends back in India scarcely mentioned the war, but rather told of his work and his progress, inquired after family members, even occasionally gave advice. To his two brothers back in India, he sent a parcel full of books of English literature.

The persistent bother of finding and preparing food for himself undermined his sense of well-being somewhat. So did the English chill, and the clothes that cinched at his fat waist. And the peculiar absence of letters from Janaki. But mostly, he still rode the crest, kept working, happily and hard, at mathematics.

In his first letter to Hardy, he had sought help in publishing his results. And now, during 1914 and 1915, letters home showed how preoccupied he was with seeing his work in print. "I . . . have written two articles till now," he wrote in June 1914. "Mr. Hardy is going to London today to read a paper on one of my results before the London Math. Society."

"I have written three papers till now. The proof sheets have come. I am writing three more papers. All will be published at the end of the vacation, i.e. in October," he wrote in August.

"I am very slowly publishing my results owing to the present war," he wrote in November 1914.

"It will take some months for me to write that paper systematically and publish it," he wrote Narayana Iyer in November 1915.

After years as a mathematician known only to himself, then only to Madras, Ramanujan plainly relished the prospect of appearing in prestigious English mathematics journals. To appear in print was the only tangible sign of recognition you could hold up to family and friends, the only way the world would know what you'd done. At one point, when he learned that Ananda Rao, a mathematics student at King's who was the son of a Madras judge and relative of R. Ramachandra Rao, was preparing an essay for a Smith's Prize, he went straight to Hardy to ask whether he, too, might try for it. *Doing* mathematics satisfied deep emotional and intellectual needs in Ramanujan. But credit, kudos, *appreciation* satisfied quite other, just as insistent, needs. Intellectually, Ramanujan was not, and never could be, "just like everyone else." But in this social, public realm, he was: he wanted recognition, needed it.

And now he got it. In 1915, no fewer than nine of his papers appeared, five of them in English journals, as against six—all but one of them in the *Journal of the Indian Mathematical Society*—during the previous twenty-seven years.

The advantages he now enjoyed in England were hardly lost upon him. He had originally promised to return to India after two years. And when he wrote his friend Subramanian in June 1915, he still talked of returning to India the following year. But, in another letter the following month, he suggested his return might be only temporary. "I think it may be necessary," he wrote, "to stay here a few years more as there is no help nor references in Madras for my work."

For a while after he came to Cambridge, apparently, Ramanujan's shyness was read as unfriendliness, and students sometimes taunted him. But by now he was a more popular, even legendary figure, his room accorded the status of a shrine. "It was a thrill to me to discover on reaching Cambridge in July 1915 that I was going to be a contemporary of Ramanujan," recalled C. D. Deshmukh, until then a student at a college in Bombay. "Ramanujan's current achievements were common talk amongst us Indian students." Ananda Rao remembered Ramanujan at tea parties and other social gatherings, mixing freely among both English and Indians. Mahalanobis, with whom he took Sunday morning strolls, recalled him as reserved in large groups, expansive in small ones. In any case, he was the toast of the Indian students—*the* mathematical genius, the man whom the English had moved heaven and earth to bring to Cambridge. Two surviving photographs from this period show him in a group; in both he occupies the very center of the composition.

Ramanujan mostly stayed in Cambridge during the long periods between terms. (The academic calendar came to only about twenty-two weeks a year.) But sometimes he'd get to London, where he visited the zoo or the British Museum. One time, he and his friend G. C. Chatterji went to see *Charley's Aunt,* a farce about Oxbridge undergraduate life. In it, young Lord Fancourt disguises himself as Charley's aunt from Brazil, the requisite chaperone for the girl Charley has invited to his rooms for lunch. Since its first performance in 1892, it had become a hardy annual of the London stage, full of wigs and fluttering eyelashes, of tea poured in hats, of silliness and comic confusion. It was, of course, just Ramanujan's speed. He laughed until he cried.

By mid-October of 1915, Ramanujan had moved from his rooms in Whewell's Court to new ones on Staircase D of Bishop's Hostel, just behind Great Court. This wasn't the original Bishop's Hostel built in 1669, whose sharply sloping roof and gabled windows loomed just outside his east window. Two red-brick structures had been built adjacent to the old building in 1878, on the site of what had once been the college stables, and Ramanujan's rooms were on the second floor of one of them. He had one large sitting room, perhaps fourteen feet wide by twenty long, with a tiny bedroom and a small cooking area—actually the "gyp" room, a combination coal cellar, pantry, and kitchen—just off it.

The layout was almost identical to his old rooms in Whewell's Court. But now, Hardy was closer yet. Kumbakonam . . . Madras . . . Chestertown Road . . . Whewell's Court . . . it was as if some relentless force drew him ever closer to Hardy. Now, in Bishop's Hostel, all that divided them was a hundred paces.

Late in 1915, Ramanujan's big paper on highly composite numbers, his most important body of work during that first year or so in Cambridge, appeared in the *Proceedings of the London Mathematical Society.* Back the previous June, Hardy had introduced his friends at the Mathematical Society to it. In November, a manuscript was ready. But revisions were required, and it didn't reach final form until March 1915. Now, at last, it was in print.

So long, and ranging over so vast a terrain, it was divided into discrete numbered sections and even had its own table of contents page to guide readers through it. Sometime earlier, Ramanujan had invited W. N. Bailey and another mathematician, S. Pollard, to see it. "He started at the beginning," Bailey would recall, "and quickly turned over the pages as he explained the ideas and the arguments very briefly. Pollard wrestled manfully with the argument and was rewarded by a severe headache. I gave up the struggle earlier."

A composite number, recall, is a number that is *not* prime. The number

21 is composite, the product of 3 and 7. So is 22, whereas 23 is prime, the product only of itself and one. Now for each composite number, you can list all the numbers that divide it. For example, 21 has the divisors 1, 3, 7, and 21. For 22, it's 1, 2, 11, and 22. Twenty-four can be divided by 1, or 2, or 3, or 4, or 6, or 8, or 12, or 24.

And this last one, 24, was the kind of number with which Ramanujan's paper dealt. Its eight divisors numbered more than those of any other composite number less than 24—and made it, in Ramanujan's terminology, "highly composite." Twenty-two has four divisors, 21 has four, 20 six. None less than 24 has even as many as seven, much less eight. A highly composite number, then, was in Hardy's phrase "as unlike a prime as a number can be." Ramanujan had explored their properties for some time; in the earliest pages of his second notebook he'd listed about a hundred highly composite numbers—the first few are 2, 4, 6, 12, 24, 36, 48, 60, 120—searching for patterns. He found them.

The prime factors from which any composite number, *N*, was built could be written in the form,

$$N = 2^{a_2} \times 3^{a_3} \times 5^{a_5} \ldots$$

where a_2, a_3, a_5, and so on are just the powers to which the prime numbers, 2, 3, 5 . . . , are raised. The highly composite number 24, for example, can be viewed as $2^3 \times 3^1$. In this notation, then, $a_2 = 3$, and $a_3 = 1$. Ramanujan found that, for any highly composite number, a_2 was always equal to or larger than a_3, a_3 was always equal to or larger than a_5, and so on. Count as high as you liked, you'd never find a highly composite number like $N = 2^3 \times 3^4 \times \ldots$ *Never.* And he showed that, with two exceptions (4 and 36), the last a_n necessary to construct a highly composite number was *always* 1. He went on to prove these and other truths through fifty-two pages of reasoning that Hardy would term "of an elementary but highly ingenious character."

The problem Ramanujan had addressed, Hardy observed, "is a very peculiar one, standing somewhat apart from the main channels of mathematical research. But there can be no question as to the extraordinary insight and ingenuity which he has shown in treating it, nor any doubt that his memoir is one of the most remarkable published in England for many years."

* * *

At Cambridge, every student had a tutor who looked after him and monitored his progress. Ramanujan's was E. W. Barnes, who would call Ramanujan perhaps the most brilliant of all the top Trinity students (which included Littlewood) to have come before him. Though later to become a bishop in the Anglican church, Barnes was now a mathematician of some standing; he had been one of Andrew Forsyth's earliest disciples, had with Hardy been among the leading advocates of Tripos reform, and had made substantial mathematical discoveries of his own. Now, in November 1915, he wrote Francis Dewsbury, registrar of the University of Madras, of Ramanujan's progress, which he termed "excellent. He is entirely justifying the hopes entertained when he came here." His two-year scholarship, soon coming to an end, ought to be "extended until, as I confidently expect, he is elected to a Fellowship at the College. Such an election I should expect in October 1917." Ramanujan, he was saying, was in line to become a Fellow of Trinity.

A few days later, Hardy also wrote Dewsbury, calling Ramanujan "beyond question the best Indian mathematician of modern times. . . . He will always be rather eccentric in his choice of subjects and methods of dealing with them. . . . But of his extraordinary gifts there can be no question; in some ways he is the most remarkable mathematician I have ever known."

In Madras, Sir Francis Spring joined the chorus, specifically requesting a two-year extension of Ramanujan's scholarship. Through the late fall and early winter of 1915–1916, Madras authorities debated whether the scholarship should be extended for one year or two. If just one, Spring wrote Dewsbury, Ramanujan was inclined not to return to India during the summer of 1916 as planned; for should his scholarship end in the spring of 1917, he'd be back in India just nine months later anyway.

The university held, however, that, if the Trinity fellowship went through, its own scholarship would overlap. So one year, with the possibility of further extensions, it was.

The Madras scholarship came to 250 pounds a year, which was supplemented by a 60-pound-per-year "exhibition" from Trinity. In 1914, the average English industrial worker made about 75 pounds per year. The threshold for paying income taxes—reached by less than 7 percent of the working population—was 160 pounds. So even with the 50 pounds he sent to support his family in India, Ramanujan was comfortably fixed—

especially given what Hardy would call Ramanujan's "almost ludicrously simple tastes."

Ramanujan had no official college duties. He could do as he pleased. He could immerse himself in mathematics without fretting over financial want, either his own or his family's. Yet something still nagged at him—his lack of a degree, the tangible public marker of academic achievement. In his case, it was the merest formality. But he *wanted* it.

Admission as a research student normally meant you already held a university diploma or certificate. But in his case, the requirement had been waived. And now, in March 1916, he received a B.A. "by research," on the basis of his long paper on highly composite numbers. He'd put up his five pounds dissertation fee. He'd paid two pounds each to his examiners. And now, a dozen years and two college failures after leaving Town High School, he had his degree.

In the early afternoon of March 18, Ramanujan posed for a photograph to mark the occasion with a group of students in their academic robes outside the Senate House. The shortest and stockiest of the lot, he stood squarely at attention, like an army recruit in boot camp, his mortarboard sitting flat atop his head. His trouser legs were a couple of inches too short. His suit bulged, its buttons straining.

Whether because the scholarship had been extended for only a single year or for fear of the U-boats then ravaging British shipping, Ramanujan didn't return to India in the spring. Over much of the next year, he continued instead to work with Hardy on a problem that would indissolubly link their two names in the annals of mathematics.

That June, Hardy followed up his letter of the previous year to Dewsbury with an official report, his delight in telling of Ramanujan's progress marred only by the war:

> In one respect Mr. Ramanujan has been most unfortunate. The war has naturally had disastrous results on the progress of mathematical research. It has distracted three-quarters of the interest that would otherwise have been taken in his work, and has made it almost impossible to bring his results to the notice of the continental mathematicians most certain to appreciate it. It has moreover deprived him of the teaching of Mr. Littlewood, one of the great benefits which his visit to England was intended to secure. All this will pass; and, in spite of it, it is already safe to say that Mr. Ramanujan has justified abundantly all the hopes that were based upon his

> work in India, and has shown that he possesses powers as remarkable in their way as those of any living mathematician.

Hardy's account of Ramanujan's work was "necessarily fragmentary and incomplete," he apologized. But

> I have said enough, I hope, to give some idea of its astonishing individuality and power. India has produced many talented mathematicians in recent years, a number of whom have come to Cambridge and attained high academical distinction. They will be the first to recognize that Mr. Ramanujan's work is of a different category.

The previous December, British, Australian, and New Zealand troops had suffered a catastrophic defeat at Gallipoli. U-boats continued to take their bloody toll of Allied ships. The machine guns chattered away in France. By the end of the year, lists of sometimes four thousand casualties per day would appear in the newspaper.

For Ramanujan, in mid-1916, things could hardly have been brighter. But he, too, would ultimately be struck down by the war.

CHAPTER SEVEN

The English Chill

[1916 to 1918]

1. HIGH TABLE

Beneath the seemingly unruffled surface of Ramanujan's life in wartime England lay signs that his nerves were tautly strung, his sensitivities balanced on a hair trigger.

Sometime probably in early 1916, he learned that his friend Gyanesh Chandra Chatterji, a twenty-one-year-old from the Punjab region who held a Government of India state scholarship to study in Cambridge, planned soon to marry. To help celebrate the good news, he invited Chatterji and his fiancée to dinner.

Back in India, Ramanujan had probably never cooked in his life, had conceivably never even stepped into a kitchen. But here, with neither wife nor mother to serve him and unwilling to trust to the vegetarian purity of the college kitchen, he'd had to learn. Sometimes, on Sundays, he had Indian friends over for *rasam,* a thin peppery soup, or other South Indian fare. "Delicious," a friend later recalled. And once, S. Kasturirangar Iyengar, owner and editor of South India's preeminent English-language newspaper, the *Hindu,* visited him in Cambridge and lavished praise on the *pongal,* a lentil and rice dish, that Ramanujan served him.

By now he took no little pride in his culinary skills and, to honor Chatterji and his fiancée, set about preparing them a feast.

On the appointed day, Chatterji showed up at Ramanujan's rooms in Bishop's Hostel. With him was his fiancée Ila Rudra, a student at the local teacher's college; and, probably as chaperone, Mrinalini Chattopadhyaha, a woman from Hyderabad in her early thirties then studying ethics at Cambridge's Newnham College, who would go on to become active in the Indian labor movement and run a school for untouchables.

With his guests seated and his apartment awash in the cooking smells of South India, Ramanujan served soup. All was well.

Did his friends wish to have more? asked Ramanujan after a time. They did indeed.

Then: a third helping? Chatterji apparently accepted. But this time, his fiancée and the other woman declined. The dinner conversation continued. . . .

Where is Ramanujan? Abruptly, they looked around and realized they were alone in Ramanujan's apartment. Their host had vanished.

For an hour, Chatterji and the others waited. Finally, he walked down the single flight of stairs, out the door, through the gate into Great Court, and across the cobbled courtyard to the porter's lodge, where he inquired after Ramanujan. Why, yes, Chatterji was told, Mr. Ramanujan had called for a taxi.

Perplexed, he returned to Ramanujan's rooms and sat there with the others, waiting. Until ten o'clock they waited, the time by which guests had to leave. Still no Ramanujan.

And no Ramanujan, either, the next morning, when Chatterji checked on him. For four days, he heard nothing, grew increasingly fearful. Then, on the fifth day he received a telegram from Oxford, about eighty miles away. Could Chatterji wire him five pounds? The amount was today's equivalent of three or four hundred dollars.

Next day, Ramanujan was back in Cambridge. "I felt hurt and insulted when the ladies didn't take the food I served," he told Chatterji. "I did not want to come [back] in while they were in the house." He had needed to get as far away, as fast as he could; with the money in his pocket, that was Oxford.

It was the same impulse, in the face of what he viewed as humiliation, that had driven him to Vizagapatnam ten years earlier. It was the rash, precipitous gesture of a man stretched thin.

To all appearances, Ramanujan had made a splendid adjustment to a foreign country and an alien life. Mathematically, he had lived up to the fondest hopes of Hardy, Littlewood, Neville, and his other English friends. Socially, too, he had seemed to adapt, at least at first. Neville would tell of his delight in cracking jokes and discussing philosophy and politics, would speak approvingly of his simplicity and "instinctive perfection of manners."

But inside, Ramanujan was like a checking account from which funds are only withdrawn, never deposited. Doing mathematics took vast personal energy. So did adjusting to his new life in England, as anyone will attest who has ever tried to penetrate a foreign culture. Together they drained his physical and emotional reserves. Eventually, the account must run dry.

Sometime in 1916, perhaps around the time of the Chatterji dinner, it did. A constellation of forces had conspired to stretch thin his nerves, weaken him physically, isolate him socially. Indeed, the meal he prepared for Chatterji and his fiancée may have loomed far larger in his mind than it ever did in theirs. For Ramanujan normally went without mealtime company, unlike Hardy and the other Trinity fellows.

The college fellows in their black academic robes solemnly trooped into the great candlelit Hall. A webbed understructure of wooden arches graced the high ceiling. Dryden and Tennyson, Newton, Thackeray, Francis Bacon, and other Trinity notables glowered down from their portraits on the walls. Silver adorned bare wood tables; the tablecloths of peacetime were gone now for fear the light they reflected might beckon German zeppelins. Once all the fellows were seated, the senior among them, sitting at one end of the long table, recited a Latin grace. And the meal began.

This was High Table, so called because the tables at which the fellows ate were set on a platform built up about four inches from the floor. At Trinity, as elsewhere in Cambridge, it was the focal point of the college's social life, and long-standing notions of conversational good form governed it.

For one thing, you never got too serious. "If one has done a hard day's thinking one does not want to *work* at conversation," Littlewood would say. "Dinner conversation is in fact easy and relaxed. No subject is definitely barred, but we do not talk shop in mixed company, and, Heaven be praised, we abstain from the important and boring subject of politics." It was a place not of great profundity but of wit, wine, and release—release from the high tension of translating ancient Greek, or writing about the fall of Constantinople, or proving a new theorem in the theory of numbers. Here, the Important receded into distant memory. Here, trifles had their day.

You could see it in the little book kept over the years that recorded suggestions and criticisms of meals served at High Table. Hardy had been among several to declare in 1909 that Berlin pudding and Hirschhorn fritters ought never appear again on the menu. Later, he and Littlewood defended plum pudding with wine and camperdown sauce against those who would see it blacklisted; over the days and weeks, alliances were built around the issue, compromises forged. At another point, Hardy was embroiled in a battle over whether fruit pies should be served hot, subscribing to the view that, as another fellow put it, "a man who will eat a *hot* fruit pie is unfit for decent society."

It was good fun, evenings of camaraderie that eased days of hard solitary work. Littlewood got his share of it over the years, before he left for the army. So did Neville. So, of course, did Hardy, for whom High Table, with its light, frothy repartee, was his natural terrain. During the war, officers attached to the hospital unit or quartered at the college showed up. So may have a few Indians. But not Ramanujan. For him, dinner in Hall was a slice of Trinity life from which he was excluded.

"Mutton, sir? Beef, sir?" the waiters would bustle around the noisy hall. *That* was the problem.

Ramanujan was a vegetarian unusually strict in his orthodoxy—if not for South Indians generally then at least for those in England. In crossing the seas he had defied Brahminical strictures. He had forsaken his tuft. He mostly wore shoes and Western clothes. But as he had promised his mother, he clung fiercely to the proscriptions most central to Brahminic life, on food.

The story is told of a hungry Brahmin who requests food from a man he meets along the road. He knows nothing of the man's caste, or character, but he is too hungry to ask. Belly full, he sets out again and, after a few minutes, reaches the house of a Brahmin, who puts him up for the night. That evening, he spies a gold statue in the house, aches to have it, and under cover of night, spirits it away. All the next day, his guilt mounts. Finally, shaken by remorse, he returns to the house to give it back. "Oh, I knew you would steal something," says the host. "You see, I saw you take food from that man yesterday, and I knew he was a thief."

It was this spirit that was built into Brahminic food prohibitions—that from whom you take your food *matters*. In accepting food from just anyone, who knows what sins he has committed in this or a former life?

Brahmins varied, naturally, in the details of their observance. Some forswore onions and garlic on the theory that these foods raised sexual appetites, while others extended the prohibition even to cabbage and potatoes. Remembered details differ on what Ramanujan would and would not eat. One friend has him eating eggs, another not. Some said he refused onions, or even tomatoes. But all recall the absolute rigidity with which he clung to his observance.

Early in his stay, he had once or twice ordered fried potatoes from the college kitchen. *Fried in lard,* joked another Indian, also a Tamil Brahmin, trying to get his goat. That was all it took. Whether true or not, Ramanujan never again ordered anything from the college kitchen.

Instead, he cooked in the tiny, small-windowed alcove, equipped with electricity and a little gas stove, just off his sitting room. When he could get the ingredients, he ate what he'd eaten back in India—rice, yogurt, fruits, *rasam,* and sambhar, a thick, spicy, potato-laced vegetable stew. His friend Mahalanobis recalled him standing over that little stove, stirring vegetables over the fire. But it was something he did much more often alone.

From his window, Ramanujan could look out over the roof of the adjacent building to the spired steeple atop the college Hall. There, at High Table during the long winter evenings of 1916, the candles flickered, the conversation hummed. But of all that, Ramanujan never shared.

2. AN INDIAN IN ENGLAND

Ramanujan was not the first Indian to come to England for an education and feel isolated from the alien world around him. Mohandas K. Gandhi, the future apostle of nonviolence and leader of the Indian independence movement, arrived in England in 1887, the year of Ramanujan's birth. He wrote later:

> I would continually think of my home and country. My mother's love always haunted me. At night the tears would stream down my cheeks, and home memories of all sorts made sleep out of the question. It was impossible to share my misery with anyone. And even if I could have done so where was the use? I knew of nothing that would soothe me. Everything was strange—the people, their ways, and even their dwellings. I was a complete novice in the matter of English etiquette and continually had to be on my guard.

Between 1892 and 1906, Cambridge University admitted about twenty Indians per year, during Ramanujan's time a little more. At any one moment, something like a thousand Indians were scattered among English colleges. In any case, their numbers and the magnitude of their problems in adjusting to English life were enough to spur government studies of them in 1907, in 1922, and again in later years. And the recurring theme of these studies, which is ubiquitous, too, in stories told by Indians over more than a century, was the maddening reserve, the unfathomable distance, of the ordinary Englishman.

Sometimes, to be sure, Indians experienced downright racial prejudice; it was common enough, for example, that ugly rumors of it reached Ramanujan's mother in India. But more often, it was the peculiar shyness of the English that left Indians feeling so adrift. "We think it must be admitted," concluded a report by the Lytton Committee on Indian Students in 1922, "that the British and the Indian student each has his racial characteristics which imposes an initial barrier to intimacy." Whereas the Indian, for his part, tended to be oversensitive to any hint of patronage, "the British has a reserve which causes him to be slow in making friends even with his own countrymen, and he is apt to regard with suspicion the first attempts of any stranger to cultivate his acquaintance."

Observed a student in a much later study: "The initial difficulty is to break the extraordinary reserve of the English people, their correct but cold behavior, formal, unemotional, courteous and decent to a degree, but detached, both to take sides or involve themselves."

A still later student complained of having to stock a ready supply of pleases and thank-yous. To him, it was just crude barter: you give me something, I give back a thank-you. Far worse, though, was English indifference. Some Indians learned to accept, even embrace, it as "respect for privacy," but most simply saw it as unfeeling and cold. The conversational task of the Englishman, it could seem, was to be scrupulously correct yet remain untouched and unmoved, taking care always to convey the clear impression that, at bottom, he just didn't give a damn. One Indian student would dream of dying, his body going undiscovered for weeks, reeking and rotting all the while. Neville would read the fact "that none of us, as far as I know, ever pressed [Ramanujan] for the true reason for his initial refusal to come to England" in 1913 as typical of "the reticence of his English friends."

Nowhere were these traits more prevalent, of course, than in the patrician class. In the House of Lords just before World War I, one of their number perceived a polite "detachment almost amounting to indifference." Another complained of the "profound weariness and boredom" its members affected. One later visitor to England, the Indian Nirad Chaudhuri, wrote how "One evening, when dining at a club, I tried in my innocence to open a conversation across the table, and I admired the skill with which the intrusion was fended off without the slightest suggestion of discourtesy." Fended off, though, it was.

Back home, Indians recalled, people would come up to you, sit down, start talking, and in five minutes know all about you—whether you were married, had children, where you were from, what kind of work you did. Whereas in England, returning Indians advised their countrymen, formal introductions were de rigueur. One story, set in India, told of a swimmer whose cries for help sent everyone rushing to his aid. Everyone, that is, save the lone Englishman, who sat where he was, apparently unmoved. "Oh," he replied when asked later why he'd not helped, "were we introduced?"

Cambridge boasted its own brand of aloofness. A book aimed at Indian students in England told how even college porters went about their duties "without the least concern about our new comer and with an air of indifference." And a student of Ramanujan's tutor, E. W. Barnes, once set Barnes apart from most other Cambridge dons who, he said, gave "the impression that they were not greatly interested."

Laurence Young, the son of two mathematical contemporaries of Hardy and Littlewood at Cambridge, told how when Littlewood, as an old man, visited him in Wisconsin he would row him out on to the lake to view the sunset. Littlewood never said a word, and Young, in time, surmised that he was bored. But when one day Young suggested that the water might be too rough for their excursion, "his face dropped . . . and I quickly looked again at the lake and pronounced it smooth. This is typical of Cambridge—what you admire, you merely do not speak of." Littlewood himself would remark that "When you make your speech at a Trinity Fellowship Election, do not expect them to break into irrepressible applause; no one will blink an eyelid."

It was a wall erected around one's feelings, a great silence of the emotions. In Cambridge, the emphasis was on ideas, events, things, work, games—*anything*, it seemed, but the deeply personal.

At first Ramanujan may have marveled at these peculiarities of the English, much as Western tourists do the sight of untended cattle on the streets of Delhi or Madras. But he was not one to eagerly embrace foreign ways; rather, he was apt to stand apart from them. Stubborn, self-driven, self-willed, he was every bit the product of his country and its customs. Unlike some Madrasi intellectuals, he'd never lived a Westernized life in India; he was too much, and too recently, a son of Kumbakonam. He maintained an Indian's deep-felt deference to the wishes of his parents. He'd cried when his *kutumi* was cut off. In England, he remained steadfastly vegetarian, kept a poster of an Indian deity in his room, and each morning performed the Brahmin's fastidious morning ritual. He changed into a new, ritually pure dhoti, applied the *namam* caste mark on his forehead, performed his devotions, then wiped it off. Only then, if he were going out, would he don Western clothes.

Even if he didn't always let his mind wander back home, there were times when the small, familiar things of South Indian life insinuated their way into his awareness. The smells of his mother's cooking on Sarangapani Sannidhi Street, or of burning cow dung in the streets of Madras. The bright colors of religious festivals parading down the streets of Kumbakonam, accompanied by the strumming and jingling of the musicians. The reds and oranges of sari-clad women along the banks of the Cauvery, white dhotis setting off the dark brown skin of laborers in the fields. The vibrant greens of the vegetables, the coconuts and bananas and mangos sold down by the market near the river. And always, the bright blue sky and high overhead sun.

Among the English, Ramanujan could not long forget his foreignness. His musical accent was alien to their ears. His skin was darker by many shades than theirs—which, the winter chill bringing color to their cheeks, was more a rosy pink than anything you could call "white." Everywhere were churches and chapels, Christian crosses, and Jesus Christ. Here, in this strange land, families scattered, children paying parents nothing like the respect that was a law of social life in India. Meanwhile, all Cambridge resounded to the sound of tramping feet bound for an alien war.

Even the prevalence of body odors among the English mystified him—until, the story goes, one day he was enlightened about it at a tea party. A woman was complaining that the problem with the working classes was that they failed to bathe enough, sometimes not even once a week. Seeing disgust writ large on Ramanujan's face, she moved to reassure him that the Englishmen *he* met were sure to bathe daily. "You mean," he asked, "you bathe only *once* a day?"

Ramanujan was not the kind of chameleonlike figure who does well at the tough job of reshaping himself to fit a foreign culture; he was not flexible enough, could not sink down effortlessly into his new English life.

Nor could he long be immune to that succession of subtle, slight rebuffs the ordinary Englishmen dispensed, with scarcely a thought, every day. He would have needed to experience few incidents of aloofness and reserve to damp his sunny openness and send him running back to the cozy den of his mathematical research.

And that, it seems, is what he did. He withdrew. "I remember him during those years, though I never spoke to him," recalled B. M. Wilson later. "He was in fact very rarely seen." For long stretches, he scarcely left his room. Back in India, he liked to work in the cool of the night, the better to escape the midday heat. Now, even without the heat, he worked at night, alone.

Ramanujan was not the first foreigner to retreat into his shell in a new country; indeed, his was the typical response, not the exceptional one. One later study of Asian and African students in Britain observed that a sense of exclusion "from the life of the community . . . constituted one of the most serious problems with which they were confronted . . . [and had] a serious psychological effect" upon them. Another study, this time of Indian students in particular, reported that while 83 percent of them saw friends more or less every day back in India, just 17 percent did while in England.

During the Easter Term of 1916, Hardy was a member of the Trinity Sunday Essay Society. So was E. H. Neville. So was Bertrand Russell. Ramanujan was not. While Hardy played tennis or whatever cricket could still be had in wartime England, Ramanujan remained steadfastly sedentary. When some of his friends joined the Majlis, the Indian students' debating society (where, it has been said, "succeeding generations of scions of Indian aristocracy picked up their nationalism and radicalism"), Ramanujan did not. When Ramanujan's paper on a type of Diophantine equation was read to the Cambridge Philosophical Society on October 30, 1916, it was Hardy who read it, not Ramanujan. When Hardy read a joint paper at a meeting of the London Mathematical Society on January 18, 1917, Littlewood was there to hear it, and so was Neville, and so was Bromwich—but not Ramanujan.

There, in Bishop's Hostel, Ramanujan was a shut-in.

From the small west window of his sleeping alcove, the Gothic windows and stone walls of New Court loomed just a few feet away. The east window got whatever morning sunlight wasn't blocked by the building across the way. But there usually was not much sun to block. The English skies were notoriously overcast. And around Christmas, Cambridge—

which is as far north as Labrador—would be mostly dark by four in the afternoon. Even the long summer days were, by Madras standards, sunless; a typical July brought rain a dozen days a month.

The war only made things worse. Cambridge and surrounding East Anglia, which projected out into the North Sea toward the Continent, were particularly exposed to zeppelin raids. In part to protect King's College Chapel, streets were kept dark at night. Cambridge, someone would remember, "was wrapped in a medieval gloom for some three years and men had to grope their way about the streets as best they could." If the England streetscape was normally gray and drear, now it was black, fairly pressing you back inside.

In South India, the boundaries between inside and outside were not so fixed and immutable as they were in England, where you were forever trying to escape the chill. Walls and windows were more permeable. Insects, smells, and sounds brought outside inside. Chipmunks and lizards scampered through window shutters. Whereas in Cambridge, amidst the stony, solid permanence of its five-hundred-year-old walls, there was an ever-present sense of demarcation and division.

So that in the winter, especially, Ramanujan's apartment could feel like a prison. A plush-lined prison, perhaps, but a prison nonetheless. And while *driven* back into it by the English reserve, the winter chill, and the dark streets and wartime gloom, he was *lured* back into it by the delight he got from his work with Hardy.

3. "A SINGULARLY HAPPY COLLABORATION"

Mathematician Norbert Wiener would one day note how, in one sense, number theory blurs the border between pure and applied mathematics. In search of concrete applications of pure math, one normally turns to physics, say, or thermodynamics, or chemistry. But the number theorist has a multitude of real-life problems before him always—in the number system itself, a bottomless reservoir of raw data. It is in number theory, wrote Wiener, where "concrete cases arise with the greatest frequency and where very precise problems which are easy to formulate may demand the mathematician's greatest power and skill to resolve."

In 1916, one such problem lay in the area of number theory known as "partitions."

On the surface, the problem was so simple it went back to almost the first days of grammar school: $2 + 2 = 4$. But that's just one way to add up numbers to get 4. There are others. Like 1+3. Or 1+1+2. Or 1+1+1+1. And, lastly, to be scrupulously complete about it, just plain 4 itself. These (aside from mere rearrangements of the same numbers) are the *only* ways of adding up integers to get 4. Count them up, and you get five different ways, or "partitions." Mathematicians say that the number of partitions of 4 is 5. Or,

$$p(4) = 5$$

More generally $p(n)$, read "*p* of *n*," represents the number of partitions of any number, *n*, and is known as the partition function. What, the mathematician wonders, can be said about it? How could we evaluate it for any *n*?

In principle, it's not difficult to go through all the possibilities for a given number and add them up. *In principle*. The problem is that the number of partitions rises very fast. The number of partitions of 3, $p(3)$, is just 3 (3, 1+1+1, 1+2). But by the time you get up to 10, $p(10) = 42$. And $p(50) = 204{,}226$. Two hundred four thousand, two hundred twenty-six different ways of adding up the integers. Just listing them all, one every five seconds, would take two weeks. And that's just for $n = 50$. So the question is, can you find a formula for $p(n)$ that sidesteps the awful arithmetic and spits out the number of partitions for any number you please?

As usual, it was Euler who made the first real dent in the problem. Working from an area of mathematics which later became known as elliptic modular functions, Euler detoured around the problem to construct what is known as a generating function. In theory, a generating function supplies not just a particular answer to a particular problem but all the answers to all related problems; it *generates* answers as fast as the numbers can be plugged in. With it, you step off into new mathematical terrain where the object of attention seems no longer to be $p(n)$, but some new function, $f(x)$. Except that with $f(x)$ in hand you can go back and get what you really want, $p(n)$. In this case, Euler's generating function gave rise to a "power series," which is just a series of terms each of successively higher powers and each multiplied by something, some coefficient:

$$f(x) = 1 + \sum p(n)x^n$$

which means:

$$f(x) = 1 + p(1)\,x^1 + p(2)\,x^2 + p(3)\,x^3 + \ldots$$

It was those coefficients, $p(1)$ and so on, that potentially supplied the answers. Because if you could actually evaluate the power series, the coefficients would turn out to be not any old numbers, but just the $p(n)$ desired. In Euler's scheme, if you wanted, say, $p(50)$—which is 204,226—you'd just head out to the fiftieth term of the series, where you'd note the coefficient of x^{50}. So if you had done it right, that part of the series would read:

$$\ldots + 204{,}226\,x^{50} + \ldots$$

Now, this was not just black magic. It was *natural* that the generating function for partitions be a power series, because powers combine by addition. Thus $4 \times 8 = 32$ can be written as:

$$2^2 \times 2^3 = 2^5$$

Their exponents *add up*. And adding things up is just what you do in partitions. ("A moment's consideration shows that every partition of n contributes just 1 to the coefficient of x^n," Hardy would observe in tracing the algebraic links between power series and partitions.) Squint, then, and even without following the logic in detail, you can roughly discern the route between the two seemingly unconnected realms.

This, at any rate, was how Euler set up the problem. But Euler didn't solve it, didn't say how to use his generating function to actually churn out the desired power series. He offered a strategy for arriving at $p(n)$, but little more. As Hardy and Ramanujan would write in their big paper in the *Proceedings of the London Mathematical Society*, "we have been unable to discover in the literature of the subject any allusion whatever to the question of the order of magnitude of $p(n)$." No one, in other words, had a clue.

The seeds for the attack Ramanujan and Hardy were to take came in Ramanujan's first letter to Hardy in January 1913. There, on page 7, he gave a particular theta series—a power series of a certain kind, containing only terms with squared powers, like x^1, x^4, and x^9—and claimed that to determine its coefficients one had but to evaluate the mathematical expression he furnished and select the integer closest to it.

This was not quite right. "The function," Hardy would write, "is a genuine approximation to the coefficient, though not at all so close as Ramanujan imagined." And yet, he went on, "Ramanujan's false statement was one of the most fruitful he ever made, since it ended by leading us to all our joint work on partitions"; indeed, Ramanujan's function from 1913 bore a strong family resemblance to the problem's ultimate solution, which came only in 1937.

In the course of their work, Ramanujan and Hardy were led to what would come to be known as the circle method. The circle method made use of Cauchy's theorem, which might seem at first as inappropriate for the job as anything you could imagine. Cauchy's theorem lay in the domain of "analysis," the broad area of mathematics that includes calculus and deals with "continuous" rather than "discrete" quantities; how long you've been pregnant is a continuous quantity, how many children you have a discrete one. Partitions were a discrete quantity; you couldn't have 6.719 partitions just as you couldn't have 6.719 children. It had to be 6, exactly, or 7.

But Cauchy's theorem had a history of being applied to such problems by now, and Hardy could be thanked for it; he was foremost among twentieth-century mathematicians in pioneering the "analytic theory of numbers," which took the powerful tools developed over more than two centuries for continuous quantities and, through feats of mathematical legerdemain, applied them to the integers of number theory. In an account of his work with Ramanujan, Hardy noted that "the idea [of using Cauchy's theorem had become] an extremely obvious one: it is the idea which has dominated nine-tenths of modern research in the analytic theory of numbers; and it may seem strange that it should never have been applied to this particular problem before." It hadn't, he went on, in part because of "the extreme complexity of the behavior of the generating function $f(x)$ near a point of the unit circle."

Hardy referred to the fact that the integral at the heart of Cauchy's theorem couldn't be evaluated as it stood, because the "contour" over which it was to be integrated held impermissible points, where its value wasn't defined. So they would have to find an alternative path *close to* the forbidden "unit circle," systematically dissect it, make approximations as they went along. This was the new strategy.

Basing a result on a series of approximations seemed to guarantee that it would itself be an approximation. But should they expect anything better? Hadn't prime numbers resisted all efforts at exact calculation, forcing mathematicians to be content with no more than a rough estimation that grew relatively more accurate as *n* increased? Hadn't all Ramanujan's attempts to make the prime number theorem exact run afoul of mathematical pitfalls? Surely partitions would prove similarly intractable, making them happy with almost any decent approximation.

To see how far off they were—to check the worth of their approximations by how close they came to the precise result—they put Major MacMahon to work.

The son of a brigadier general, Percy Alexander MacMahon was a sixty-one-year-old mathematician who had served in the Royal Artillery—including a stint in Madras back in the 1870s—before seriously taking up mathematics. "With his moustache, his 'British Empah' demeanor and worst of all his military background," it was once written of him, "MacMahon was hardly the type to be chosen by Central Casting for the role of the Great Mathematician." But after leaving the army, he had gone on to become a professor at Woolwich, the army school, and since 1904 had been associated with Cambridge's St. John's College. His expertise lay in combinatorics, a sort of glorified dice-throwing, and in it he had made contributions original enough to be named a Fellow of the Royal Society in 1890. MacMahon was a whirlwind of a calculator. Sometimes, in fact, he would take on Ramanujan in friendly bouts of mental calculation—and regularly thrash him.

But now, MacMahon was putting his calculating skills to good use and, using a simple formula that led directly from Euler's earliest work in partitions, had arduously hand-calculated, in impossible, endless streams of numbers, the values for the first two hundred $p(n)$. By the roughest of analogies, he was adding up 29 + 29 + 29 + . . . 1,000,001 times to get 29,000,029, while Hardy and Ramanujan sought a way to simply multiply 29 by 1,000,001. But they didn't have it yet, and MacMahon's tedious efforts supplied a benchmark against which to test it when they did.

When they tried it early in their work, with their approximation strategy still relatively primitive, they were already encouragingly close; for $p(50)$ and $p(80)$, the error stood at just 5 percent. Further refinements, with correspondingly smaller errors, reinforced their conviction that they were on the right track.

Then, in December 1916, the two men, perhaps having gone their separate ways until the start of Lent term in January, came a big step forward. Ramanujan wrote Hardy a postcard, squeezing into the tight space one more contribution to their continuing mathematical dialogue:

> It therefore appears that *in order that $p(n)$ may be the nearest integer to the approximate sum, S need not be taken beyond* $\beta\sqrt{n}/\log n$ *and cannot be taken below* $\alpha\sqrt{n}/\log n$ I hope you can easily prove these. Then the problem is completely solved. Major MacMahon was kind enough to send me a type-written copy of the 200 numbers. The approximation gives the exact number. I think you knew this already from him.

Ramanujan alluded to a way of making the number of terms of the series they used to approximate $p(n)$ itself depend on n. This, as Littlewood wrote later, was "a very great step, and involved new and deep function-theory methods that Ramanujan obviously could not have discovered by himself." What it seemed to mean, as Hardy and Ramanujan would write, was that "we may reasonably hope, at any rate, to find a formula in which the error is of order less than that of any exponential of the type e^{an}; of the order of a power of n, for example, or even bounded."

A *bounded* error, an error they could fix within set limits: they would have been happy with that. But their results were better still: When

> we proceeded to test this hypothesis by means of the numerical data most kindly provided for us by Major MacMahon, we found a correspondence between the real and the approximate values of such astonishing accuracy as to lead us to hope for even more. Taking $n = 100$, we found that the first six terms of our formula gave
>
> $$\begin{array}{r} 190568944.783 \\ +348.872 \\ -2.598 \\ +.685 \\ -.318 \\ +.064 \\ \hline 190569291.996 \end{array}$$
>
> while $$p(100) = 190569292;$$
>
> so that the error after six terms is only .004.

Similar precision applied to $p(200)$. Their method was supplying an answer whose error was not just "bounded"—which could, after all, mean bounded but *large*—but small enough to round off to the nearest integer. "These results," wrote Hardy and Ramanujan, "suggest very forcibly that it is possible to obtain a formula for $p(n)$, which not only exhibits its order of magnitude and structure, but may be used to calculate its *exact* value for any value of n."

This was the shocker. Nothing in the history of this problem, nothing in the work with primes, had prepared them for anything like it. What they had done with partitions was just what Ramanujan had thought he'd done with primes but, Hardy showed him, hadn't. What was more, the uncanny accuracy of their results attested to the power of the approximating technique they had used to get them. The circle method it would be called, from how it let you draw oh-so-near, but never actually touch, the forbidden circular path. So subtle and inspired were the approximations it permitted that it went beyond approximation to promise exactitude. Two decades later, sure enough, Hans Rademacher came up with the missing piece of the puzzle and made the formula exact.

Did Ramanujan suspect there was an exact solution all along? That's what the eminent Norwegian number theorist Atle Selberg has suggested. Selberg, in fact, argues that Hardy's insistence on certain methods of classical analysis actually impeded their efforts; and that lacking faith in Ramanujan's intuition he discouraged a search for the kind of exact solution Rademacher produced twenty years later.

In any case, their partitions solution was big news, the circle method they'd used to come up with it a stunning success. In late 1916, Hardy dashed off an early account, under his own name but offered "as the joint work of the distinguished Indian mathematician, Mr. S. Ramanujan, and myself," to the *Quatrième Congrès des Mathématiciens Scandinaves* in Stockholm. Early the following year, a brief joint paper appeared in *Comptes Rendus,* as "Une Formule Asymptotique Pour le Nombre Des Partitions De n." And a one-paragraph reference to the French journal article appeared in *Proceedings of the London Mathematical Society* in March. The forty-page paper setting out their work in full detail didn't appear until 1918.

Ramanujan and Hardy: as a mathematical team, they would remind Pennsylvania State University mathematician George Andrews of the story of the two men, one blind and the other lacking legs, who together could do what no normal man could. They were a formidable pair. On the strength of their work on partitions alone, which by itself justified Ramanujan's trip to England, their names would be linked forever in the history of mathematics.

Cambridge mathematician Béla Bollobás has observed that while Hardy furnished the technical skills needed to attack the problem,

> I believe Hardy was not the only mathematician who could have done it. Probably Mordell could have done it. Polya could have done it. I'm sure there are quite a few people who could have played Hardy's role. But Ramanujan's role in that particular partnership I don't think could have been played at the time by anybody else.

Whatever the proper assignment of credit, "We owe the theorem," Littlewood would write, "to a singularly happy collaboration of two men, of quite unlike gifts, in which each contributed the best, most characteristic, and most fortunate work that was in him. Ramanujan's genius did have this one opportunity worthy of it."

For Ramanujan, it was all deliciously addictive. A decade before, his discovery of Carr had left him so single-mindedly devoted to mathematics that he no longer could function as an ordinary college student. Now, something like that was at work again, only worse. For now it was not only his own delight in mathematics that spurred him on, but the encouragement he got from Hardy. Hardy, this embodiment of all that was highest and best in the Western mathematical tradition, with his immense technical prowess and rich knowledge of the whole mathematical world of England and the Continent, was all Ramanujan could want in a colleague and mentor. And that *he* saw such breathtaking originality in him could do nothing to restrain Ramanujan's eagerness to get on with their work.

"A character so remarkably free from the petty meannesses of human life . . . the most generous of men." That's what C. P. Snow once said of Hardy. Another time, he called him "freer from the emotion [of envy] than any man I have ever known." Indeed, though Hardy deemed Ramanujan's natural mathematical ability superior to his own, no hint survives of so much as a wisp of envy tainting his relationship with him. Despite any private twinges, he seems always to have been Ramanujan's unalloyed friend and supporter. All his life he championed him, hailed his gifts. Recognizing Ramanujan's genius, he wanted only to push it toward its limits.

And that, if anything, was just the problem.

Hardy was in many ways the best and truest friend Ramanujan ever had. He was considerate, loyal, and kind to him. And yet in at least one way, unintentionally, he probably did Ramanujan harm. For in the ardent hopes he had for him, in his unbridled wish to see that he lived up to his potential—in keeping him to the mark, *driving* him—Hardy only fed Ramanujan's addiction.

He couldn't keep Ramanujan from deepening the hole he was digging for himself. He may even have helped him dig it deeper.

4. DEEPENING THE HOLE

Hardy was an aristocrat of the intellect, raised to value high achievement and dismiss anything less. As a mathematician, he was all Ramanujan could want. But he was also a formidable and distant figure who demanded always, and only, the best from him. "His sense of excellence was absolute; anything less was not worthwhile," remembered an Oxford economist, Lionel Charles Robbins, who knew Hardy later. "What Dr. Johnson said of Burke applied exactly to Hardy: his presence called forth all one's powers."

J. C. Burkill, a Cambridge undergraduate beginning in 1919 who later himself became a prominent mathematician, remembers feeling always intimidated by Hardy—always "below him," as he puts it. Whereas Littlewood, say, came across as thoroughly human and accessible, chatting away amiably in Hall, Hardy was busy being brilliant. "When he was conversing," says Burkill, "he felt he had to be on a high plane." The great Hungarian mathematician George Polya told how Hardy once expressed disapproval of Polya's failure to pursue a promising mathematical idea. The two of them, with a third mathematician, were visiting a zoo. At one point, Polya recalled, a caged bear "sniffed at the lock, hit it with his paw, then he growled a little, turned around and walked away. 'He is like Polya,' remarked Hardy. 'He has excellent ideas, but does not carry them out.' "

You didn't turn to G. H. Hardy for an unjudging shoulder to cry on. Once, when the mathematician Louis J. Mordell wrote him, complaining that his papers were getting picked apart by the editors of a mathematical journal because of minor stylistic faults, Hardy refused to indulge him. "I know I have spent over three hours over the journal proofs of a note of yours," he wrote back, "and have made over thirty corrections on a page. All 'trivialities'—so trivial that you have never noticed them, or at any rate commented on them: but a morning's work gone west." The remainder of Hardy's eleven-page letter showed similar irritation. *Don't be so easy on yourself,* it said.

Hardy, in short, was a stern taskmaster. His was a personality of expectations, of high performance. From him, Ramanujan could get encouragement and, in those ways in which Hardy could express it, friendship—but little in the way of pure, uncritical nurturing.

Hardy's reply to Ramanujan's first letter from India showed that same holding-to-account—*Prove it*—and he never let up once Ramanujan was in England. At one point, writing to Ramanujan, who was then in the hospital, about their current work, his eagerness for the mathematical fray plainly warred with his concern for his friend's health: "If I get out any more I will write to you again. I wish you were better and back here—there would be some splendid problems to work at. I don't know if you feel well enough to think about such difficult things yet." Then, a postscript: "At present you must do what the doctors say. However you might be able to think about these things a little: they are very exciting." As maddeningly equivocal as the letter was, Ramanujan would have been obtuse indeed to miss its message: *The work awaits you.*

And so, if any part of Ramanujan wanted to relax, pursue his pet philosophical notions, investigate the psychic theories of the English physicist Oliver Lodge that so intrigued him, take the train into London to visit the zoo, as he did once or twice, or otherwise stray from mathematics, it got scant support from Hardy.

Hardy's urgings, to be sure, found fertile soil in Ramanujan's own obsessive bent. Later, he would write Hardy that while the hospital in which he was then a patient was uncomfortably cold,

> the bath rooms are nice and warm. I shall go to the bath room with pen and paper every day for about an hour or so and send you two or three papers very soon. This thought did not strike me before. Else I would have written something already. In a week or so I may perhaps have a complaint against me from the doctor that I am having a bath every day. But I assure you beforehand that I am not going to bathe but to write something.

Ramanujan, sick and in the hospital, was *apologizing* to Hardy for having failed to do more mathematics!

His marriage in 1909 had induced Ramanujan to stitch himself back into the wider social world. Now, in England, the threads of connection were once again severed. By early 1917, he was a man on a mission, propelled toward his destiny, oblivious to all but mathematics. After three years in Cambridge, his life was Hardy, the four walls of his room, and work. For thirty hours at a stretch he'd sometimes work, then sleep for twenty. Regularity, balance, and rest disappeared from his life.

He was not the first man to sacrifice his health on the altar of mathematics. Jacobi gave this retort to a friend who worried that excessive devotion to his work might make him sick: "Certainly I have sometimes endangered my health by overwork, but what of it? Only cabbages have no nerves, no worries. And what do they get out of their perfect wellbeing?" And before Jacobi, there was Newton himself. "Never careful of his bodily health," E. T. Bell wrote of him, "Newton seems to have forgotten that he had a body which required food and sleep when he gave himself up to the composition of his masterpiece. Meals were ignored or forgotten."

In Hardy, Ramanujan had the intellectual companionship he'd long missed. In Cambridge, with its stately stone chapels and quiet courts and great libraries, he had before him all the riches of Western civilization. Yet now he went without much else that had sustained him, perhaps without his realizing it, in India. He went without family. Without dark, familiar faces and open, sunny Indian smiles. Without the sound of friendly Tamil. Now those hidden props to identity and self-esteem, so easy to take for granted or dismiss, had been pulled out from under him. Now there was no one even to prepare meals for him. No one, as Janaki and his mother had done, to place food into his hand as he worked. No one to remind him to sleep. No one to cool his fevered brain with a touch, or a sexual embrace. No one to counsel moderation, to urge him, as it were, to come in out of the rain. Ramanujan was like a balky thoroughbred with no one to groom or feed it. And by early 1917, there had been no one since the day he'd stepped aboard the *Nevasa* three years before.

Intellectually, he had come home. But Ramanujan was more than a mind; he was a body, a complex of muscle and tissue, hormones and neurochemicals. And his body had needs that his mind scarcely knew.

A generation before, in 1890, a small book aimed at Indian students, entitled *Four Years in an English University,* was published in Madras. Its author was S. Satthianadhan, a professor of logic and moral philosophy at Presidency College in Madras. Apparently based on Satthianadhan's experiences as a student at Corpus Christi College, Cambridge, the book purported to tell Indian students what life at Cambridge was *really* like, and what Indians could learn from it. And one point it conveyed with all the force of fresh revelation was the English emphasis on sports and recreation.

> A Cantab [from the medieval Latin for Cambridge] never fails to take his two hours' exercise per diem in one way or another. Seldom does one find a student in his rooms in the afternoon, however passionately fond of study he may be. One who does so and keeps to his books the whole day long will be looked upon as an abnormal character and be snubbed by the other students of the College. *Mens sana in corpore sano,*—a sound mind in a sound body,—is a maxim of universal and practical application. These young Englishmen, who pay as much attention to their bodily as to their mental development, are in no way worse off as students. These men, who can walk twelve miles a day or row sixteen, without being tired in the least are just as hard-working as the German students; and it is these strong, healthy, muscular young men who turn out Wranglers and First Class Classics.
>
> If there is one lesson which our students in India must learn from English students, it is this—to pay as great an attention to their bodily as to their mental development.

Ramanujan, it need hardly be stated, did not. He had no interest in sports; if Hardy tried to interest him in cricket it didn't take. He had always been fat, largely oblivious to his body, almost pathologically sedentary. Still, he was not yet Satthianadhan's vision of the typical Indian student: "A study-worn, consumptive-looking individual, without any energy, appearing twice as old as he really is, fit rather to be an inmate of the hospital than a frequenter of the lecture room."

It was as hospital "inmates" that not a few Indian students in England ultimately found themselves. When the Lytton Committee on Indian Students went around the country in 1921 and 1922 it heard one Oxford man observe that the health of Indian students tended to break down toward the end of their second year, almost as if in obedience to natural law. A Darwinian streak that ran through much committee testimony suggested why: in their own country, Indians had adapted to their environment, in particular the hot sun and spicy food. Now, in England, the thinking went, they were faced with alien conditions. The less fit failed to adapt, and so were struck down.

Whatever the scientific validity to such a view, it was true that national identities were less blurred in those days, making for more to adapt *to*; India was more inviolably India, England more England. There were not, as there are today, Indian restaurants and food stores littering the streets of London and Cambridge. As for the climate, there was no air-conditioning or central heating to moderate its extremes. You either endured it or got out. That was why well-heeled Englishmen journeyed to Italy or Spain during the winter; or why, in India during the worst of the summer heat, they visited hill stations like Simla in the North, or Ootacamund in the South. Those who could not escape, meanwhile, faced the full force of the heat, eased only by feeble fans or, midst the damp and chill of English winters, were left to the mercies of ineffectual coal fires.

Indians come to study in England, of course, could not escape the alien and hostile climate and so, it was believed, risked their health. The director of the University of Edinburgh's Indian Student Hostel told the committee that the health of Indian students there was generally good, "except perhaps," he said, "in the case of those coming from the southern parts of India who suffered rather from the severity of the climate and were inclined to develop tuberculosis and disease of the chest." The president of the Royal College of Physicians in Edinburgh commented that among Indian students tuberculosis was common, it being aggravated by the climate and, in the words of the report, "the fact that the more religious of the Indian students insisted on keeping to the diet of their cult—a regime unsuited to the conditions under which they found themselves."

Ramanujan did indeed keep to the diet of his "cult," and from even his first days in England, before the war, doing so had posed problems for him. In a letter to a friend soon after his arrival, he complained about "the difficulty of getting proper food. Had it not been for the good milk and fruits here I would have suffered more. Now I have determined to cook one or two things myself and have written to my native place to send some necessary things for it." In other letters, he asked that certain provisions be sent him or no longer sent him, gave accounts of shipments gone astray or arriving in poor condition, expressed thanks for those that did arrive intact. Monthly, Narayana Iyer sent him powdered rice in tin-lined boxes. Others sent him spices or pickled fruits and vegetables.

Back in India, Janaki later recalled, Ramanujan would sometimes abruptly stop eating, or else hurry through it, to pursue a mathematical thought; meals were something to be dispensed with. But now, with no one to cook for him, they had become an awful bother. Preparing from scratch South Indian meals—assembling the ingredients, soaking and grinding lentils, cutting vegetables, boiling rice, and so on, all the way through the final cooking—took *time*. And time, while working on what to him were the most seductively challenging problems in the world, he resented having to give up. So, as Neville's wife, Alice, for one, would report, Ramanujan sometimes cooked only once a day, or sometimes only once every other day, and then at weird hours in the early morning.

The simplest "solution," of course, was a species of asceticism. Back in Madras, he had enjoyed mango, or banana, or jackfruit with his rice and yogurt. And there was brinjal, prepared that special way his mother did it, and . . . But the great days of good South Indian meals were gone. He had written Subramanian in 1915, "I am not in need of anything as I have gained a perfect control over my taste and can live on mere rice with a little salt and lemon juice for an indefinite time."

For a time, certainly, he could. But not indefinitely. Ramanujan's strict vegetarianism had cost him, in High Table, a social outlet that might have helped keep him on a more even keel. Now, during wartime, it risked his health, too.

5. "ALL US BIG STEAMERS"

The streets were dark. The mood was black. The war dragged on.

When Ramanujan became a research student at Trinity College in 1914, he was entered in the same thick Admissions Book as all other Trinity men. He signed his name on page 8 of a volume begun just the previous year, then filled in columns running all across the page with his place of birth, father's name, school, and other such routine information.

But examine that book today, and you find that on the next page after Ramanujan's, the stately march of densely filled columns comes to a chilling halt. Suddenly, for the first time, gaps appear in the record, blank spaces. The name of John de Vere Loder appears, but not his signature. Islay Makimmon Campbell's admission is listed; but he never signed either. Both were young men who, though admitted to Trinity, had taken their places at the front and never reached Cambridge. The next page is worse, like a mouthful of teeth half of which are missing, or a bombed city with every other house reduced to rubble.

Beside many names appears further information: Charles Jervoise Dudley Smith, "killed 16 June 1915." Donald Holman, "killed 8 Aug 1918." John Brerton Howard, "died of wounds 6 Sept 1918." For twenty pages, it goes on like that, white spaces in the Trinity Admissions Book like white marble headstones of the men who died in Flanders and in France.

The university stayed open, but it was only a ghost of what it had been. Trinity was depopulated, its enrollment plummeting from almost six hundred before the war to forty-seven in October 1916. By that time, as one young officer training in Cambridge wrote, "the pulse of the University had almost stopped beating; there were very few undergraduates in residence except boys under military age, Asiatics, and the physically unfit ('infants, Indians and invalids')." It was true: those taking the mathematical Tripos in 1916, for example, included names like Tripathi, Mahindra, Prosad, and Saravanamutti.

The war, entered upon with such patriotic fervor, did not end soon, but grew grimmer in its ceaseless toll. The early optimism faded. Deadly static trench warfare replaced sweeping battlefield maneuvers. To gain a hundred yards of machine-gun-swept mud might cost a thousand lives. The sad, stupid reality of the conflict, fed by the awful casualty reports, fell like a pox-ridden blanket across Britain. It must have particularly galled Hardy that when the First Eastern General Hospital moved from Nevile's Court in March 1915, it was to a temporary hospital built, behind King's and Clare Colleges, on a cricket ground. All its eleven acres ultimately quartered the sick and wounded.

The war touched everything. "Military Field Boots, Made to Order," read an advertisement in the *Cambridge Magazine*. "Best Quality Marching Boots Kept in Stock." Food prices rose. Shortages set in. By May 1916, the crisp brown stock gracing the cover of the *Gazette* of the First Eastern General Hospital had given way to flimsy blue paper. By October, with the hospital having already treated thirty-three thousand casualties, the cover was the consistency of toilet paper.

"Never have we had to chronicle so terrible a list of losses as this week's number [issue] contains," wrote the editors of the *Cambridge Magazine* in its issue of October 14, 1916. "Scarcely a day has passed but names familiar to many still in residence have appeared in the official lists." Then, college by college, it recorded the toll: Hopgood, Hudson, Johnson, Keeling, Knight . . .

"It is our pride and sad privilege," said Sir Joseph Larmor two weeks later in addressing the London Mathematical Society after two years as its president, "to recall the names of the cultivators of our science who, in response to their country's appeal in time of national peril, have already laid down their lives on her behalf. In E. K. Wakeford, Scholar of Trinity College, Cambridge, not a few of us had recognized a future leader in geometrical science."

Ramanujan's tutor, E. W. Barnes, cast a colder, more bitter light on such losses. In a sermon toward war's end, he declared: "Of my pupils at Cambridge at least one-half, and practically all the best, have been killed or maimed for life; the work that I did [teaching mathematics over the years] has been for the most part wasted."

Sometimes it seemed everybody died. The proprietor of a grocery back in Hardy's hometown, Robert Collins, died.

So did artillery battery commander W. Graham, who five years before back in India, had been among those in the Madras administration to contribute to the debate on Ramanujan's merits.

Later, at Winchester College, a War Cloister honoring Wykehamists who fell during the war would rise on a site near where Hardy had played soccer in the 1890s. It was engraved with the names of five hundred of them.

What Barbara Wootton, writing in *In a World I Never Made,* best remembered was

> an endless succession of memorial services in college chapels or Cambridge churches. In a way, what distressed me most . . . was the unashamed public grief of the fathers of the young men. . . . According to the standards of our family circle, tears were to be shed only by children. Very occasionally, women perhaps might weep—but grown men never. . . . The sight of distinguished professors and famous men whom I had been brought up to regard with awe openly crying in church disturbed me profoundly.

"Things here are sad and sorrowful," Trinity vice-master Henry Jackson wrote on January 25, 1917. He was cheered a little, ardent militarist that he was, by the tramp of recruits drilling in Nevile's Court, officers issuing commands, men marching in formation. But: "Many friends are dead. Few are left here. I am anxious about the home people. Gout and rheumatism punish me. Deafness incapacitates me. And there is always the war, the war."

* * *

In February 1915, Germany had imposed a submarine blockade of the British Isles. In October 1916, it was intensified; now ships would be sunk without warning.

BRITISH SHIPS SUNK
BY SUBMARINES

10,000 TONS LOST

read an unexceptional headline in *The Times* of London on February 17, 1917. The same edition carried word that ports in America, which had not yet joined the war, were under virtual blockade. Four million bushels of export wheat were locked up in silos in Minneapolis because East Coast ports were clogged with grain that couldn't be loaded.

On the eve of the war, Britain depended on imports to the tune of two-fifths of her meat, four-fifths of her wheat. Rudyard Kipling had written:

> *For the bread that you eat and the biscuits you nibble,*
> *The sweets that you suck and the joints that you carve,*
> *They are brought to you daily by all us Big Steamers—*
> *And if any one hinders our coming you'll starve!*

Now many of the Big Steamers weren't getting through. And some Englishmen began to starve. There was, it was said, one sure way to tell rich from poor in England. You had only to look at them; the poor were shorter, less physically developed, bore the stigmata of undernourishment. Now poor nutrition became more widespread. Bertrand Russell's second wife Dora, writing in *The Tamarisk Tree,* recalled how "people began to feel the effects of the loss of essential fats and other necessities. When I saw my two clergyman uncles in the latter stages of the war, I was shocked to see how these rosy, well-covered men had shrunk."

Food was rationed. Prices rose; by 1916, the food bill for a working-class family had climbed 65 percent compared to before the war. By early 1917, severe shortages had set in. "The usual week-end potato and coal scenes took place in London yesterday," the *Observer* noted on April 8th.

> At Wrexham a big farm-wagon laden with potatoes already weighed into shillingsworths was brought into the square by the agriculturalists who at once proceeded to sell them to all comers. The wagon was surrounded by hundreds of clamouring people, chiefly women, who scrambled on to the vehicle in the eagerness to buy. Several women fainted in the struggle, and the police were sent for to restore order.

Even High Table felt the shortages. "We are virtuously obedient to the controller of food," wrote Jackson on March 26, 1917: "fish and potatoes but no meat on Tuesday and Friday; meat but no potatoes the rest of the week: bread rolls half their old size: portions strangely dwarfed." But if for Englishmen it was hard getting potatoes and sugar, what of the specialty foods Ramanujan craved that were hard to come by even in peacetime? In June 1914, he had written that the easy availability of "good milk and fruits" had helped ease his food problems. Now, all fruits and vegetables were difficult to procure.

The war had reached Ramanujan.

The food shortages coupled with his irregular eating habits could hardly have fortified him against any diseases to which isolation, overwork, and climate may have predisposed him. In his third year in England, he embodied the distinction some physicians today make between being well and merely not being sick. He was an illness waiting to happen.

W. C. Wingfield, medical superintendent of one famous English sanatorium, would observe that the disease with which he was most familiar, tuberculosis, was brought on, or aggravated, by *faulty modes of life*, which he defined as "overwork, overplay, overworry, undernourishment, lack of necessary sunshine and fresh air, or chronic intemperance in any form."

Except, perhaps, for "overplay," Ramanujan was guilty of all of them.

6. THE DANISH PHENOMENON

Both during his life and afterward, it was a matter of some mystery just what laid Ramanujan low. But *something*, in the spring of 1917, did. In May, Hardy wrote the University of Madras with news that Ramanujan was sick—afflicted with, it was thought, some incurable disease.

Was this the time to send him back to India? The idea was broached. But many Indian physicians were on war duty; getting him adequate medical care, it was thought, might be impossible. Then, too, he might never reach Indian soil; the U-boats made ocean travel perilous. The future Dora Russell in August 1917 accompanied her father, a high Admiralty official, to New York. "My father had an able personal assistant," she wrote, "but in view of the submarine menace he felt that he could not press her to accompany him." So he asked Dora instead. "We travelled in convoy and were never allowed to be one moment without our lifebelts beside us." The war had denied Ramanujan access to mathematicians with whom he had been brought to England to work. It had undermined his nutrition, perhaps priming him for disease. Now, sick, it helped keep him in England.

It was an anxious time. He was admitted to a "nursing home"—actually a small private hospital catering to Trinity patients—on Thompson's Lane, overlooking the Cam across from Magdalene College, and within a stone's throw of the Neville house. He was very ill. The prognosis was so poor that Hardy asked the master of Trinity for help in getting word to Ramachandra Rao in India by special dispatch. Later, when Ramanujan seemed a little better, he asked Subramanian to contact Ramachandra Rao and allay any fears the earlier report had raised.

By that time, Ramanujan was out of the hospital, perhaps back in Bishop's Hostel. There is some evidence that Hardy himself nursed him for a while. But whoever did, it couldn't have been easy. Because Hardy's letter to Subramanian already bore the stamp of what was to complicate all efforts to restore Ramanujan's health: "It is very difficult to get him to take proper care of himself," Hardy wrote. Ramanujan, he was saying, was a terrible patient.

Not that this was anything new. Back about 1910, when Ramanujan got sick while living in Madras, his friend Radhakrishna Iyer put him up for a while. "As a patient," he recalled later, "Ramanujan was not exemplary; he was obstinate and would not drink hot water and insisted on eating grapes which were sour and bad for him." Radhakrishna called in a doctor who examined him and, mercifully, ordered that Ramanujan be sent back to his parents in Kumbakonam.

He had not changed in seven years and, all through his care, was a plague on his caregivers. "A difficult patient always inclined to revolt against medical treatment," is the way two of his Indian biographers would describe him. "Difficult to manage," is what Hardy had to say. "Whenever he started with a doctor he was full of confidence, faith and hope," says Béla Bollobás, a Trinity College mathematician who was a friend of Littlewood and who has taken a special interest in Ramanujan's life. "[But] when he realized the doctor couldn't help him, he went to the other extreme and he saw that this doctor was no good at all. 'I fell into his trap, how, how can I get away from him?' "

Ramanujan was picky about his food, wouldn't do what he was told, forever complained about his aches and pains. Self-willed as ever, he had no *faith* in medicine. "There are very few doctors who would care to have Ramanujan in their nursing homes," it would be said of him, "and fewer still who would bother with humoring Ramanujan's palate."

Ramanujan's pigheadedness, along with the uncertainty surrounding his diagnosis, led him to see at least eight doctors and enter at least five English hospitals and sanatoriums over the next two years. Probably around October, he became a patient at the Mendip Hills Sanatorium at Hill Grove, near the city of Wells, in Somerset. There he came under the care of Dr. Chowry-Muthu, an Indian doctor who, as it happens, had accompanied Ramanujan on the *Nevasa* three years before. Dr. Muthu was a tuberculosis specialist.

The earliest diagnosis of Ramanujan's ailment was a gastric ulcer, support for which waxed and waned all through his treatment; at one time, exploratory surgery was considered.

One doctor held that Ramanujan's hydrocele operation back in India had, in fact, been to remove a malignant growth, and that the cancer was now spreading; but since Ramanujan's condition didn't deteriorate, most doctors dismissed the idea.

Blood poisoning was another possibility; an idea heard later was that, perhaps overly impressed with their presumed sterility, Ramanujan would eat canned vegetables whose labels gave assurance of their purely vegetarian origin. Bypassing proper pots and pans, the theory went, Ramanujan would cook them right in the cans, over the gas flame in his room, perhaps contracting lead poisoning from the soldered lids.

It was tuberculosis, though, for which Ramanujan was treated at Mendip Hills and for which he would most consistently be treated in the coming years, and tuberculosis—consumption, phthisis, the White Plague—which remains the most likely candidate today.

TB accounted for as many as one in three deaths in European cities during the midnineteenth century and, by the early twentieth, after half a century of decline, still caused one in eight deaths in Britain. In 1882, Robert Koch showed that the disease stemmed from infection by a particular microorganism, the slow-reproducing bacillus *Mycobacterium tuberculosis*. The bacillus could infiltrate most any bodily tissue, causing lesions, little granulous pockets, in the spine, the eyes, the bones, the kidneys, or the lymph nodes, but most typically in the lungs. Fever, night sweats that left the patient drenched by morning, cough, difficulty breathing, spitting of blood, and weight loss were typical symptoms—but *not* invariant ones. Diagnosis was always tricky. Often, the disease followed a peculiar course. Sometimes it took an abrupt, violent turn; this was the feared "galloping consumption." Often it lay low for years, only to recur. Sometimes seemingly advanced cases recovered spontaneously.

To get tuberculosis you must, in the first place, be infected with the tubercle bacillus. But many people, probably most, in India as well as in England, got infected, yet never came down sick; their immune systems successfully warded off the attack. Heredity clearly played a role in who got sick. But so, too, almost certainly, did what today might be called "lifestyle" factors. While consensus on their influence eludes the research community, the evidence gives strong credence to Dr. Wingfield's impression that "overwork, overplay, overworry, undernourishment [and] lack of necessary sunshine and fresh air" help transform otherwise failed bacterial attacks into successful ones.

For one thing, a formidable body of research now points to close ties between the nervous system and the immune system, and between stress and illness. The spouses of breast cancer victims, aching with worry and despair, come down sick more than controls. Neurotransmitter receptors occupy sites on cells of the immune system, making for a natural communications link between the two systems. Stress—overwork? worry? loneliness?—the evidence powerfully suggests, can weaken the immune system and offer a ripe field for disease. One study found that Filipino sailors serving in the American navy—typically away from home for years at a time, and for much longer than other sailors—were more TB-prone than other sailors. Researchers concluded that "emotional stress associated with separation from family and friends"—simple loneliness—may well have contributed.

Could Ramanujan's vegetarianism, made harder to nutritionally maintain by chaotic eating habits and food shortages, have set him up for the disease? Here, again, evidence is suggestive, with the Danish phenomenon giving the notion a powerful boost.

In 1908, Dr. H. Timbrell Bulstrode submitted his *Report on Sanatoria for Consumption and Certain Other Aspects of the Tuberculosis Question* to Parliament. In it, he included a chart showing a comfortingly steady decline in the tuberculosis death rate in England and Wales, from close to 300 deaths for every 100,000 population around 1850 to 120 by 1904. Similar declines were being recorded during this period in Switzerland and Germany, in Denmark and the United States.

Then came the Great War.

With it came a wild spasmodic leap in the death rate. In Prussia, tuberculosis deaths shot up from about 150 per hundred thousand to 250. The same in Belgium, Italy, and the other belligerent countries. England recorded a less dramatic, if equally unambiguous jump, 17 percent between 1913 and 1917—erasing, in two years, twenty years of steady decline.

But why did *Denmark* also show a precipitous 30 percent rise? Denmark was neutral, untouched by war or civil strife. Moreover, why did the Danish figures resume their downward course beginning in 1917, two years before the other countries?

As it happens, Denmark shipped large quantities of meat and dairy products to England and other belligerents during the first two years of the war. Prices rose dramatically at home, but wages failed to keep pace. Danish per capita consumption plummeted. But with the resumption of unrestricted submarine warfare around 1917, Danish food piled up in Denmark. Consumption on the home front rose—and with it, apparently, TB resistance. "There is ample evidence, accepted by virtually all authorities," wrote the English physician R. Y. Keers in explaining the war's impact on tuberculosis deaths, "that the main factor in the production of this increased mortality was malnutrition—a view supported by the figures from Denmark."

What was the crucial missing nutrient that meat or dairy products supplied? Quite possibly vitamin D—a deficiency of which may have left Ramanujan, among thousands of others, more vulnerable to the tubercle bacillus.

It was not until 1920 that vitamin D's role in the prevention of rickets, a bone disease, was discovered. Much recent research, however, links it also to the immune system. And in what has been called a piece of "admirable epidemiological detective work," Welsh physician P. D. O. Davies in 1985 showed how vitamin D deficiency could explain, through its immune system effects, the much higher TB rate among immigrants to Britain from the Indian subcontinent.

Davies noted that the disease was *thirty* times higher in this group than among native Britons; that studies showed more than a third of them had vitamin D deficiency; that vitamin D clearly played a role in marshaling monocytes and macrophages—key immune system players. "A Possible Link Between Vitamin D Deficiency and Impaired Host Defence to *Mycobacterium Tuberculosis,*" Davies titled his paper in *Tubercle,* the British tuberculosis journal. "Can this complex [biochemical] pathway really explain the incidence of tuberculosis in dark-skinned immigrants to the United Kingdom and the cure of skin tuberculosis by vitamin D treatment?" asked another medical researcher, Graham A. W. Rook, in evaluating the logic of Davies's argument. He concluded: "It probably can."

Prime sources of vitamin D include egg yolks, organ meats, and fatty fishes. Indeed, cod liver oil was used as early as the eighteenth century to treat tuberculosis. With the possible exception of egg—the evidence here is contradictory—Ramanujan ate none of them.

Another source of vitamin D today is milk specifically fortified with it. But fortified milk wasn't available in Ramanujan's day.

There is another prime source of vitamin D, one that explains why Indians back in India, for example, unlike those in Britain, have normal levels of the vitamin: the sun. The sun gives off not just visible light but ultraviolet radiation, and ultraviolet rays activate cholesterol in the skin to make vitamin D.

Ramanujan got little sun. In Cambridge, high up near the Arctic Circle, there wasn't much to begin with. And the English cloud cover blocked most of the rest. Then, too, Ramanujan didn't leave his rooms much, often working at night and sleeping by day. Even working by a sunny window would have done no good: ordinary window glass absorbs the ultraviolet rays that make vitamin D.

In the fifth edition of his classic *Principles and Practice of Medicine,* published in 1902, the great physician William Osler, referring to the sheer capriciousness with which infection by the tubercle bacillus ended in disease, observed:

> There are tissue-soils in which the bacilli are, in all probability, killed at once—*the seed has fallen by the way-side.* There are others in which a lodgment is gained and more or less damage done, but finally the day is with the conservative protecting force—*the seed has fallen upon stony ground.* There are tissue-soils in which the bacilli grow luxuriantly; caseation and softening, not limitation and sclerosis prevail and the day is with the invader—*the seed has fallen upon good ground.*

Both the common medical wisdom of his own day and scientific evidence from our own suggest that Ramanujan, during his first three years in England, had become fertile ground indeed for the growth of *M. tuberculosis.*

In the years between 1899 and 1913, Mendip Hills Sanatorium claimed hundreds of "cures." And five other sanatoriums within a twenty-mile radius apparently also found the country just across Bristol Channel and the Severn River from Wales conducive to recovery. Still, geography and cure rate notwithstanding, something didn't take for Ramanujan at Mendip Hills; he was there but briefly, and soon, in about November, he was transferred to Matlock House Sanatorium in Derbyshire. There, off and on for the best part of a year, he was treated by at least three doctors and ran up bills of at least £240, the equivalent today of perhaps $20,000.

Early in his stay, he wrote Hardy:

> I have been here a month and I have not been allowed fire even for a single day. I have been shivering from cold many a time and have not been to take my meals sometimes. In the beginning I was told that I could not possibly have any except the welcome fire I had for an hour or two when I entered this place. After a fortnight of stay they told me that they received a letter from you about one and promised me fire on those days in which I do some serious mathematical work. That day hasn't come yet and I am left in this dreadfully cold open room.

The Matlock staff wasn't being flinty eyed or cruel in denying Ramanujan the warmth of a fire; it was just that cold open rooms like the one Ramanujan endured were, during this period, the most accepted treatment for tuberculosis.

Physicians facing TB today take their pick from a variety of drug regimens to suit considerations of cost and possible side effects. In developed countries at least, the disease has been largely wiped out. But it wasn't until the 1950s that the first effective drug, streptomycin, came along. Until then, tuberculosis remained the great White Plague, resistant to all that medical science could throw at it. Koch, discoverer of the bacillus, had trumpeted a vaccine—which didn't work. There was surgery, involving the opening, treatment, and draining of tubercle abscesses. There was the artificial pneumothorax vogue, in which the tuberculous lung was collapsed, by injecting it with a gas, and allowed to "rest," presumably leading to healing; one of Ramanujan's doctors, the prominent London chest disease specialist Harold Batty Shaw of Brompton Hospital, would a few years later be among those to remark on this treatment's rise to popularity. Gold salt injection had its proponents. So did patent medicines like "Tuberculozyne," taken with milk three times a day, and "Crimson Cross Fever and Influenza Powder for the Cure of Consumption."

But most widely accepted of all during Ramanujan's time was the open-air treatment pioneered in the late nineteenth century, and calling for bed rest, typically in open lodges exposed to the fresh air, lots of food, and measured amounts of light exercise.

"Open-air treatment for tuberculosis had preceded the bacteriological revolution in origin and was little influenced by it," Linda Bryder has observed in *Below the Magic Mountain*, her study of tuberculosis treatment in England. Tuberculosis was bred in urban slums? The great outdoors were linked to health? Then open air, far out in the countryside, was what the lungs of the hapless consumptive needed. Such, anyway, was the treatment rationale.

Launched in Germany, the open-air strategy heavily influenced British practice, especially in the years around the turn of the century. Indeed, one of the most influential of the early German sanatoriums, at Nordrach, in the Black Forest, lent the weight of its name to one of those near Mendip Hills. It was known as Nordrach-upon-Mendip. There was a Nordrach-on-Dee, too.

It was no longer possible, Bulstrode observed in his 1908 report, to meaningfully distinguish sanatoriums from ordinary hospitals, "seeing that practically all hospitals for consumption have now made some attempt towards carrying out the 'open air' treatment." And by 1913, there were at least fifty-two British sanatoriums embracing it. Mendip Hills was one of them. So was Matlock, though as a more recent convert; it had earlier served as a hydropathic establishment specializing in heat therapy.

By Ramanujan's time, being left to the mercies of the cold was the treatment of choice. "Wide-open windows and cure on outside balconies were the order, bleak and cold as the night or day might be," Rene and Jean Dubos wrote in *The White Plague*. "The art of wrapping oneself in blankets became an essential part of the cure, almost a ritual." Architects competed to find new ways of bringing fresh, cold air to patients. Some sanatoriums had open-air chapels. More typically, patients lay on beds lined up in long, wide open corridors; *Liegehallen*, the Germans called them. Sometimes, sanatoriums were wholly unheated. Even children were not exempt; one school for consumptive children in Northumberland recorded a temperature of thirty degrees Fahrenheit in November 1915. At another sanatorium, just after the war, a girl described what amounted to a hut on top of a hill in which her consumptive sister stayed. "There was no door, no glass in the windows; the window and the rain blowed [sic] through."

In misery at Matlock, Ramanujan was not alone.

But there were other reasons why Ramanujan was—and had been, and would be—unhappy during the period of his care.

In September 1917, Hardy had written Subramanian of Ramanujan's seemingly improved prospects. In his letter, he'd noted that "it was only a few months ago—when he was for a time in a Nursing Hospital here—that we discovered that he was not writing to his people nor, apparently, hearing from them. He was very reserved about it, and it appeared to us that there must have been some quarrel."

There is a certain shock of revelation here, a sense almost of having been caught napping. *We discovered . . . apparently . . . it appeared . . . some quarrel.* The vagueness is distinctly un-Hardylike. And in the way Hardy attributes his ignorance to Ramanujan's reserve, there is even a trace of defensiveness.

A little later, another hint: writing Hardy from Matlock, Ramanujan seems to be acquainting him for the first time with some of the most rudimentary facts of his personal life:

> It is true that I promised my mother that I was going home at the end of 2 years; I wrote them several letters 1½ years ago that I was coming over there for the long vacation; but I had many letters of protest from my mother to the effect that I ought not to come to India till I took my M.A. degree. So I gave up the idea of going there.
>
> It is not true that I am getting letters from my wife or brother-in-law or anybody. I had only a few formal letters from my wife just explaining to me why she had to leave my home. . . .
>
> The initial S. in my name stands for Srinivasa which is my father's name. I haven't got a surname, really speaking.

Ramanujan had not returned home in 1916 because his mother had implored him not to? His wife was not writing him? She had left his family's home? And Ramanujan had no real first name?

It was all news to Hardy.

There was, indeed, trouble back home; "quarrel" was scarcely the word for it. And it had reached across the waters to upset Ramanujan in England. Making matters worse, Hardy had known nothing of it, and so could hardly have done much to ease his distress.

Sometime in early 1917 something had gone badly wrong with Ramanujan's body; he had come down sick. But by the end of the year, it was not only his body that was troubled, but his mind. By then, certainly, he was not a happy man. Happy men do not try to kill themselves.

7. TROUBLE BACK HOME

Ramanujan got no letters from Janaki not because she didn't write them but because his mother intercepted them.

One time, with a package destined for Ramanujan prepared for pickup and her mother-in-law out of the house, Janaki slipped into it a brief note. But Komalatammal returned early, spied the note, opened it, read it, dismissed it as childish or silly or stupid, and refused to send it. Janaki was upset. But what could she say? Or do? She was a girl of seventeen, and her mother-in-law was, well . . . her mother-in-law. So she said nothing, consoling herself that that was just how things were.

That *was* just how things were in the Indian extended family. Mother-in-law and wife troubles were a given in many Indian households, perhaps most. They were the stuff of jokes—*and,* as usual, the butt of moralistic opprobrium from Western observers, whose sensibilities were offended by so much of what they saw in India. The institution of child-brides, wrote Herbert Compton in 1904, was abomination enough, but the girl's customary fate in her new home was, if anything, even worse.

> It is pitiable for the child-wife, torn from a home that contained all she knew of happiness, to be obliged to submit herself to the temper, caprice, and often tyranny of her husband, but when to this is added the despotism and cruelty of several elderly women, who often avail themselves of her helplessness, and if she fails to find favour in her husband's eyes, almost invariably take their cue of unkind conduct from him, her lot may be better imagined than described. She has absolutely no place to go to for comfort and sympathy if it is not to be found in her new home. There is no escape, and no matter what her sufferings, her parents' home is closed to her. An appeal to them meets with a rigid command to submit herself to her husband.

Some years before, the *Hindu* had gone to the trouble of defending Indian culture against such criticism. "The tyrannical mother-in-law," it argued, "is not the rule in Hindu society; and even she is not so black as she is painted. Nor is she so persistent and unchanging in her cruelty to the girl-wife." In its very defensiveness, of course, the editorial spoke some kernel of truth.

An Indian marriage was a mating—or a clash—of families. The wife, a newcomer to her husband's family, was apt to be deemed an interloper, a threat to the household sway long held by her mother-in-law. Besides, she was just a child. Her mother-in-law, who had undergone the same trials when she was a bride, was there to shape her, just as the hard-bitten drill sergeant does new Marines. But here "boot camp," as it were, extended over years—until the wife bore her own children and then, in the course of time, became a mother-in-law herself. Only then might she take a dominant place in the household. In the meantime, her husband normally acted toward her, at least publicly, with a species of indifference. The wife's lot was to defer, unquestioningly, to her mother-in-law and the other older women of the house. In some homes, she would not even hand something to her husband without giving it to her mother-in-law first, who in turn would pass it to him.

Plainly there were seeds of domestic tension in such a situation. And in the case of Ramanujan's family, they grew in more than usually fertile soil. For one thing, Komalatammal's identification with her son—his birth long sought and the object of fervent prayers—was particularly strong. And while young Indian men were often more attached to mother than to wife, Ramanujan was more than usually so; he looked like her, thought like her, shared the ardency of her temperament. Then, too, Komalatammal, close to fifty by now, was a formidable figure of a woman, by one account "insanely jealous" of her son, while Janaki was a meek and callow teenager.

Seven decades later, Janaki, eighty-eight years old, was a stooped old woman, living in Triplicane, Madras, with her forty-five-year-old adopted son, Narayanan, his wife and three children. Their modest house stood behind a low wall with an iron fence, a few feet back from a busy street accented by coconut palms heavy with fruit. Inside, wrapped in a burgundy sari, Janaki sat on a bare wood bench against a wall, where she had only to look up to see a bronze bust of her late husband, garlanded with flowers, the gift of his admirers from around the world. Her skin was glossy, stretched over bones barren of fat. Hunched and frail, she got around the house only by painfully pushing a wooden chair, which functioned as a walker. Nearly deaf, she could hear Narayanan only when he shouted through a rolled-up magazine into her ear. As she replied, in loud staccato bursts, to questions asked her, her face would sometimes grow contorted with the effort of simply listening and speaking. At other times, it would break out into a broad, captivating smile.

According to some who knew her, Janaki was more confident and assertive now than in years past. And yet, such was a daughter-in-law's place as she had been brought up to accept it that even now, nearing ninety and known to be bitter about her treatment at Komalatammal's hands, she took pains to show respect for her long-dead mother-in-law. Through Narayanan, she expressed gratitude to her for the opportunity to marry Ramanujan. And she asked that certain difficulties in their relationship be couched in properly respectful circumlocutions; that it be said, for example, that they were simply "not able to see eye to eye"; and that she fled the household at one point merely because she "wanted to change the atmosphere."

Those close to Janaki, however, suggested some of the basis for her resentment. While Ramanujan remained in India, Komalatammal apparently kept them from sleeping together as husband and wife. Once he had left for England, Janaki was given only the coarsest material for saris. She got no money of her own, but depended on her mother-in-law for the merest trifle. She was made to trek across town to the banks of the Cauvery for water with never a word of thanks. She was the butt of her mother-in-law's abusive language. Finally, of course, her letters to Ramanujan were intercepted, as his were to her. At one point, apparently, Ramanujan wrote his mother asking that she have Janaki join him in England. His mother, not telling Janaki, wrote back that it was out of the question.

Komalatammal's side of the conflict does not come down to us, except that, by some accounts, she blamed Janaki, on the basis of her horoscope, for Ramanujan's ill health; had he married someone else, she was certain, he would not have gotten sick. It may be, however, that she bore toward Janaki no special animosity at all, but merely wrote her off as the child she still was. Almost forty years her senior and used to having her way around the house—*and* with Ramanujan—she could scarcely be expected to give Janaki much voice or autonomy until she grew up. It demands little imagination, then, to see her dismissing with scarcely a wave of her hand Janaki's pleas and protestations.

Whatever the precise family dynamics, it was a scene rife with hard feelings and harsh words. Things were so bad that, at one point, even Ramanujan's half-blind father, who had at first opposed the marriage of Janaki to his son and scarcely ever figures in accounts of family life, stood up for her against his wife.

Finally, Janaki found an excuse to get away. Her only brother, Srinivasa Iyengar, then working in Karachi, in what is now Pakistan, was getting married. The wedding would be back in her hometown of Rajendram, where she and Ramanujan had been wed. Obviously she would have to go. Through mutual friends in Kumbakonam, her parents had known something was amiss, and it may have been their idea to use the wedding as pretext to get Janaki out of Komalatammal's clutches.

A little later, Janaki, now at her brother's house in Karachi, wrote Ramanujan, and this time the letter got through. Could he send her some money for a new sari and for a wedding gift for her brother? Dutifully, Ramanujan sent the money. But by now bitter at the long silence from his wife and knowing only what he heard from his mother, he let no warmth or feeling slip into his reply.

The trouble at home had overflowed its banks, distorting Ramanujan's relationship with his whole family. His letters home first dropped off, then stopped altogether. In 1914, Ramanujan had written home three or four times a month. By 1916, sometimes two or three months passed before he wrote. During 1917, the family heard from him not at all.

For a long time, perhaps shamed at feeling abandoned by his own family, Ramanujan kept silent about it. But finally, while in the Cambridge nursing home, he could hold it in no longer. He told Hardy. And he told his friend Chatterji. Visiting him one day, Chatterji found him looking unhappy. "What's the matter?" he asked.

"Oh, my house has not written to me," Ramanujan replied, using a common South Indian idiom for "my wife."

"Well," joked Chatterji, though familiar with the idiom, "houses don't write."

The "quarrel" disturbed Ramanujan's equanimity, made for a tangled snag in his emotional lifeline back to India. That was bad enough. But that Hardy knew so little of his personal side at this late date, three and a half years after he came to England, testified to something more—and, perhaps, worse. Presumably, Hardy was his best friend in England. At least before he got sick, the two had seen each other almost every day.

And yet, of Ramanujan's lack of letters from home—and, very likely, of the strain of his adjustment to England, and of the winds of loneliness that sometimes blew through him—Hardy had, at least until recently, known nothing.

Why, if they were bosom friends, hadn't Ramanujan been able to tell him long before?

The fact is, they were *not* intimate friends. Ramanujan was cut off from India. He was cut off from the English. And, by a chasm of personality, culture, and circumstances, he was cut off from Hardy as well.

8. THE NELSON MONUMENT

For one thing, as keen an interest as Hardy took in Ramanujan, he had other things on his mind. He was an international mathematical figure. He was involved in many areas of mathematics other than those Ramanujan pursued; of forty-five papers he wrote from 1915 to 1918, only four were collaborations with Ramanujan, though others were influenced by their joint work. Hardy was active in the London Mathematical Society, attended its meetings, served as officer, sometimes took on seemingly petty "journalistic" chores for it; in January 1917, for example, he undertook to draft a leaflet advising authors on the writing of their papers. In the Cambridge Philosophical Society, too, he was active. In 1917, he campaigned for splitting the society's *Proceedings* into two separate journals, thereby presumably upgrading the pure mathematics one of them would carry.

Hardy was deeply involved in the world outside mathematics, too, in particular against the war. When Tresilian Nicholas, a young Fellow of Trinity briefly back from war service, showed up in Cambridge in 1915, he found himself seated next to Hardy in Hall. Surprised Nicholas knew nothing of some recent college business, Hardy asked him, "Whatever have you been doing?" "When I said I had been on war service in the Mediterranean," Nicholas recalled, "he gave me a look of extreme disapproval and talked to his other neighbor for the rest of the dinner."

From its onset, the war had divided Cambridge. G. E. Moore, philosophical guru of the Apostles and Hardy's "father" from fifteen years before, agonized over his stance on it. And Hardy, as Paul Levy writes in his biography of Moore, "soon became one of the people whose opinions on the war most interested [him]." For two years Hardy's views appeared regularly in his diary. "Hardy just back [from vacation]," Moore wrote on September 25, 1914, "thinks we ought to make peace as soon as France and Belgium are safe." (They never were.)

During this period, Trinity was torn by the Bertrand Russell affair. The leading mathematical philosopher of his day, Russell had already become the impassioned antiwar campaigner he would remain on into Vietnam days. A pacifist, he was unpopular among conservative senior fellows who now, with the junior fellows at the front or otherwise involved in the war, ran Trinity. In April 1916, a schoolteacher named Everett, a conscientious objector granted exemption from combatant service, was called up for service in the noncombatant corps; he refused, was court-martialed, and sentenced to two years hard labor. Russell, active in the Non-Conscription Fellowship, came to his defense in a leaflet, and went on record as its author. He was convicted for making statements prejudicial to recruitment and discipline, and stiffly fined.

On July 11, 1916, Trinity dismissed Russell from his lectureship. "Trinity in Disgrace," ran a headline in the *Cambridge Magazine* lamenting the college's action. Hardy, who later chronicled the affair in *Bertrand Russell and Trinity,* was among those fellows—Littlewood, Barnes, and Neville were others—who protested the action. (Hardy's "little book," one review said of it, "is a reminder of a way of life where the participants did their best to hurt each other by day and dined together by night.")

When, on May 5, 1917, *Cambridge Magazine* carried a small ad for the Cambridge Branch of the Union of Democratic Control—billed as "an Association for the expression of independent opinion concerning foreign policy and the settlement after the War"—readers were informed they might seek "further information from G. H. Hardy, Trinity College." Earlier, Hardy had helped keep at least one mathematician out of the war by building a case for the importance of his work to the national interest. And later, in 1918, he would protest the firing, because of his antiwar views, of an otherwise competent university librarian. "It is not, so far as I know, that Dingwall has *done* anything," Hardy would write. "It is purely and simply that he holds views which are held to be obnoxious . . . Russell's case (bitterly as I resent the Council's action) was quite different: he *had* done things, right or wrong, but at any rate perfectly tangible and definite."

Hardy, in short, was *busy,* busy with matters that siphoned off time and energy from his relationship with Ramanujan. As closely as the two men often worked, Ramanujan was, inevitably, less the beacon of Hardy's life than Hardy was of Ramanujan's.

But even if Hardy weren't so busy, an immense personal and cultural gap stood in the way of real intimacy between the two men.

Some years later, the English mathematician Alan Turing would complain of Hardy's lack of even superficial friendliness. It was 1936, Hardy was spending the year at Princeton, and Turing found him "very standoffish or possibly shy. I met him in Maurice Pryce's rooms the day I arrived, and he didn't say a word to me." Hardy loosened up later, but as Turing's biographer, Andrew Hodges, observes, "although 'friendly,' the relationship was not one that overcame a generation and multiple layers of reserve"—this though Hardy "saw the world through such very similar eyes."

Hardy and Ramanujan, who saw the world through such very dissimilar eyes, had far more to overcome.

Whatever its psychosexual roots, Hardy had lowered about himself a lovely, lacy veil of personal defenses that was even more formidable than that of the ordinary Englishman's. An Indian admirer of Hardy would remark on his "parental solicitude" toward Ramanujan. It was an apt choice of words; their relationship was marked by distance, not comradely intimacy. Ten years older than he, Hardy remained always the parent, a kind and obliging parent, perhaps, but forbidding, demanding, and remote, too.

If you stand in the middle of London's Trafalgar Square, which commemorates Admiral Nelson's defeat of the French in 1805, at the base of the Nelson Column you'll see, up close, sculpted depictions of his various naval campaigns. But as you lift your gaze to the top of the great 167-foot-tall fluted column, Nelson himself is just a cloaked figure in a three-cornered admiral's cap, too high to make out even the bare outline of his features. One day, offered the hypothetical choice between being commemorated by such a statue, glorious but distant, and a lower, more approachable one, Hardy would choose the former. He *preferred* the safety of barriers, privacy, distance.

The ultimate barrier to their relationship, of course, was, as Hardy would write, that "Ramanujan was an Indian, and I suppose that it is always a little difficult for an Englishman and an Indian to understand one another properly." It was Kipling's verse all over again: "East is East and West is West and never the twain shall meet." For the English, India was impenetrable, scarcely possible to understand. But *ensuring* that the cultural gap blocked a closer friendship between them was that Hardy scarcely tried to bridge it.

Hardy, C. P. Snow once said of him, "would have been the first to disclaim that he possessed deep insight into any particular human being." The untidy contours of human personality were not his home turf. Yes, he might ask Ramanujan about his knowledge of art or philosophy—the European kinds of things that one might brilliantly discuss with advantage at High Table. And he knew something of Ramanujan's tastes in literature and politics. But he talked to him scarcely at all about his family, or South India, or the caste system, or the Hindu gods. He didn't probe, he didn't pry. "I rely, for the facts of Ramanujan's life, on Seshu Iyer and Ramachandra Rao," Hardy began an account of him later—*not* facts he'd gleaned from Ramanujan himself.

Even so safely neutral a matter as Ramanujan's mathematical influences in India never profited from Hardy's questioning. "Here I must admit that I am to blame," Hardy would write, "since there is a good deal which we should like to know now and which I could have discovered quite easily. I saw Ramanujan almost every day, and could have cleared up most of the obscurity by a little cross-examination." But he never did, never stepped past the mathematics of the moment, "hardly asked him a single question of this kind. . . .

> I am sorry about this now, but it does not really matter very much, and it was entirely natural. [Ramanujan] was a mathematician anxious to get on with the job. And after all I too was a mathematician, and a mathematician meeting Ramanujan had more interesting things to think about than historical research. It seemed ridiculous to worry about how he had found this or that known theorem, when he was showing me half a dozen new ones almost every day.

Mathematics, then, was the common ground of their relationship—perhaps the *only* one other than their mutual pleasure in having found one another. Like many an Englishman, Hardy hid behind his reserve, disdaining any too-presumptuous an intrusion into Ramanujan's private life. He was not ideally suited to draw out a lonely Indian, to ease his adjustment to an alien culture, to shelter him from the English chill.

9. RAMANUJAN, MATHEMATICS, AND GOD

Exemplifying the distance between the two men was Hardy's refusal to view Ramanujan, in matters of religious belief, as any different from the usual run of atheists, agnostics, and skeptics he knew among Cambridge intellectuals; or indeed, to see his mind as flavored by the East at all.

In the 1930s, E. T. Bell would remark that Ramanujan had broken the rules by which mathematicians evaluate their own. "When a truly great [algorist, or formalist] like the Hindu Ramanujan arrives unexpectedly out of nowhere, even expert analysts hail him as a gift from Heaven," he wrote, crediting him with "all but supernatural insight" into hidden connections between seemingly unrelated formulas.

Supernatural insight.

A gift from Heaven.

It is uncanny how often otherwise dogged rationalists have, over the years, turned to the language of the shaman and the priest to convey something of Ramanujan's gifts. Hardy was the first Western mathematician to thoroughly examine Ramanujan's notebooks, but over the next seventy-five years many others would, too. And repeatedly they have been reduced to inchoate expressions of wonder and awe in the face of his powers, have stumbled about, groping for words, in trying to convey the mystery of Ramanujan.

"We have no idea how he did the marvelous things he did, what led him to them, or anything else," says mathematician Richard Askey, a Ramanujan scholar at the University of Wisconsin in Madison. Says Bruce Berndt, after years of working through Ramanujan's notebooks: "I still don't understand it all. I may be able to prove it, but I don't know where it comes from and where it fits into the rest of mathematics." He adds at another point, "The enigma of Ramanujan's creative process is still covered by a curtain that has barely been drawn."

Something of this same enigmatic flavor makes its way into Littlewood's account of Ramanujan's work with partitions. Attempting to trace the progress of Ramanujan's thinking, he ultimately throws up his hands, frustrated and perplexed:

> There is, indeed, a touch of real mystery [here]. If only we *knew* [the result in advance], we might be forced, by slow stages, to the correct form of Ψ_q. But why was Ramanujan so certain there *was* one? *Theoretical* insight, to be the explanation, had to be of an order hardly to be credited. Yet it is hard

to see what numerical instances could have been available to suggest so strong a result. And unless the form of Ψ_q was known already, *no* numerical evidence could suggest anything of the kind—there seems no escape, at least, from the conclusion that the discovery of the correct form was a single stroke of insight.

Ramanujan, in the language of the Polish emigré mathematician Mark Kac, was a "magician," rather than an "ordinary genius."

> An ordinary genius is a fellow that you and I would be just as good as, if we were only many times better. There is no mystery as to how his mind works. Once we understand what he has done, we feel certain that we, too, could have done it. It is different with the magicians. They are, to use mathematical jargon, in the orthogonal complement of where we are and the working of their minds is for all intents and purposes incomprehensible. Even after we understand what they have done, the process by which they have done it is completely dark.

Mystery, magic, and dark, hidden workings inaccessible to ordinary thought; it is these that Ramanujan's work invariably conjures up, a sense of reason butting hard up against its limits.

But at reason's limits *does* something else take over? Do we here flirt with spiritual or supernatural forces outside our understanding?

It's an unlikely, anachronistic, even heretical notion today, with science and Western rationalism everywhere triumphant. But there is scant reason to doubt that Ramanujan himself thought so. South Indians of otherwise presumably rationalist bent—accountants, lawyers, mathematicians, professors—recall him as wholly at ease in the spiritual world to which his mother and grandmother introduced him as a child, and ready to see in it a source of mathematical inspiration.

T. K. Rajagopolan, a former accountant general of Madras, would tell of Ramanujan's insistence that after seeing in dreams the drops of blood that, according to tradition, heralded the presence of the god Narasimha, the male consort of the goddess Namagiri, "scrolls containing the most complicated mathematics used to unfold before his eyes."

R. Radhakrishna Iyer, a classmate at Pachaiyappa's College, recalled one day asking Ramanujan about his research only to have him reply, in Radhakrishna's words, "that Lord Narasimha had appeared to him in a dream and told him that the time had not come for making public the fruits of his research."

And it was on a day about a year before he left for England, in 1912 or early 1913, that Ramanujan, showing his work to mathematics professor R. Srinivasan, made the statement in which he pictured equations as products of the mind of God.

Ramanujan's friends invariably noted his interest in astrology and his penchant for interpreting dreams. His boyhood friend Anantharaman would tell how once, when his brother had a dream, Ramanujan read it as foretelling a death in the street behind their house. (There was.)

K. Gopalachary, a friend of Ramanujan from Madras days, said that Ramanujan even attributed his early interest in mathematics to a dream—a dream about, of all things, a street peddler hawking pills.

Ramanujan's belief in hidden forces and the powers of the supernatural was never, at least back in India, something about which he felt the need to apologize or keep quiet. It was no mere matter of private conviction to him, consigned to the periphery of his life, or some pet theory about which he merely liked to speculate. Time and again, he acted on it.

Thus, in the hectic year before he left for England, he found time to prepare astrological projections, fixing auspicious times for religious functions for relatives and friends. He was convinced, from studying the lines on his palm, that he would die before he was thirty-five, and told his friends as much. Anantharaman would record that he attributed to a temple near Trichinopoly the power to cure mental ailments and advised sufferers to go there. And while still at Pachaiyappa's College, Ramanujan dreamed of a family whose child was near death, went to its parents, and in obedience to the dream, asked them to move the child to another house. "The death of a person can occur only in a certain space-time-junction point," he said.

One evening before he left for England, Ramanujan returned to his house in Triplicane in an electric streetcar (which had been introduced to Madras in 1895, years before anywhere else in India). The driver, enjoying himself, was alternating sudden stops with sharp accelerations, jerking his passengers around unmercifully. Said Ramanujan, sitting with a friend on the long bench behind the driver: "That man imagines he has the power to go slow or fast at his pleasure. He forgets that he gets the power through the current that flows in the overhead wires. . . . That," he said, invoking the Hindu term for the illusion, or vanity, thought to deflect humans from God, "is the way *maya* works in this world."

Ramanujan, then, was steeped in the belief system of his culture. So many stories, from so many quarters, taking so many forms, over so many years, add up to nothing like the skeptical rationalist Hardy imagined, nothing like a man mechanically adhering to the quaint orthodoxies of his family and his caste.

Yet that is indeed what Hardy insisted. "I do not believe in the immemorial wisdom of the East," he would one day declare in a lecture devoted to Ramanujan,

> and the picture which I want to present to you is that of a man who had his peculiarities like other distinguished men, but a man in whose society one could take pleasure, with whom one could drink tea and discuss politics or mathematics; the picture in short, not of a wonder from the East, or an inspired idiot, or a psychological freak, but of a rational human being who happened to be a great mathematician.

Ramanujan's religion, Hardy insisted, "was a matter of observance and not of intellectual conviction, and I remember well him telling me (much to my surprise) that all religions seemed to him more or less equally true."

He was sure, Hardy wrote on another occasion, that "Ramanujan was no mystic and that religion, except in a strictly material sense, played no important part in his life."

For years, C. P. Snow would keep a large photo of Hardy in his otherwise almost bare study and otherwise respected him in every way. But from Hardy's insistence that Ramanujan "did not believe much in theological doctrine, except for a vague pantheistic benevolence, any more than Hardy did himself," Snow took care to distance himself. "In this respect," he wrote, "I should not trust his insight far."

Still, taking their cue from Hardy, most Western observers, and some Indians, have wholly detached Ramanujan's mystical streak from his mysterious ability to forge new mathematical linkages. Hailing the one, they've dismissed the other, and written off his credulousness—his weakness for astrology, his arrant superstitions, his devotion to Namagiri—as an unfortunate eccentricity peripheral to his mathematical inventiveness but which has somehow to be stomached for the sake of it.

For Ramanujan, though, the split was not so sharp. The man whom Hardy met in England in April 1914 was a man still of South India, who had grown up on the Indian gods and the relaxed fluidity of Hindu belief. In him, the natural and the supernatural, Jacobi and Namagiri, Number and God, found a common home, stood in something like an easy intimacy.

Did Ramanujan's religious belief bestow on him his mathematical gift? Certainly not, since otherwise all those with kindred beliefs would share it. Nor did he necessarily gain mathematical insights through anything like the means he thought he did and to which he assigned credit. Nor, to state the obvious, does the fact that Ramanujan believed what he believed mean that what he believed is so. On the other hand, in how the mystical streak in him sat side by side, apparently at perfect ease, with raw mathematical ability may testify to a peculiar flexibility of mind, a special receptivity to loose conceptual linkages and tenuous associations.

Despite his emphasis on rigor, G. H. Hardy was not blind to the virtues of vague, intuitive mental processes in mathematics. Bromwich, he would write, for example, "would have had a happier life, and been a greater mathematician, if his mind had worked with less precision. As it was, even the best of his work is a little wanting in imagination. For mastery of technique in a wide variety of subjects, it would be difficult to find his superior, but he lacked the power of 'thinking vaguely.' " And some such extraordinarily developed ability to "think vaguely" may have been among Ramanujan's special gifts.

Ramanujan's belief in the Hindu gods, it stands repeating, did not *explain* his mathematical genius. But his openness to supernatural influences hinted at a mind endowed with slippery, flexible, and elastic notions of cause and effect that left him receptive to what those equipped with more purely logical gifts could not see; that found union in what others saw as unrelated; that embraced before prematurely dismissing. His was a mind, perhaps, whose critical faculty was weak compared to its creative and synthetical.

It is the critical faculty, of course, that keeps most people *safe*—keeps them from rashly embracing foolishness and falsehood. In Ramanujan, it had never developed quite as fully as the creative—thus giving him the credulousness, the appealing innocence, upon which all who knew him unfailingly remarked. Without that protective screen, as it were, he risked falling prey to the silly and the false—as many, over the years, would view his belief in palmistry, astrology, and all the rest of the esoterica to which he subscribed.

And yet, without that screen did he thus remain more open to the mathematical Light?

Hardy was not, to say the least, at home in this mental universe.

"I have always thought of a mathematician as in the first instance an *observer,*" he said in a Cambridge lecture in 1928, "a man who gazes at a distant range of mountains and notes down his observations."

> His object is simply to distinguish clearly and notify to others as many different peaks as he can. There are some peaks which he can distinguish easily, while others are less clear. He sees A sharply, while of B he can obtain only transitory glimpses. At last he makes out a ridge which leads from A, and following it to its end he discovers that it culminates in B. B is now fixed in his vision, and from this point he can proceed to further discoveries. In other cases perhaps he can distinguish a ridge which vanishes in the distances, and conjectures that it leads to a peak in the clouds or below the horizon.

But about the veiled process by which one might come to discern those peaks in the first place, Hardy remained largely silent. Indeed, rarely in a long life of doing mathematics and writing about it did he choose to discuss the creative process behind it, not even in his *Mathematician's Apology,* otherwise so rich with his insights into the mathematician's world. Always it was the *product* of that process, the theorem itself, that interested him. He might want to establish, through proof, whether or not it was true. Or, perhaps, to evaluate its place in the mathematical firmament; in the *Apology,* for example, he speaks of theorems almost as an art critic might the works in a gallery show, evaluating them by this or that yardstick of mathematical beauty.

Almost the only time he did write about mathematical creativity came many years later, near the end of his life, in a review of a book by Jacques Hadamard, *The Psychology of Invention in the Mathematical Field.* What philosophers or poets could say about the creative process in mathematics, Hardy felt, was next to nothing. But Hadamard was a mathematician, and a great one. He had run up the highest score ever recorded on the entrance examination to the Ecole Polytechnique, France's premier school of science. He had, with the Belgian Charles J. de la Vallée-Poussin, proved the prime number theorem. What *he* had to say about "invention in the mathematical field" was worth listening to.

Hardy agreed with Hadamard that

> unconscious activity often plays a decisive part in discovery; that periods of ineffective effort are often followed, after intervals of rest or distraction, by moments of sudden illumination; that these flashes of inspiration are ex-

> plicable only as the result of activities of which the agent has been unaware—the evidence for all this seems overwhelming.

But beyond this, Hardy seemed to say in every word of his review, he thought it best not to venture. Too soon are we thrust upon the Unknowable; it was better to meekly sidestep the issue than mire our explanations in foolishness. Indeed, he lauded Hadamard for being "wisely diffident and tentative in his conclusions."

How unconscious activities "are related to those of a more normal [sic] kind, to fully conscious work or reflection on the fringe of consciousness, how they function and what is the proper language in which to describe them, are terribly difficult questions," Hardy wrote. When Hadamard did offer a tentative explanation for what Hardy called "the most puzzling question" of all—how you seize one from among the welter of ideas your unconscious serves up—Hardy made explicit how he felt: "It may be so," he wrote, "though I cannot say that I find it very convincing; but I am no psychologist, and my distaste for all forms of mysticism may be prejudicing me unduly."

At one point, Hardy admitted that, quoting Hadamard, " 'the unconscious is not merely automatic, it has tact and delicacy,' even that 'it knows better how to divine than the conscious self, since it succeeds where that has failed.' But I do not *like* this kind of language," Hardy went on; to him it verged on nonsense.

"It is something of a relief," he wrote in the most revealing admission of all, "to pass to the later chapters, which are full of interesting and less controversial matter."

This soft, ineffable region of unconscious processes, of vague, hazy connections and suddenly appearing insights, of loose ties and nameless links, was not, Hardy's remarks suggest, any place where he himself liked to dwell. He was uncomfortable discussing it, uncomfortable thinking about it. That mathematics was a "creative" activity was not the question; it was among the most creative. But as to its source—there he didn't care to delve. Certainly, then, any resort to Eastern mumbo jumbo to explain Ramanujan's mathematical gifts would not have drawn from him a warm reception.

But it was Hardy, the dedicated atheist, who represented the extreme position, and Ramanujan who was more in line with the large body of belief and conviction, within the Western tradition as well as that of the

East, that perceived links between creativity and intuition on the one hand and spiritual forces on the other.

The Greeks, for example, invoked the muses—goddesses to whom poets looked for inspiration. Both the English and French languages speak of "divining" the truth. Hadamard himself noted that the sheer inaccessibility of unconscious thought had, to thinkers over the centuries, endowed it with higher powers.

> That unconsciousness may be something not exclusively originating in ourselves and even participating in Divinity seems already to have been admitted by Aristotle. In Leibniz's opinion, it sets the man in communication with the whole universe, in which nothing could occur without its repercussion in each of us; and something analogous is to be found in Schelling; again, Divinity is invoked by Fichte; etc.
>
> Even more recently, a whole philosophical doctrine has been built on that principle in the first place by Myers, then by William James himself . . . , [in which] the unconscious would set man in connection with a world other than the one which is accessible to our senses and with some kinds of spiritual beings.

Just as India was not alone in attributing creative insights to divine influence, Ramanujan was not alone among mathematicians in holding strong religious beliefs. Newton was an unquestioning believer, felt humility in the face of the universe's wonders, studied theology on his own. Euler, in E. T. Bell's words, "never discarded a particle of his Calvinistic faith," and grew more religious as he grew older. Cauchy was forever trying to convert other mathematicians to Roman Catholicism. Hermite had a strong mystical bent. Even Descartes, that father of Enlightenment rationality, answered the call of the spirit: "His religious beliefs were unaffectedly simple in spite of his rational skepticism," writes Bell. "He compared his religion, indeed, to the nurse from whom he had received it, and declared that he found it as comforting to lean upon one as on the other." (Adds Bell, wisely: "A rational mind is sometimes the queerest mixture of rationality and irrationality on earth.")

Even among mathematicians not religiously minded, one finds evidence of at least respectful allusion to the dark terrain between faith and reason. Gauss, for example, once proved a theorem, as he wrote, "not by dint of painful effort but so to speak by the grace of God." James Hopwood Jeans, Hardy's Cambridge classmate and a famous applied mathematician, wrote: "From the intrinsic evidence of his creation, the Great

Architect of the Universe now begins to appear as a pure mathematician." Even Littlewood, commenting on an incident in which "my pencil wrote down" the solution to a particularly bedeviling problem, could write, "If we may reject divine bounty, it happened exactly as if my subconsciousness knew the thing all the time." Littlewood's, of course, was the usual safe, ironic Cambridge skepticism, perhaps no more than a stylistic device—just as the other statements may be seen as no more than metaphor. Yet each contained the barest breath of ambivalence or humility in the face of the mysterious origins of human creativity.

Hardy, though, did not admit to such ambivalence. For him, the whole spiritual realm was just so much bunkum. He *knew*—this was *his* faith—that wherever Ramanujan's genius came from, there was something straightforward to explain it. He would write:

> I have often been asked whether Ramanujan had any special secret; whether his methods differed in kind from those of other mathematicians; whether there was anything really abnormal in his mode of thought. I cannot answer these questions with any confidence or conviction; but I do not believe it. My belief is that all mathematicians think, at bottom, in the same kind of way, and that Ramanujan was no exception.

Ramanujan's mathematics, he was saying, was the product of the reasoned working of a reasoning mind, and nothing more needed to be said.

Someone would later observe that "Hardy's deep reverence for mathematics and for all things of the mind was precisely of the same kind as impels other people to the worship of God; the only enigma about Hardy was that this never seemed to occur to him." And at least for public consumption, it never did. Had Ramanujan scoured the British Isles, he could have found no one less sympathetic to his spiritual side, no one who, in this one realm, could appreciate him less.

In defending his belief in Ramanujan's lack of genuine religious feeling, Hardy would argue that "if a strict Brahmin like Ramanujan told me, as he certainly did, that he had no definite beliefs, then it is 100 to 1 that he meant what he said." No matter that Ramanujan's statement was, in the context of the relaxed tolerance of Hinduism, by no means incompatible with strong religious feeling. Hardy concludes that "if Ramanujan's friends assumed that he accepted the conventional doctrines of [Hinduism] . . . and he did not disillusion them, he was practicing a quite harmless, and probably necessary, economy of truth."

A harmless economy of truth was indeed what Ramanujan was perpetrating but, almost certainly, it was *Hardy* who was its object. Ramanujan probably wasn't long in England before Hardy let him know, perhaps without even realizing it, that invocations of Namagiri were not apt to be well received. Faced with a man once described as "an atheist evangelical," and hardly wishing to provoke his benefactor and friend on what was such touchy ground, Ramanujan simply never revealed to him the richness and extent of his inner spiritual life.

And *that* was the problem: with Hardy, Ramanujan could not let his hair down—had to dissemble, could not be himself. There remained between the two men a great, unbridgeable gap. Even after working with him for several years, the conclusion is inescapable, Hardy never really *knew* Ramanujan—and thus could be no real buffer against the profound loneliness Ramanujan felt in England.

10. SINGULARITIES AT X=1

Three-quarters of a century later, in 1989, the black dean of an all-black women's college in Georgia recalled her stint as an administrator at a mostly white college up north. "The black students I knew at Haverford may have done well," she said, "but I never got the feeling they were happy"—because they couldn't be themselves, couldn't be black. And just such a split applied to Ramanujan in Cambridge. He had, under Hardy's guiding hand, done well; but he was not happy.

He may not have even realized it himself; Ramanujan was not someone closely attuned to his own feelings. It would have been perfectly possible for him to exult in mathematics, to derive enormous satisfaction from his intellectual dialogue with Hardy, to be to all outward appearances content—indeed, to *call* himself content—and yet, for other needs he could scarcely name and only dimly sense, feel incomplete.

Seasoned travelers invariably note how adjusting to another culture demands flexibility, a willingness to assume the coloration of the new country. The experiences of expatriates, immigrants, and visiting students all attest to the same thing. But flexibility was one trait Ramanujan did not possess in great measure. It took enormous energy to cast off the old and take on the new, not to mention the will to do so. And Ramanujan put all his energy into mathematics. He did not flex as the raw English winds blew around him.

So he broke. First in body, then in mind.

* * *

In 1916, Ramanujan's tutor Barnes had written the University of Madras that, given Ramanujan's achievements, it seemed likely he would be elected a Fellow of Trinity College the following October. But October 1917 came and went without Ramanujan's election. At the time, the college was wracked with dissension over the Bertrand Russell affair, and Ramanujan's champion, Hardy, was squarely in the out-of-favor camp. Then, too, it seems certain, in light of future events, simple racism was a factor; Ramanujan, after all, was a black man.

The disappointment left Ramanujan's mood darker, the whole structure of his personality that much shakier.

It was around this time that he entered Matlock, which could hardly have lifted his spirits. English sanatoriums were typically presided over by stern, patriarchal figures, strict disciplinarians who ruled with an iron hand. And Matlock was in the mold. A friend would later recall that Ramanujan was "cowed down by Dr. Ram, who seems to have told him, 'As long as you are a patient and not well you are not free and the doctor has control over your movements.' "

Matlock was typical of English sanatoriums in at least one other respect: it was out-of-the-way geographically. "There is much evidence to suggest," writes Bryder in her study of British tuberculosis care, "that inmates of sanatoria felt estranged from the outside world. Not only did their geographical isolation make visits difficult and therefore infrequent, but social attitudes accentuated that isolation." At Matlock, located about 150 miles northwest of London, in the Peak District of Derbyshire, Ramanujan enjoyed no steady stream of visitors. Getting there was not easy. The following year, a friend who did make the trip, A. S. Ramalingam, would write of the "cold weary journey" on the night train he'd had to endure. Hardy was almost certainly thinking of the Matlock period when he wrote, a few months after Ramalingam's visit, of Ramanujan's "long illness and the spells of comparative solitude" influencing his mental state.

At Matlock, Ramanujan was cold and miserable much of the time. The Trinity fellowship rejection rankled. And he was too sick to be productive mathematically, which also distressed him. The doctors didn't suit him. He couldn't get food to eat and the food he could get he didn't like; sometimes he craved the hot *dosai*, a kind of pancake, that Anantharaman's mother would cook up for him back in Kumbakonam. Finally, he was getting little in the way of nurturing and emotional support from *home*. And he wasn't getting it from Hardy, either.

He grew profoundly depressed. At one point, he had nightmares in which he was visited by images of his own abdomen as a kind of mathematical appendage with "singularities," points in space marked by indefinable mathematical surges like those he and Hardy had explored in their partitions work. Intense pain might show up at $x = 1$, half as much pain at $x = -1$, and so on. The nightmares recurred.

Ramanujan was at a low ebb, balanced precariously on the edge of mental instability.

Undeterred by the Trinity rebuff and hoping to boost Ramanujan's morale, Hardy set about trying to get his friend the recognition he felt he deserved. On December 6, 1917, Ramanujan was elected to the London Mathematical Society. Then, two weeks later, on December 18, Hardy and eleven other mathematicians—Hobson and Baker were among them, as were Bromwich, Littlewood, Forsyth, and Alfred North Whitehead, Bertrand Russell's collaborator on *Principia Mathematica*—together put him up for an honor more esteemed by far than any fellowship of a Cambridge college: they signed the Certificate of a Candidate for Election that nominated him to become a Fellow of the Royal Society.

The Royal Society was Britain's preeminent scientific body, going back to 1660 when Christopher Wren and Robert Boyles helped found it. There were, at about the time Hardy put up Ramanujan, 39 foreign members, including the Russian Ivan Pavlov, the American Albert Michelson (of Michelson-Morley experiment fame), and 6 other Nobel Prize winners. The Royal Society counted in all 464 members in physics, chemistry, biology, mathematics, and every branch of science. Being an F.R.S. meant that forevermore those three little letters would be appended to your name, appear on your own scientific papers, and on letters addressed to you. It was the ultimate mark of scientific distinction. Younger scientists lusted after it, older scientists lamented their lack of it.

C. P. Snow tells the story of H. G. Wells, author of *War of the Worlds*, *The Time Machine*, many other scientific romances, and serious works of history and social comment as well. But while famous and the recipient of numerous honors, "there was," as Snow would write of him, "precisely one honor he longed for. It went back to his youth, when he day-dreamed about being a scientist. He wanted to be an F.R.S. And this desire, instead of becoming weaker as he got older, became more obsessive." He never received it, because though he had studied science in school and had, in Snow's words, "been the prophet of twentieth century science more effectively than any man alive," he had not himself actually made original research contributions.

It was this signal honor Hardy sought for his friend and for which he set out Ramanujan's "Qualifications" in his distinctive calligraphic hand:

> Distinguished as a pure mathematician, particularly for his investigations in elliptic functions and the theory of numbers. Author of the following papers amongst others: "Modular Equations and Approximations to Pi," *Quarterly Journal*, vol. 45; "New Expressions for Riemann's Functions $\xi(s)$ and $\Xi(t)$," *ibid*, vol. 46; "Highly Composite Numbers," *Proc. London Math. Soc.*, vol. 14 . . . Joint author with G. H. Hardy, F.R.S., of the following papers: "Une formule asymptotique pour le nombre des partitions de *n*," *Comptes Rendus*, 2 Jan. 1917 . . .

Thus it continued, listing Ramanujan's papers, and ending with perhaps the most important of all—"Asymptotic Formulae in Combinatory Analysis," the big partitions paper still awaiting publication in the *Proceedings of the London Mathematical Society*.

On January 24, 1918, the names of Ramanujan and 103 other candidates were read out at a meeting of the society. If past experience applied, only a few of them would be elected.

There was no question in Hardy's mind, or Littlewood's, or anyone else's, that Ramanujan merited the honor. Still, few candidates made it the first time out, and by normal practice his nomination was premature. Hardy had been thirty-three years old when elected in 1910. Littlewood himself had made it only the previous February, also at age thirty-three—more than a decade, and dozens of notable mathematical papers, beyond his early glory as a Senior Wrangler. Ramanujan, still twenty-nine at the time of his nomination, had contributed to European mathematics for just a few years and still had a modest publication record, at least in number.

But Hardy's concern for Ramanujan's health moved him to press his claim with unusual urgency. J. J. Thomson, discoverer of the electron, winner of the Nobel Prize in 1906, and then president of the Royal Society, had asked him to outline the circumstances surrounding Ramanujan's candidature. "If he had not been ill I would have deferred putting him up a year or so," Hardy admitted: "not that there is any question of the strength of his claim, but merely to let things take their ordinary course. As it is, I felt no time must be lost."

Ramanujan might not make it to the next election, he was saying, and the society would have to live forever with its failure to honor him. "I am nervous about trying to rush him," Hardy continued,

> and I am aware that for the time being I am not an ideal supporter. And I realize that the R.S. has many other things to consider. But there is no doubt that (especially after his disappointment in the Fellowships) any striking recognition now might be a tremendous thing for him. It would make him feel that he was a success, and that it was worth while going on trying. It is this much more than the fear of the R.S. losing him entirely which seems to me important.
>
> I write on the hypothesis that his claims are such as, in the long run in any case, could not be denied. This is to me quite obvious. There is an absolute *gulf* between him and all other mathematical candidates.

Hardy's letter suggested another reason for Ramanujan's mental anguish: *it would make him feel that he was a success*. For all Hardy's encouragement, Ramanujan had come to understand how great a price he'd paid for his isolation in India. From the moment he'd stepped off the boat he'd been confronted, through Hardy, with how much he hadn't known, learned, or appreciated before—with function theory, with Cauchy's integral theorem, with so much else that was common knowledge in the West and which, by rights, he should have learned ten years before.

"It is perhaps useless to speculate as to his history had he been introduced to modern ideas and methods at sixteen instead of twenty-six," Hardy would write after Ramanujan's death. "It is not extravagant to suppose that he might have become the greatest mathematician of his time."

Of Ramanujan's work in India, Hardy would observe that it was inevitable that much had been anticipated, since Ramanujan labored under "an impossible handicap, a poor and solitary Hindu pitting his brains against the accumulated wisdom of Europe."

And this was what he had to say of the period between when Ramanujan left Government College and when he joined the Madras Port Trust: "The years between eighteen and twenty-five are the critical years in a mathematician's career, and the damage had been done. Ramanujan's genius never had again its chance of full development."

Given his relentless honesty, could Hardy have failed to convey such sentiments to Ramanujan? And could Ramanujan have failed to have been wounded by them?

Ramanujan may have known nothing of Hardy's efforts to have him named an F.R.S.; but if he did know of them, by now he doubtless assumed they would come to nothing, could lead only to the kind of humiliating blow he had suffered when Trinity turned him down in October.

On February 11, ending a period of silence that had extended over more than a year, Ramanujan wrote his family in India. About this same time, perhaps a little earlier, during a period when he was briefly away from Matlock, he tried to kill himself.

Today in London, you can buy T-shirts, posters, mugs, and other souvenirs emblazoned with the great labyrinthine grid that represents the London Underground, with its dozen or more distinct lines and hundreds of stations. One day in January or February of 1918, it was at a station somewhere in this network, then smaller and newer, that Ramanujan threw himself onto the tracks in front of an approaching train.

What happened next would be easy enough to read as a miracle. A guard spotted him do it and pulled a switch, bringing the train screeching to a stop a few feet in front of him. Ramanujan was alive, though bloodied enough to leave his shins deeply scarred.

He was arrested and hauled off to Scotland Yard. Called to the scene, Hardy, marshaling all his charm and academic stature, made a show of how there, before the police, stood the great Mr. Srinivasa Ramanujan, a Fellow of the Royal Society, and how a Fellow of the Royal Society simply could not be arrested.

In fact, Ramanujan was *not* an F.R.S. He would hardly have been immune from arrest in any case, and the police were not fooled for a minute. But they investigated, learned Ramanujan was indeed reputed to be an eminent mathematician, and decided to let him go. "We in Scotland Yard did not want to spoil [his] life," the officer in charge of the case said later.

Just what triggered Ramanujan's desperate bound onto the tracks doesn't come down to us. Certainly, though, if the mere refusal of his dinner guests to accept a third helping could foment such a storm of shame in him that he had to get up and leave, deeper humiliation might spark action more precipitous still. And in 1917 he had certainly experienced his share of it. Rejected by Trinity. Seemingly abandoned by his wife. Left sick and dependent in the sanatorium, helpless even to command the food he wanted. Unable to produce the work he felt his friends expected of him. Confronted by the knowledge that much of his past work had been rediscovery and, viewed blackly enough, a waste of time.

And for the feelings all this stirred in him, there was no safety valve, at least not among his English friends. Littlewood was gone. So was Neville; he'd lost his Trinity fellowship in 1917, probably due to his antiwar views, and was off in London. Hardy, meanwhile, was no one with whom he felt relaxed enough to bare his soul.

More than likely, he did not deliberately set out to kill himself. It was doubtless a rash act, spurred by some new humiliation, unknown to us today, piled on top of all the others. Once again, Ramanujan had acted impetuously when overcome with shame.

Late in February, back at Matlock, Ramanujan learned of his election, on the eighteenth, to the Cambridge Philosophical Society. This was not so much an honor as just one small legitimization of his stature in the scientific world.

About ten days later, he received a telegram from Hardy, sent from Piccadilly in London.

He read the telegram once.

He read it a second time.

He read it again, the words still congealing in his mind, making no sense: Hardy was advising him of what he already knew, that he had been elected to the Philosophical Society. Which was fine, of course. But he already knew as much. What was Hardy's point in wiring him?

Once more he read the telegram, and this time, finally, a new word exploded from the page. It was not the Philosophical Society to which Hardy was advising him he had been named a fellow, but the *Royal* Society.

Of 104 candidates for election that year, he was 1 of just 15 elected. "My words are not adequate to express my thanks to you," he wrote Hardy. "I did not even dream of the possibility of my election." In May, he would become S. Ramanujan, F.R.S.

India, which soon heard the news, was thrilled. On March 22, the Madras members of the Indian Mathematical Society wrote Hardy in thanks "for the aid and guidance you have been extending to Mr. S. Ramanujan in his work." In a postscript, P. V. Seshu Iyer added his personal thanks for the "care you have been bestowing on him during these months when his health has not been good."

By May, Ramanujan's health was still poor. He wrote the Royal Society from Matlock on the seventeenth that he was, just then, too sick to travel to London for his formal admission into the society.

Around this time, he heard from A. S. Ramalingam, the South Indian engineer he'd met, just off the boat from India, at the Cromwell Road reception center. Soon after the declaration of war, Ramalingam had joined the army and in early 1916 had begun work at a shipyard in Jarrow, in the north of England almost in Scotland. Since 1914, the two men had not been in touch. But now, perhaps learning of the Royal Society election, he wrote Ramanujan through Hardy. Hearing nothing in reply, he wrote Hardy directly, learned of Ramanujan's condition, and wrote Ramanujan at Matlock. Ever the engineer, Ramalingam was nothing if not persevering.

So, apparently, was his whole family. In one of his letters home he had mentioned food rationing. That was all it took. For two months, they had been showering him with South Indian food. Finally, he cabled home, "Stop sending food." By now he had parcel upon parcel of it, all piled up. Maybe, he wrote Ramanujan now, he might like to share in the booty?

Ramanujan wrote back asking for some ghee, the special butter, clarified by boiling to resemble oil, and some spicy Madras-style foods. These Ramalingam promptly sent. One thing led to another, and soon, on Sunday, June 16, Ramalingam was visiting him at Matlock, where he stayed until after lunch on Tuesday. For three days they talked—of the war, of Christian missions, of conditions in India, and of much else. Ramalingam had heard Ramanujan's mental state was impaired, but saw no evidence for it now.

His physical state, though, was another story. "I was shocked and horrified," he wrote Hardy after returning to Jarrow a few days later, "to find him in the thin, weak and emaciated state I found him in." His illness, the suicide attempt, his food problems at Matlock, had taken their toll.

Ramalingam's letter went on for twelve large lined pages documenting Ramanujan's condition, recording the views of his doctors, detailing his meal schedule, and suggesting improvements to his care. But in all its length, there is nothing like "As you are doubtless aware . . . ," or "As Ramanujan may have told you . . . ," or "As you may have learned from Ramanujan when you saw him last. . . ." Nothing he'd learned from Ramanujan during the three days gave Ramalingam reason to abridge his letter in the slightest, or to intimate that anything in it might already be known to Hardy. Rather, it reads like a middle manager's report intended for a high executive presumed to lack all detailed knowledge of the subject.

Ramalingam's letter, long as it was, had a clear focus. "It is with regard to food that I have to write somewhat harshly and tersely and at a good length," Ramalingam wrote. Ramanujan, he as much as said, was killing himself by slow starvation.

Diet was a big part of any sanatorium "cure." Tuberculosis patients were called "consumptives" because the disease *consumed* them. Almost invariably, weight loss went with it, and one prevalent notion held that fattening up the patient could slow or reverse the disease's course.

For breakfast, Matlock tried to feed Ramanujan scrambled eggs on toast, with tea. For lunch, rice, chilies, and mustard fried in butter, cucumber and lemon and, sometimes, green peas. Whatever it was, it was awful, the cooks regularly botching things. Ramanujan wrote Ramalingam the day after his friend left, complaining that the new cook had ruined the *appalappu*, rice flour fried in oil or ghee; she'd burned some, while leaving others raw. "The curried rice," he complained, spelling out the word in Tamil, "was just like *akshata*"—raw rice, used only on ritual occasions, and not meant to be eaten. Boiled rice: the poor woman couldn't even get *that* right.

Despite its priority claim on rationed foods, the sanatorium couldn't seem to procure some foods Ramanujan liked. Like bananas. Or cheese, for the macaroni and cheese he enjoyed; Ramalingam later sent him some. Butter, meanwhile, was so expensive in the wake of wartime price hikes that the Matlock administration grumbled about frying his potatoes in it.

Something, Ramalingam could see, had to give. While at the sanatorium, he asked if he could cook something for his friend. "No, you can't go to the kitchen," he was told. Could he write out a recipe and have the cook follow it? This suggestion, too, met resistance.

Sanatoriums, of course, were used to food complaints and typically dismissed them as manifestations of disease. Tuberculosis patients were supposed to be especially finicky. One whose complaint reached the Ministry of Health, for example, was written off as suffering from "the warped temperament occasionally found associated with his malady which makes him impatient and discontented with institution discipline."

The real problem, Ramalingam could see, was Ramanujan's stubbornness. "He is thinking of his vegetarianism even at the expense of his health and life," he wrote Hardy. "But one cannot but think of him as cranky and headstrong when he refuses cream and, say, plums on it." Writing to Ramanujan the same day, he minced no words:

> I will have to be a bit harsh with you. Both from my talk with Dr. Ram and after my second thoughts, I am impressed with your being so particular about your palate. Well, you will have to choose between . . . controlling your palate and killing yourself. You must try and get yourself to like porridge or oatmeal, cream, etc. My friends have strongly advised me not to let you indulge in pickles and chillies. . . .
>
> I am not going to the extremes and asking you to take beef tea or Bovril [trade name of a concentrated beef extract] though considering your life, my asking you to take such a thing is quite excusable, nay desirable, and even unavoidable. Be reasonable and don't be bigoted.

If there was a time to relax his vegetarianism, this was it. Orthodox Jews, who also proscribe certain foods, may relax the laws of kashrut when otherwise health would be impaired. But Brahminical practice makes no such dispensation, and Ramanujan was not about to make one either.

Sometime before, probably even before entering Matlock, Ramanujan was enjoying one of his occasional spells out of hospital and staying at a lodging house catering to Indians in London. At breakfast, he drank a commercial beverage, Ovaltine, represented to him as vegetarian. Later that day, idly examining the can, he saw listed among its ingredients some trace of animal product. Mortified, he *had to get out of there.* Abruptly, he packed his bags and was out the door; it was Vizagapatnam and the Chatterji dinner all over again.

This time, as he neared Liverpool Street Station to catch the train for Cambridge, tons of bombs from silent, high-flying German zeppelins came raining down on the city. This was likely the raid of October 19, 1917, in which twenty-seven people were killed. One bomb fell behind a row of working-class cottages, destroying three of them, killing four women and eight children. Yet Ramanujan felt he was being *personally* punished for drinking the Ovaltine. Later, writing the landlady to explain his precipitous departure, he pictured the raid, in the words of a friend who heard the story later, as "punishment meted out to him by God for having partaken of anything non-vegetarian."

If anything, Ramanujan had become even more finicky. At one point he'd been on a diet largely of bread and milk; now he would no longer eat them. Nor would he eat Matlock's porridge. He demanded only South Indian food.

"Be reasonable," Ramalingam had written him. But any temptation to be "reasonable" ran hard up against the unsavory reality of English cooking. Even those of Ramanujan's Indian friends less particular than he, or who had forsaken vegetarianism altogether, could have told Ramanujan what all Europe knew—that English food was heavy, ill prepared, maddeningly bland, and monotonous. When Gandhi, also a vegetarian, came to England a generation earlier, he wrote that "even the dishes that I could eat were tasteless and insipid." As for Ramanujan, Neville would recall: "I have known him ask with unaffected apologies if he might make his meal of bread and jam because the vegetables offered to him were novel and unpalatable."

After his Matlock visit, something Dr. Kincaid had said gnawed at Ramalingam's composure. Ramanujan, the doctor assured him, could eat anything he liked, even spicy pickles and chilies. Yet it was Ramalingam's understanding that tuberculosis patients should not get foods like that—unless, that is, they were dying. He wrote Hardy:

> When the patient has gone too far to be remedied, and when it is a question of only a few weeks or months, it matters little [what he eats], as long as the patient feels happy and comfortable in the last days of his life. In permitting me to give anything to Ramanujan to eat, is this the object of Dr. Kincaid?
>
> It is anguishing and breaking one's heart to feel that Ramanujan, with his wonderful capabilities and valuable contributions, should be given up for such as hopeless. War with all its horrors might have made us callous to the whole-sale slaughter and loss of lives but surely should Ramanujan be given up?

11. SLIPPED FROM MEMORY

In the fall of 1918, Ramanujan's name was once again put in for a Trinity fellowship. Hardy, apparently too closely identified with Ramanujan by now and long embroiled in Trinity politics, was this time discouraged from putting up his name. So Littlewood (who was for a time back in Cambridge, done in by overwork and a concussion) did instead. The racial issue flared. One foe of Ramanujan's candidacy, Littlewood wrote later, "went about openly saying that he wasn't going to have a black man as Fellow."

Word of the suicide attempt had gotten around, and Ramanujan's opponents seized on it. Didn't the college bylaws expressly require good mental health? And wasn't a suicide attempt evidence that Ramanujan lacked it? "Grave doubts were being felt about his mental state," Littlewood's friend and former Tripos coach, R. A. Herman, told him.

So Littlewood, who was himself too ill to show up in person, wrote a report taking on the rumors and furnishing two medical certificates to the effect that Ramanujan was indeed mentally sound.

"This evidently shocked the moderate members of the Election Committee," Littlewood wrote later, "and it was moved and passed that they be not read." The election could now proceed on the merits of Ramanujan's case. Of course, Ramanujan had a most emphatic argument in his favor—the three letters that now graced his name. For a Fellow of the Royal Society to be denied a Trinity fellowship would be a scandal. "You can't reject an F.R.S.," Littlewood told Herman, who opposed Ramanujan. "Yes," Herman replied, "we thought that was a dirty trick." Littlewood, we may imagine, just smiled.

By now, Ramanujan was out of Matlock. Ramalingam had written Hardy of Ramanujan's wish to leave it and go to London, where he could get Indian food more easily and, perhaps, receive visitors more conveniently. Ramalingam didn't think that was a good idea. What about the air raids? He suggested instead that Ramanujan go to southern Italy, or the south of France, where the climate might do him good, or at least some place in England less prone to attack.

But Ramanujan got his way and now, learning of the Trinity fellowship, was a patient at a small hospital overlooking a perfectly proportioned little square in the heart of London, on which George Bernard Shaw had lived in the 1890s and Virginia Woolf for four years until 1911. Fitzroy House was a grand five-story townhouse with an oval winding staircase punched up through its center and a massive front door reminiscent of a castle gate. While there, Ramanujan saw several specialists, but his diagnosis remained as uncertain as ever. Bouts of high fever still came in irregular bursts. He had pain no one could trace. He was on tap for a tooth extraction around this time, and one doctor even attributed all his pain to that.

"My heartfelt thanks for your kind telegram," Ramanujan wrote Hardy from Fitzroy House when he got the good news. "After your success in getting me elected by the Royal Society my election at Trinity probably became very much less difficult this year."

He wrote that letter on a Friday, probably October 18. The following Monday, still elated, he wrote Hardy: "Please tell Mr. Littlewood and Major MacMahon that I thanked them very much. Had it not been for your pains and their encouragement, I would be neither the fellow of the one nor that of the other." Then he asked for some details about his fellowship.

That was paragraph one. Paragraph two began: "I have considered more or less exhaustively about the congruency of $p(n)$ and in general that of $p_r(n)$ . . . by four different methods"—the first of which bore on a result that, he added, "you are publishing now. . . ."

As Hardy had foreseen, the honors accorded him, especially the F.R.S., had lifted Ramanujan's spirits—leading to what Neville would term "a brief period of brilliant invention" beginning about the spring of 1918. The paper Ramanujan mentioned, delivered to the Cambridge Philosophical Society two weeks before, represented its fullest flowering.

"A recent paper by Mr. Hardy and myself," it began, referring to their joint work on partitions published earlier that year

> contains a table calculated by Major MacMahon, of the values of $p(n)$, the number of unrestricted partitions of n, for all values of n from 1 to 200. On studying the numbers in this table I observed a number of curious congruence properties, apparently satisfied by $p(n)$. Thus
>
> (1) $p(4), p(9), p(14), p(19), \ldots \equiv 0 \pmod 5$,
> (2) $p(5), p(12), p(19), p(26), \ldots \equiv 0 \pmod 7$,

and so on, right down the page.

The paper was entitled "Some Properties of $p(n)$, the Number of Partitions of n," and what made it important was that until now most "properties" of the partition function had eluded discovery. The number of partitions, $p(n)$, recall, refers to how many ways you can add up numbers to get n. But of even such basic facts as, for example, whether the partition function was odd or even for a particular n, mathematicians remained ignorant.

The table to which Ramanujan referred was a long, dry list of numbers, the product of laborious hand-calculation, that MacMahon had prepared two years before; these were the values, all the way up to $p(200)$, that Ramanujan and Hardy had used as benchmarks against which to check their general formula for $p(n)$. But then Ramanujan had looked deep into

MacMahon's sterile list and made one of those leaps of the imagination that would astonish mathematicians over the years.

To say what he saw, Ramanujan used the language of congruences, which expresses facts about divisibility. Two numbers are congruent when they can be divided by the same number, and leave the same remainder (which may be 0). Take, for example, this straightforward case of division:

$$14/7 = 2$$

In the special language of congruences, you say

$$14 \equiv 0 \pmod{7},$$

which means that you can divide 14 by 7, the "modulus," with zero left over. And

$$15 \equiv 1 \pmod{7}$$

means that 15, divided by the modulus 7, leaves a remainder of 1. The number 22 is congruent to 15 because it, too, when divided by 7 leaves 1.

"Congruences are of great practical importance in everyday life," Hardy with coauthor E. M. Wright would point out in *An Introduction to the Theory of Numbers*. "For example, 'today is Saturday' is a congruence property (mod 7) of the number of days which have passed since some fixed date. . . . Lecture lists or railway guides are tables of congruences; in the lecture list the relevant moduli are 365, 7, and 24."

What Ramanujan had found in MacMahon's table were certain persistent and intriguing patterns best expressed in this simple language. For example, he found that

$$p(5m + 4) \equiv 0 \pmod{5}$$

Make m anything you like and, whatever you picked, the number of partitions would, Ramanujan showed, always be exactly divisible by 5. For example, let $m=0$. Then $5m+4$ is just equal to 4. How many partitions are there of 4? The answer is 5. Is 5 exactly divisible by 5? Why, yes. You could make $m = 1{,}000{,}000$ and ask what $p(5{,}000{,}004)$ was and, with not a clue as to what this astronomical number might be, you could say with absolute confidence that it could be divided by 5. Ramanujan also came up with a similar identity that said $p(7m+5)$ was divisible by 7. Now, in his Cambridge Philosophical Society paper, he proved these results, and conjectured others, one of which would succumb to proof later that year.

On the same day it got Ramanujan's congruences paper, the Philosophical Society received another from him, one that brought Ramanujan's work in England full circle and testified to both the genius and lost potential in his mathematical life.

This second paper went back to two striking identities he had discovered sometime before 1913 and later shown to Hardy. (An identity is an equation true for all values of the variable. So that whereas $x-2=3$ is an ordinary equation, true only for $x=5$, $(x-2)\ (x+2) = x^2-4$ is true for every value of x.) One of them was

$$1+\frac{q}{1-q}+\frac{q^4}{(1-q)(1-q^2)}+\frac{q^9}{(1-q)(1-q^2)(1-q^3)}+\dots$$
$$=\frac{1}{(1-q)(1-q^4)(1-q^6)(1-q^9)(1-q^{11})(1-q^{14})(1-q^{16})(1-q^{19})\dots}$$

The other took a similar form. And sometime after Ramanujan's arrival in England, probably in 1915, MacMahon saw in them something Ramanujan had not—something that bore, once again, on partitions.

Students of the additive theory of numbers, which is the formal name for the field, took an interest not only in partitions generally but in certain *classes* of them. Take the number 10. The number of its partitions—or to invoke a precision that now becomes necessary, the number of its "unrestricted" partitions—is 42. This number includes, for example,

$$1+1+1+1+1+1+1+1+1+1=10$$

and

$$1+1+1+1+2+2+2=10$$

But, one might ask, what if you excluded partitions such as these by imposing a new requirement, that the smallest difference between numbers making up the partition always be at least 2? For example,

$$8+2=10$$

and

$$6+3+1=10$$

would both qualify, as do four others, making for a total of six. All the other thirty-six partitions of 10 contain at least one pair of numbers separated by less than 2 and are thus ineligible.

That's one class of partitions. Here's another, formed by a second, quite distinct exclusionary tactic: What if you only allowed partitions satisfying a specific algebraic form? For example, what if you restricted them to those comprising parts taking the form of either $5m+1$ or $5m+4$ (where m is a positive integer)? If you do that, the partition

$$6+3+1=10$$

fails to qualify. Why? Because not all the parts, the individual numbers making up the partition, satisfy the condition. The part 6 does; it can be viewed as $5m+1$ with $m=1$. So does 1, which can be viewed as $5m+1$ with $m=0$. But what about 3? Make m anything you want and you can't get a 3 out of either $5m+1$ or $5m+4$ (which together can generate only numbers whose final digits are 1, 4, 6, or 9).

One partition that would qualify is

$$6+4=10$$

Another is

$$4+1+1+1+1+1+1=10$$

Each satisfies the algebraic requirement. In all, qualifying partitions come to six.

Now six also happens to be the number of partitions that fit the first category. Except that it doesn't "happen to be." It *always* turns out that way. Pick any number. Add up all its partitions satisfying the "minimal difference of 2" requirement. Then add up all its partitions satisfying the "$5m+1$ or $5m+4$" requirement. Compare the numbers. They're the same, every time.

This is what Ramanujan's identity, suitably interpreted, revealed: two seemingly distinct mathematical subworlds fused together in a single unifying relationship. In volume 2 of his book *Combinatory Analysis*, which came out in 1916, MacMahon devoted to it and the other identity a whole chapter, "Ramanujan's Identities." He verified it by reams of hand-calculation as far as any reasonable person could, "so that there is practically no reason to doubt its truth; but," he added, "it has not yet been established" through formal proof.

But it *had* been, in a paper twenty years before.

One day in 1916 or 1917 Ramanujan was rummaging through the 1894 volume of the *Proceedings of the London Mathematical Society* when there, near the bottom of page 318, he saw it. It was entitled "Second Memoir on the Expansion of Certain Infinite Products," and two of those infinite products were just the identities he thought *he* had discovered. Before MacMahon's book came out, Hardy had shown Ramanujan's identities around. Did anyone know proofs for these marvelous theorems? Could anyone furnish any? No one could. Yet here, in black and white, like a specter from the past, was evidence that someone already had.

"I can remember very well his surprise, and the admiration which he expressed" for the older work, Hardy would say of Ramanujan. As for any loss, or even bittersweet ambivalence, that he may have also felt, Hardy said nothing.

The man who anticipated Ramanujan was Leonard James Rogers, a remarkable character if ever there was one. Born in 1862, in Oxford, where his father was an economist, he had not only done well on the Oxford equivalent of the Cambridge mathematical Tripos, but had earned a bachelor of music degree in 1884. He was a fine pianist, an exceptional mimic, liked to affect a broad Yorkshire accent. He knitted. He skated. He gardened. He was a natural linguist. "Surely," it was said of him,

> no position in diplomacy would have been unattainable to one endowed with his easy mastery of languages, his quick intelligence, his sparkling wit, his fine presence, his athletic grace, his courtly charm that no woman could resist. Yet of what the world counts success he achieved practically nothing.

Rogers was, in spirit, a gifted amateur who, despite his abilities, never pursued his mathematical career with the single-minded devotion so necessary, then as now, to establish a big name: "He did things, and did them well, because he liked doing them, but he had nothing of the professional outlook, and his knowledge of other people's work in mathematics was vague. He had very little ambition or desire for recognition."

Even in the years before Ramanujan rediscovered him, he wasn't quite the mathematical nonentity his obituaries presented him as; for years a mathematics professor at what became the University of Leeds, he had quite a few published papers to his credit. Still, he possessed something of a knack for making notable contributions that were promptly forgotten, at least in the short run. Just three years earlier, for example, the *Proceedings of the London Mathematical Society* had run this correction:

> Prof. Rogers, in his paper "On the Quinquisectional Equation" (*Proc. London Math Soc.*, Vol XXXII, pp. 199–207), gives the equation in explicit form. I regret very much that Prof. Rogers's work, which I must have seen at the time of [its] publication, had entirely slipped from my memory, so that I was led to state that the problem had not been dealt with since Cayley's paper in Vol. XII.

In the case of what later became known as the Rogers-Ramanujan Identities, Hardy suggested later, the Rogers originals appeared "as corollaries of a series of general theorems, and possibly for this reason, they seem to have escaped notice, in spite of their obvious interest and beauty." Then, too, the proofs were tortuous; Rogers's aesthetic sensibilities didn't extend to the written word.

Still, there they were.

A correspondence ensued. In the book to which he'd put the finishing touches in April 1916, MacMahon had said Ramanujan's identities were yet unproved. Now, with the falsehood enshrined in black and white, MacMahon wrote Rogers, as Rogers recalled a little later, "regretting that he had overlooked my work before it was too late." In October 1917, Rogers wrote MacMahon with a new, simpler proof. In April 1918, probably while at Matlock, Ramanujan wrote Hardy with his own.

On October 28, 1918, the two proofs, along with Ramanujan's important congruences paper, were read at the annual general meeting of the Philosophical Society.

Two weeks later, the war ended.

The Bolshevik Revolution had led to a collapse of the Russian armies facing the Germans. Germany had withdrawn all but a million men from the eastern front, rushed them across central Europe and sent them hurtling west toward Paris. Once again, as in 1914, German armies breached the Marne. Once again they came within sight of Paris. But this time, there was a difference. A million fresh American soldiers had arrived in France. The Germans, exhausted by years of war, rolled back, and soon had to sue for peace. The armistice was declared on November 11, 1918.

Very shortly, in the words of the *Cambridge Review*, "there was a very creditable pre-war bonfire in the Market Place, fed chiefly with packing cases round which there was dancing; but not all the dancers performed with two sound legs." By one o'clock that afternoon, some Girton College women, normally sequestered in their own campus on the fringes of town, had joined the afternoon revels. At five that evening, many went to the Thanksgiving Service at King's College Chapel. "As we stood in the dim candlelight of that wonderful chapel," one of them wrote, "it seemed as if Earth and Heaven were no longer divided, and as if Time and Eternity were one."

On November 26, Hardy wrote Francis Dewsbury in Madras with word on Ramanujan. "I think it is now time," he wrote, "that the question of his temporary return to India and of his future, generally, should be reconsidered."

CHAPTER EIGHT

"In Somewhat Indifferent Health"

[From 1918 on]

1. "ALL THE WORLD SEEMED YOUNG AGAIN"

Ramanujan seemed better and, Hardy wrote Dewsbury, "on the road to a real recovery." He had gained almost fifteen pounds. His temperature had steadied. The doctors now favored blood poisoning as the source of his ills, and this, it seemed, had "dried up."

Was it not time, then, that Ramanujan return to India? The reasons against doing so had disappeared. The sea lanes were safe. He had achieved all he had set out to in England. His Trinity fellowship imposed no residency requirement. He need not stick around while his Royal Society candidacy was up in the air; it no longer was. So, the thinking went, with Ramanujan on the mend why retard his recovery by keeping him in England any longer?

Of course, more was going on behind the scenes. "He has apparently been approached (with a view to return) directly by several friends," Hardy wrote. "It is possible, I think, that the suggestion has not been made in the most tactful way possible; at any rate, it seems to have turned him rather against the idea of going"; something they said had pushed one of Ramanujan's numerous buttons. Mindful of Ramanujan's sensitivities, and of no mind to trespass on them, Hardy advised Dewsbury that "the suggestion would best be made more or less officially and by letter simultaneously to him and to me." Offer Ramanujan a university position that left him free to do research and occasionally visit England

and he, Hardy, would favor his return—and Ramanujan, he felt sure, would, too.

Hardy's letter bore a reminder that the emotional malaise that had catapulted Ramanujan onto the subway tracks earlier that year was not wholly cured. "He will return to India with a scientific standing and reputation such as no Indian has enjoyed before, and I am confident that India will regard him as the treasure he is. His natural simplicity and modesty has never been affected in the least by success—indeed," he added, "all that is wanted is to get him to realize that he really is a success."

Ramanujan *still* didn't think he was? In November 1918? With close to twenty major papers appearing in the past four years? With an F.R.S. appended to his name? And a fellowship at Trinity College? All this did not convince him? In public, Ramanujan affected a South Indian brand of aw-shucks modesty. But inside, he still wanted something more. Wanted it from Hardy? Was Hardy, on whom the war had made Ramanujan so dependent, in private more niggardly in his encouragement, more aloof and detached, than he was in public?

In any case, the wheels were being set in motion for Ramanujan's return to India the following year. But was he going back because he was really better? Or because he was worse, his chances for recovery in England seen as remote? That, at least, is what one of his Indian biographers implied later. "Mr. Ramanujan's disease had assumed serious proportions by the Christmas of 1918," wrote P. V. Seshu Iyer, referring to a time only a month removed from Hardy's letter to Dewsbury, "and caused such grave anxiety to his doctors in England, that, hoping to do him good, they advised him to return to his native home in India."

Whatever was physically wrong with Ramanujan, his progress, or decline, was glacially slow—making it easy to read slight day-to-day fluctuations in his condition any way you liked. So it may have been no great change in his health to which his return to India was really due but, more simply, the end of the war.

The weeks and months just after the armistice were a time of endings and beginnings, of vast relief all across the blood-drenched soil of Europe. Streetlights blazed again. On December 7, Britons were restored the right to make cakes and pastries and to smear them with chocolate to their hearts' content. Rationing continued, as it would for some products into

Sophia and Isaac Hardy, G. H. Hardy's parents. She was upright, pious, and stern. He was a "White Knight," who "never uttered an unkind word." *Master and Fellows of Trinity College, Cambridge*

G. H. Hardy, a little before he heard from Ramanujan, at the time he became a Fellow of the Royal Society. "His face was beautiful," C. P. Snow once wrote of him, "with high cheek bones, thin nose, spiritual and austere." *Master and Fellows of Trinity College, Cambridge*

Hardy's sister, Gertrude. She called her brother "Harold," his middle name. Neither of them ever married. Both spent their lives in academic settings. Both were enchanted by intellect and contemptuous of religion. *Master and Fellows of Trinity College, Cambridge*

The "Mt. Pleasant" cottages, in one of which Hardy lived as a child, across the road from Cranleigh School. The separate entrance nook, at the extreme right of this recent photo, is new.

Cranleigh School, conceived as a "middle class school," in a recent photo. Hardy's father was drawing teacher. His mother directed the preparatory school across the road. Hardy himself attended for a few years, but by the time he was thirteen had left for a more academically challenging school.

Flint and Stone. A court at Winchester School, a traditional English public school going back to the fourteenth century. Hardy hated the place and apparently never returned for a visit.

J. E. Littlewood, "Senior Wrangler," 1905. That honor went to the man scoring highest on the notorious Mathematical Tripos exam, and in English academic circles was about like being named All-American, Rhodes Scholar, and Bachelor of the Year all at once. It was Littlewood to whom Hardy turned when Ramanujan's mystifying letter arrived from India. *Master and Fellows of Trinity College, Cambridge*

E. H. Neville, Hardy's emissary to Ramanujan. He arrived in Madras in the winter of 1913 to give a series of lectures. But Hardy had given him an additional task—to bring Ramanujan to England. *Master and Fellows of Trinity College, Cambridge*

Long the first destination in England of most Indian students: 21 Cromwell Road, in the South Kensington district of London, shown here in a recent photo. Ramanujan was there in mid-April 1914.

Ramanujan's introduction to the English home: 113 Chestertown Road, in a recent photo. E. H. Neville and his wife bought it in 1913, and Ramanujan lived here for two months in 1914.

New Court, Trinity College, where Hardy lived during the time he knew Ramanujan. *Master and Fellows of Trinity College, Cambridge*

The Wren Library, Trinity College, in a recent photo taken from the Backs. Hardy lived on the second floor of the building to the right, rooms extending over the gate into New Court.

Bishop's Hostel. From about 1915 to 1917, Ramanujan lived in the building closest to the camera. From the window with the strong reflection on the second floor (directly over the ghostly figure below) Ramanujan could see the spired steeple atop the college dining hall. But his rigid vegetarianism meant that he never ate there with other Trinity College scholars. *Master and Fellows of Trinity College, Cambridge*

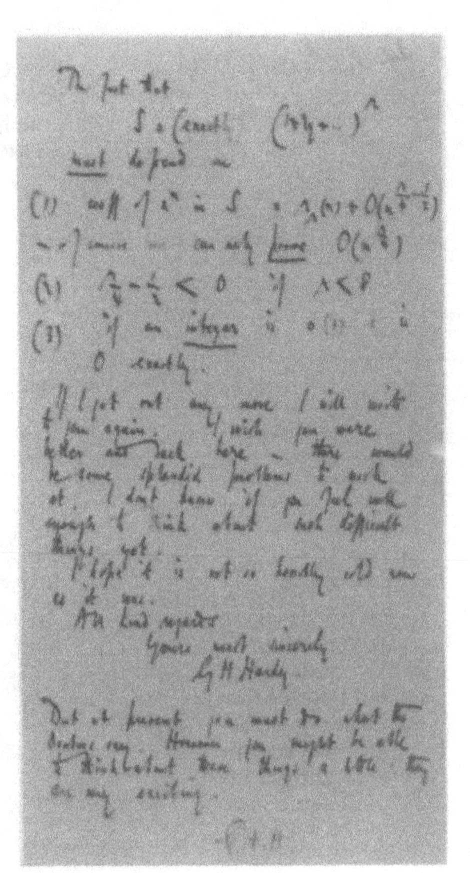

The fact that
[illegible]
must be found in
(1) [illegible]
[illegible]
(2) [illegible] < 0 if [illegible]
(3) if an integer is o(1) it is 0 exactly.

If I get out any more I will write to you again. I wish you were better and back here — there would be some splendid problems to work at. I don't know if you feel well enough to think about such difficult things yet.

I hope it is not so beastly cold now as it was.

All kind regards
Yours most sincerely
G H Hardy

But at present you must do what the doctors say. However you might be able to think about these things a little: they are very exciting.

G.H.H

The last page of a letter Hardy wrote to Ramanujan in February 1918, while Ramanujan was sick and in the hospital, about their current work. *Master and Fellows of Trinity College, Cambridge*

A. S. Ramalingam, an engineer whom Ramanujan met in the days after he arrived in England. Three years later, Ramalingam visited him at Matlock, a tuberculosis sanatorium, where his friend's emaciated condition and finicky eating habits left him worried. *Ragami's Collections, Madras, South India*

The Indian stamp issued in 1962 to honor Ramanujan.

"Gometra," the house off Huntington Road, in the Chetput section of Madras, where Ramanujan died.
Ragami's Collections, Madras, South India

P. S. Chandrasekar, Ramanujan's physician back in India. Ramanujan's death, he wrote in his diary the following day, was "a tragedy too deep for tears."
Ragami's Collections, Madras, South India

Hardy in his prime. "To sit that way," someone once said of him, "you have to have been educated in a public school." (He was.) *Master and Fellows of Trinity College, Cambridge*

A photo taken in 1941, snapped by someone from *The Picture Post*, as Hardy watched a rugby game between Cambridge and Oxford. He was 64 years old—and after a seemingly endless youth, looked it. *Master and Fellows of Trinity College, Cambridge*

1920, but a double meat ration was announced for Christmas 1918. On December 9, demobilization began.

During the war, 2162 Cambridge men had been killed, almost 3000 wounded—altogether about a third of all who served. Some 80,000 wounded had come in through the rail station over the four years. But now, within months of war's end, the university was already returning to something like its prewar proportions.

The *Nevasa,* the British India Lines ship that had carried Ramanujan to England in 1914, went back to carrying travelers and tourists. She had become a troop ship early in the war. Then, her funnel painted yellow and a broad band painted around her white hull in green, she was made over into a hospital ship, with 660 beds, which for two years bore the sick and wounded between Suez, Basra, and Bombay. Later, a troopship again, she carried American soldiers to France, at least twice having to outrun U-boats. Now, in 1919, she was a passenger ship once more.

Hardy's friend, the historian G. M. Trevelyan, returned to England "to get on with writing history books again. I had no other ambition in life. The delightful delusion that we had done with total war at least for a generation, perhaps for ever, gave a zest to domestic and personal happiness. I shall never forget the exhilaration of a Cornish holiday with my wife and girl and boy at Easter 1919, when all the world seemed young again, and the sands and rocky headlands rejoiced."

About the time of the armistice, Ramanujan had left Fitzroy Square for another nursing home, this one in the suburb of Putney, a few miles southeast of London on the south bank of the Thames.

It was called Colinette House. From the outside it was a big, boxy, undistinguished structure, in plan almost square, part of an early suburban development of freestanding brick houses built in the 1880s; Leonard Woolf had lived in one of them on his return from Ceylon a few years earlier, finding it something of a comedown from the Anglo-Indian luxury to which he had become accustomed. The interiors were more impressive, though, laced as they were with elaborate moldings, stained glass, grandly scaled rooms and handsome winding staircases. During Ramanujan's time, the eight-bedroom house at 2 Colinette Road had been made into a small nursing home presided over by one Samuel Mandeville Phillips.

Compared to Matlock, Ramanujan was much more accessible here, and Hardy (whose mother had died in Cranleigh a few weeks after the armistice) could visit him more easily; beyond the usual two hours on the train into London, Putney was just a cab ride away. Once, in the taxi from London, Hardy noticed its number, 1729. He must have thought about it a little because he entered the room where Ramanujan lay in bed and, with scarcely a hello, blurted out his disappointment with it. It was, he declared, "rather a dull number," adding that he hoped that wasn't a bad omen.

"No, Hardy," said Ramanujan. "It is a very interesting number. It is the smallest number expressible as the sum of two cubes in two different ways."

Finding numbers that were the sum of *one* pair of cubes was easy. For example, $2^3 + 3^3 = 35$. But could you get to 35 by adding some *other* pair of cubes? You couldn't. And as you tried the integers one by one, it was the same story. One pair sometimes, two pair never—never, that is, until you reached 1729, which was equal to $12^3 + 1^3$, but also $10^3 + 9^3$.

How did Ramanujan know? It was no sudden insight. Years before, he had observed this little arithmetic morsel, recorded it in his notebook and, with that easy intimacy with numbers that was his trademark, remembered it.

While Ramanujan was at Colinette House, he got some good news. All through the war, the authorities in Madras had kept up with him through Hardy, periodically extending his leave of absence from the Port Trust and his fellowships from the university. In 1917, Ramanujan's mother had come to Madras and learned that her son would be named University Professor, earning at least 400 rupees per month—six times what he made as a research scholar before he left. Now, in the wake of his F.R.S., the Madras authorities were fairly falling all over themselves to see what they could do for him. And in late December 1918 or early in the new year, he learned the university had granted him a 250-pound-per-year fellowship. This was on top of a like amount awarded him by Trinity. And it was good for six years. And it permitted him periodic trips back to England.

Ramanujan too well remembered the days in Kumbakonam when he'd lost his scholarship and had to drop out of school; and the period in Madras when he'd scraped by on Ramachandra Rao's patronage and the few rupees he made from tutoring. Besides, he was embarrassed by his comparative lack of productivity over the past year and a half. And so, on January 11, 1919, he wrote to Dewsbury:

Sir,

I beg to acknowledge the receipt of your letter of 9th December 1918, and gratefully accept the very generous help which the University offers me.

I feel, however, that after my return to India, which I expect to happen as soon as arrangements can be made, the total amount of money to which I shall be entitled will be much more than I shall require. I should hope that, after my expenses in England have been paid, £50 a year will be paid to my parents and that the surplus, after my necessary expenses are met, should be used for some educational purpose, such in particular as the reduction of school-fees for poor boys and orphans and provision of books in schools. No doubt it will be possible to make an arrangement about this after my return.

I feel very sorry that, as I have not been well, I have not been able to do so much mathematics during the last two years as before. I hope that I shall soon be able to do more and will certainly do my best to deserve the help that has been given me.

I beg to remain, Sir,
Your most obedient servant,
S. Ramanujan

About a month later, on Monday, February 24, Ramanujan was well enough to tend to the business of getting his passport. "Age: 30," the clerk recorded the information. "Profession: Research Student." Sitting for his passport photo, he held his head, with its full shock of straight black hair, cocked a little to one side, and, with luminous eyes, peered slightly up at the camera. The resulting Hollywood-handsome image was not the Ramanujan his friends knew back in India, nor the Ramanujan Hardy knew in 1914, but that of an ill, much thinner man. His shirt, buttoned to the top, was loose around the neck. Layers of fat no longer pressed against his jacket which—now ill fitting in the other direction—was two sizes too big around his neck and shoulders.

About two weeks later, on March 13, 1919, two short notes by Ramanujan appeared in the *Proceedings of the London Mathematical Society*; in them, he revealed new congruence properties of the partition function and a new link between the first and second Rogers-Ramanujan identities. That same day, lightened by notebooks and other papers he had left with Hardy, but encumbered with, among other things, at least a dozen books, a box of raisins for his younger brothers, and a big leather trunk filled with papers, he boarded the Pacific and Orient Lines ship *Nagoya*.

The *Nagoya*, of the same recent vintage as the *Nevasa* only a bit smaller, was steaming for Bombay.

The India to which Ramanujan was returning had not gone unscathed by the conflict that had bled Europe. Indeed, Madras itself had become a battleground. Immediately upon the declaration of war, the German light cruiser *Emden* had struck sea lanes in the Indian Ocean, taking prizes and destroying merchant ships. One night she appeared outside Madras. Wartime or not, the harbor lights burned brightly, illuminating the red-striped white tanks of the Burma Oil Company. Mistakenly advised of the *Emden*'s sinking, British officials were at a dinner party celebrating when the German marauder began to shell the harbor, setting fire to the oil tanks. Later, from ninety miles out at sea, the night sky was still aglow from the fires ashore.

The attack had little strategic impact, but it struck terror in the local citizenry, which may have included Ramanujan's parents and wife. Many natives, fearful of the *Emden*'s return, fled the city.

The war purged India of all but about fifteen thousand British troops. A volunteer Indian army of more than a million men had been raised, many of whom, at least in Madras, were untouchables; local recruiters promised them, "When you wear the King-Emperor's uniform, you will be able to walk through the Brahmin quarter and spit where you like."

Politically, it was a time of great new strides toward independence. In 1915, Annie Besant, English social reformer and president of Madras's Theosophical Society, began agitating for Home Rule with the publication of her daily newspaper, *New India*, and the formation of the Home Rule League. In the same year, Gandhi, already famous for the nonviolent methods of civil disobedience he had forged for the defense of Indian immigrants in South Africa, returned to India, where he began using his new social weapons to form a mass movement; five years later, he would become head of the Indian National Congress.

Change—a chafing at the bonds of the caste system, an awakening of the Indian masses, a resurgent Indian nationalism, a rediscovery of indigenous Indian ways—marked the years Ramanujan was away. All across the subcontinent, the giant was feeling its strength.

Intellectual India, too, felt it. In late December 1916, the Indian Mathematical Society held its first conference, in Madras. Lord Pentland himself opened the proceedings, being met on the steps of Presidency College by R. Ramachandra Rao, the society's fourth president. "At the present time," Lord Pentland told the conference, "a young Indian student, Mr. S. Ramanujan"—cheers erupted from the audience—"is studying at Cambridge whose career we in Southern India are watching with keen interest and high anticipation. You know the story of the discovery of his unusual talent and all here will be glad to hear how entirely he is justifying the efforts which were made to give it full scope."

Everyone knew the story by now. Back in May 1914, the *Nevasa* having borne him away barely two months before, Ramanujan was already being heralded in the Madras papers.

> Mr. S. Ramanujan of Madras, whose work in Higher Mathematics has excited the wonder of Cambridge, is now in residence at Trinity. He will read mainly with the two Fellows of the College—Mr. Hardy and Mr. Littlewood. They are going through masses of work he has already done, and are making some surprising discoveries in it!

When he received his degree, they heard about it back home. When he was named an F.R.S., Madras rolled out the red carpet for him in absentia, in the form of a meeting to honor him at Presidency College. One classmate of his from Pachaiyappa's College days, K. Chengalvarayan, later recalled how when he met old school friends during these years, "the emergence of Ramanujan into fame was usually one of the topics of talks."

Ten weeks before Ramanujan's arrival in Bombay, the Indian Mathematical Society met there for its second conference. Ramanujan's "brilliant career," his "very humble origin," and his elevation to the Royal Society were on every speaker's lips. He was returning at a bad time, though, at the height of a flu epidemic that would kill ten million people; among the mathematical society's small membership alone, it had claimed five lives already. Others who had died during Ramanujan's time abroad were E. W. Middlemast, the Presidency College mathematics professor who had given him one of his earliest recommendations; and, at the age of forty-two, Singaravelu Mudaliar, his mathematics professor at Pachaiyappa's College.

Now the lead page of the society's *Journal,* dated April 1, 1919, bore the news of Ramanujan's return. But he was coming back, it advised, "in somewhat indifferent health."

2. RETURN TO THE CAUVERY

"Where is she?" asked Ramanujan of his mother as he stepped off the ship into the maw of Bombay on March 27, 1919. *She* was Janaki. His mother was there, and his brother Lakshmi Narasimhan; the two of them had set out for Bombay on the twenty-first. But not his wife. Why fret over Janaki? sniped Komalatammal. Scarcely off the boat, Ramanujan had dropped into the family snake pit. Domestic conflict had sabotaged his last three years in England. Now they dampened his arrival in India.

In fact, the two sides of the family had been out of touch for more than a year. No one knew just where Janaki was. Perhaps back in Rajendram. Or, if she had learned of Ramanujan's arrival, maybe with her sister in Madras. So, Lakshmi Narasimhan was dispatched to write two identical letters to Janaki asking that she come meet Ramanujan in Madras.

She was back in Rajendram, it turns out, where she'd gone for her brother's wedding more than a year before. She did know of Ramanujan's return—but only thanks to the Madras papers, not because her mother-in-law had bothered to inform her. Her brother, R. Srinivasa Iyengar, had advised her not to rejoin the family in Madras. Komalatammal hated her; why place herself within the tiger's jaws once again? But then the letter came from Lakshmi Narasimhan. Ramanujan, it said, wanted her. That was all it took. She and her brother set off.

"When I go back I shall never be asked to a funeral," Ramanujan had told Neville before he left. It was one form of the taint he expected to bear, in the eyes of strictly orthodox Brahmins, for going to England. To help remove it, Komalatammal had planned to take her son directly to Rameswaram, for purification ceremonies in the great temple to which the whole family had gone for a pilgrimage in 1901. But now, in Bombay, one look at him decided her against it; Rameswaram, all the way south almost to Ceylon, was another five hundred miles and another hard day on the train beyond Madras. He was just too sick. And so, after a few nights in Bombay, Ramanujan, his mother, and brother boarded the Bombay Mail for the overnight trip to Madras.

"When I met him alighting from the railway train," Ramachandra Rao would recall of Ramanujan's arrival in Madras on April 2, "I foresaw the end." He looked awful. Making things worse, there was still no Janaki. Why? he asked his mother. This time Komalatammal said something about Janaki being off to tend her father, who was unwell.

Ramanujan was bundled on to a *jutka,* a two-wheeled, horse-drawn vehicle, which soon pulled away from the bustling area around Central Station. His old friend K. S. Viswanatha Sastri, one of those who'd seen him off from Madras five years before and whom Ramanujan had once tutored in math, followed on behind by bicycle. About three miles south of the station, they reached a large, lovely bungalow on Edward Elliots Road owned by a wealthy trial lawyer. When Viswanatha Sastri got there a little later, Ramanujan was already eating yogurt and sambhar. "If I had this in England," said Ramanujan, "I would not have gotten sick."

The weekend before his arrival in Madras, the local papers carried an account of his life prepared by Dewsbury's office. "This," Sir Francis Spring scrawled on an interoffice memo, "does not seem to go far enough in satisfying legitimate public interest in a man whose mathematical genius may yet do great things for the world and has already." He directed Narayana Iyer to rummage through the Port Trust's Ramanujan files and assemble a much fuller biography. This appeared on April 6. "Mr. S. Ramanujan is a native of Kumbakonam in the Tanjore District of Madras Presidency. He was born in 1888 [sic] of poor and, so far as English goes, illiterate parents of the Vaishnava sect of Brahmins. . . ."

Ramanujan was offered a university professorship, which he said he would accept when his health improved. Madrasi notables trooped by to visit the convalescing genius, South India's conquering hero of the intellect, who had shown the Britishers the stuff of which South Indians were made. Now and over the next year, they rushed to pick up his medical and other expenses. They offered him their homes. Top people from the *Hindu* came by. So did Ramachandra Rao, of course, and Sir Francis Spring, and Narayana Iyer.

To shield him from visitors a little, Ramanujan's doctor, M. C. Nanjunda Rao, had him moved a half mile or so south to a place called Venkata Vilas on Luz Church Road, named for a nearby Portuguese church (known locally as *Kattu Kovil,* or Jungle Temple). Now, far from the teeming Triplicane and Georgetown of his youth, he was in the very heart of cultured Madras, peopled by high-born Brahmin intellectuals, lawyers, and scholars, who lived in large compounds, luxuriant with banana trees and betel gardens. It was here, finally, on April 6, that Janaki and her brother caught up with him, followed, about a week later, by Ramanujan's father, grandmother, and younger brother, from Kumbakonam.

Later that month, on the twenty-fourth, Ramanujan wrote Dewsbury asking for an account of his expenses in England and the trip back and requesting payment of his fellowship in monthly installments. He was settling in for the long haul.

For three months Ramanujan stayed at the bungalow on Luz Church Road. And here, he and Janaki began to forge something like a real relationship. Janaki had been just thirteen when he left. Now she was eighteen. As they never really had before, they began to *talk*, perhaps now each coming to discover how Komalatammal had intercepted their letters, perhaps finding time for physical intimacy as well.

But the man Janaki, his family, and friends saw was much changed from the man they had known. There were the small superficial changes, of course. He drank coffee now, which he hadn't much before despite its popularity (more than tea) in South India. His tuft was gone. He was a bit lighter in complexion. He was much thinner; Janaki, seeing him so emaciated and coughing up phlegm, only now realized how sick he was.

But it was the change in his personality that upset his friends most. K. Narasimha Iyengar, with whom Ramanujan had lived for a time in Madras around 1912, had been there to greet him at the train platform. "I found him not a cheerful, chummy and affectionate Ramanujan," he said later, "but a thoroughly depressed, sullen and cold Ramanujan, even after seeing me, a close and affectionate friend." Recalled Anantharaman, who visited him at Luz, bearing a bunch of Kumbakonam bananas: "He was not the original Ramanujan. He could [scarcely] speak and his illness had made him peevish." He was bitter, impatient, sour. His faith was tarnished, too. Once, Anantharaman's older brother, Ganapathy, said something about gods and temples, and Ramanujan snapped back that "it was foolish talk, and they were only devils."

Janaki was not immune from his wrath and found him quick to jump on her when disturbed while working. In his passport photo, she thought, he seemed troubled, and the same look crossed his features often over the next year. Other accounts have him yelling at his brother for wasting money, or impetuously scattering things around the house, or even chewing on thermometers placed in his mouth.

The family's own record of events in Ramanujan's life, reconstructed after his death (probably by his brother Lakshmi Narasimhan) is over the next year often ambiguous, its cryptic scrawl bearing fragmentary references like "Tamil books," and "paradox about points," and "rumor that he was mad." It records the comings and goings of family members, visits to the house, the receipt of letters, and much other daily minutiae, most of it not fleshed out enough to reconstruct today. But one notation, recorded over and over during the year, leaves little ambiguity: *quarrel.* Sometimes the quarrels were between Janaki and Komalatammal, sometimes between Janaki and Ramanujan's grandmother. Sometimes the antagonists went unrecorded. Twice during April the family scribe mentioned quarrels, twice again in July, another time in August, again in September, again in October, and at other times later.

For much of the year, during which the family moved repeatedly, the household was a hellhole of simmering resentment. Once, a battle apparently broke out over whether Ramanujan should go to a sanatorium, another time over some donation or another. Ramanujan's letter from Colinette House, suggesting that some of his fellowship money go toward scholarships, was another irritant; Komalatammal preferred that her son shower more of his new affluence on the family, not strangers. Janaki was probably no more immune to money's allure than Komalatammal; years later she recalled how Ramanujan, in his final days, "said he had five thousand rupees in his savings to buy me diamond eardrops and a gold belt."

With the approach of the Madras summer and its daytime temperatures over a hundred degrees, Ramanujan's doctors advised him to head inland to escape the heat and humidity. One possibility was Coimbatore, a city about the size of Kumbakonam located in hill country most of the way west across the southern spine of the Indian subcontinent. Coimbatore would be drier, and ten degrees cooler. But Ramanujan's mother opted instead for Kodumudi, a sleepy little town of a few thousand people known for its Magudeswara temple, which was unusual in offering common worship to Brahma, Vishnu, and Siva. Kodumudi formed one vertex of a twenty-mile-wide triangle of which Namakkal and Erode, Ramanujan's birthplace, were the other two. For Komalatammal's side of the family, especially, Kodumudi was like going home.

The district comprising Kodumudi, with rainfall only half as heavy as that of Madras, was mostly brown and bare. But the town itself, occupying a site on the southern bank of the Cauvery, was green. The river, beside which a fertile levee had built up over the years, was broad here, a line of palm trees on the north shore distant enough to savor of a foreign country.

Through arrangements made by the university, Ramanujan was staying in a house located on East Agraharam Street, near the town registrar's office; *agraharam* meant the Brahmin quarter. Crossing the T of

Agraharam Street was a street that dropped down to the Cauvery at a spot where men and women came to wash clothes and bathe. And there, on August 11, 1919, at the time of the Sravanam ceremony that marked the annual changing of the sacred thread, Ramanujan openly rebelled against his mother.

Trouble had been brewing for weeks between the two of them, from even before they'd arrived in town; Ramanujan had wanted to travel first class, but his mother had insisted on second or third class. Now, Ramanujan was heading down to the river to bathe, as part of the Sravanam rites. Janaki wanted to go with him. Ramanujan said yes. Komalatammal said no.

And Ramanujan insisted, *yes*.

It was a turning point. All the hostility pent up over the past few years, the diverted letters, the endless quarrels—and now this. He spoke respectfully, but firmly: Janaki was going with him.

After that, Janaki began to occupy a larger place in her husband's life. He grew freer in her company. He told her, more than once, "If only you had come with me to England perhaps I would not have fallen ill." On Friday afternoons, he would sometimes watch her oil and wash her hair, shaking it out and letting it dry in the sun. The sight gave him great pleasure, sometimes enough to draw him away from his work. In coming months, it would be Janaki who, increasingly, would care for him during the day, give him his medication, nurse him at night. When, at one point, his mother urged him to send Janaki packing, back to her parents, Ramanujan refused.

For two months Ramanujan stayed in Kodumudi. Every Sunday the doctor visited—a C. F. Fearnside, Divisional Medical Officer. And it may only have been now, in questioning him, that Komalatammal grasped the full extent of her son's condition. At one point Fearnside suggested that Ramanujan go to Coimbatore. At another, Thanjavur, near Kumbakonam, was the recommendation. (Ramanujan would not hear of it. "They want me to go to Tan-savu-ur," he said, punning on the word's Tamil components, "the place-of-my-death.") But with the worst of summer over, the pressure was strong to go *somewhere* less out-of-the-way.

Komalatammal decreed that it be Kumbakonam.

In late August, perhaps grateful to get away from a domestic situation that had turned against her, she went on ahead to look for a house, the family home on Sarangapani Sannidhi Street apparently being deemed inappropriate for her son's care. On September 3, Ramanujan and the rest of the family left Kodumudi and, around dusk of the following day, arrived in Kumbakonam.

The surroundings of his childhood and adolescence may have mellowed Ramanujan a little because stories from this period bear a softer quality, even a nostalgic tinge. One of Ramanujan's former classmates told how "he insisted on seeing my aunt, who is an old widow on the wrong side of 70, whom he knew while at school some 20 years ago." Old friends dropped by. Balakrishna Iyer, one of those Ramanujan had approached for a job around 1910, sent him dried brinjal slices. Radhakrishna Iyer, the Pachaiyappa's College classmate who had nursed him while he was sick in 1909, recalled Ramanujan as just "a bundle of bones," lying on his cot. "Ramanujan had powerful and penetrating eyes. For a moment his eyes flashed a look of recognition," and he mumbled Radhakrishna's name.

Ramanujan had a new doctor now, P. S. Chandrasekar, brought down from Madras especially to care for him. Hardy, through the Royal Society, had approached the authorities in Madras. The surgeon general had turned to Chandrasekar, fifty, a professor of hygiene and physiology at Madras Medical College and a prominent tuberculosis specialist. At one point, Chandrasekar accompanied one of Ramanujan's oldest friends, Sarangapani (the boy who had once bested him in an arithmetic exam) to the house on Bhaktapuri Street where Ramanujan lay sick. For more than an hour Chandrasekar examined him. It was tuberculosis, he was sure, notions of blood poisoning notwithstanding. "I have a friend who loves me more than all of you, who does not want to leave me at all," Ramanujan told Sarangapani later. "It's this tuberculosis fever."

When Ramanujan boarded the *Nagoya* to return to India early that spring, hopes for recovery, at least in Hardy's estimation, had been high. But the kind of sanatoriums treating him for the past two years had one essential failing: they didn't cure tuberculosis. Of those discharged from English sanatoriums, one study found, between one- and two-thirds were dead within five years, most of them within the first two. In Ramanujan, Chandrasekar now concluded, the disease had reached an advanced stage. His fate lay more with God than in anything medicine could do.

Since coming down sick two and a half years before, Ramanujan had been through the wringer. In England, he'd seen many doctors, heard countless conflicting diagnoses, been shuffled around to numerous hospitals and nursing homes. Now in India, it was the same thing. Two places in Madras. Kodumudi. Perhaps briefly in Coimbatore. Kumbakonam. All in nine months. More doctors. More examinations. More diagnoses. Through it all, his condition grew gradually worse. Now, in Kumbakonam, the downward slide accelerated.

Just before Christmas of 1919, while writing Hardy with details of Ramanujan's accounts, Dewsbury added: "He is still in very bad health and difficult to deal with, living up country with his family. Mr. Ramachandra Rao is doing what he can for him, but Mr. Ramanujan himself will not consent to live in a suitable environment under proper treatment. It is a great pity." By one account, he refused to be further treated, by another he had simply given up the will to live. It was like pulling teeth even to get him to go to Madras, where Chandrasekar wanted him. It was the coolest, most pleasant time of the year; the heat that had originally driven him to Kodumudi had given way to daytime temperatures in the eighties. Finally, Ramanujan relented.

Sometime probably a little after the first of the year, Ramanujan, his mother, wife, and brother-in-law, Srinivasa, set off for Madras—but, for the moment, got only as far as the railway station. They had missed the train, the next one not due for another six or seven hours. Rather than subject Ramanujan to another bumpy, two-mile ride to Bhaktapuri Street, only to turn around a few hours later, they camped out at the station on the edge of town. Periodically, Janaki and Komalatammal returned to the house for food or other provisions.

The train ultimately came, of course, and they reached Madras. On January fifteenth, when Dewsbury wrote Hardy again, he gave Ramanujan's address as "Kudsia," Harrington Road, Chetput, Madras.

3. THE FINAL PROBLEM

Ramanujan was living in one of a group of substantial stucco houses built for the British around the turn of the century in a western suburb of the city. Given names like Sydenham, Ravenscroft, and Lismoyle, and sometimes supplied with tennis courts, they were popular with middle-rank Indian Civil Service officers and company officials. Several dozen of them had gone up along a lane perpendicular to Harrington Road, now called Fifteenth Avenue, but during Ramanujan's time unnamed. "Kudsia" was one of them, though apparently Ramanujan didn't stay there long. For a time, he was installed in a considerably larger one, "Crynant," in a cul-de-sac at the end of the dusty lane. Ultimately, he would be moved to another, known as "Gometra."

It was from one of these houses that on January 12, 1920, Ramanujan wrote Hardy for the first time in almost a year.

> I am extremely sorry for not writing you a single letter up to now. . . . I discovered very interesting functions recently which I call "Mock" theta functions. Unlike the "False" theta functions (studied partially by Prof. Rogers in his interesting paper) they enter into mathematics as beautifully as the ordinary theta functions. I am sending you with this letter some examples.

In how it would excite the active interest of mathematicians up until the present, it was a letter much like the one Ramanujan had written Hardy seven years earlier almost to the day. By one estimate, it represented "one of the most original pieces of mathematics, and in some ways the very best, which Ramanujan did."

As with so many other subjects that interested Ramanujan, theta functions could be represented as infinite series. Strict rules governed the formation of these series, and so long as you observed them, they always had particularly intriguing properties. For example, they were "quasi doubly periodic" and they were "entire functions." Over the years since Jacobi had first studied them, theta functions had come to exert a profound impact on fields ranging from mathematical physics to number theory. Books devoted to them typically wound up bridging vast terrain, yielding connections to fields to which at first they might seem wholly unrelated.

But theta functions were fragile. The properties that made them so intriguing to mathematicians vanished if, in trying to construct them, you departed even slightly from their canonical form—for example, by merely changing the signs of some of the terms; at a glance, they might seem indistinguishable from theta functions themselves, yet they lost most of their interesting properties. True, certain "false theta functions" explored by Ramanujan's intellectual cousin, L. J. Rogers, in a paper a few years before, held some interest; later study would apply them, for example, to partitions. But compared to the real, or "classical," theta functions that mathematicians found so rich with meaning, they were pale substitutes.

Now, in his letter to Hardy, Ramanujan was telling how a new class of functions could be constructed that were not theta functions but bore crucial similarities to them and were just as interesting. They were not "false" theta functions, like those Rogers had investigated, but what Ramanujan termed "mock" theta functions.

Classical theta functions could be put into a variant "Eulerian" form, in which terms like $(1-q)$ or $(1-q^3)$ appeared in the denominator. That meant that when $q=1$, say, or $q^3=1$, the functions were mathematically undefined and had to be explored in alternative ways (in a way loosely analogous to how Ramanujan and Hardy had explored the partition function through their "circle method"). Offering two theta functions as examples, Ramanujan noted that at such "singularities"—at the points demanding special study—"we know how beautifully the asymptotic form of the function can be expressed in a very neat and closed exponential form."

All this was spadework, with which Hardy was familiar. But now Ramanujan ranged beyond it:

> Now a very interesting question arises. Is the converse of the statements concerning the [two theta functions] true? That is to say: Suppose there is a function in the Eulerian form and suppose that all or an infinity of points are exponential singularities, and also suppose . . .

and on he went like this, fashioning his function before Hardy's eyes. When it had satisfied a host of such special conditions it was, he said, a "mock theta function." Then he went on to offer a whole slew of them—four third-order mock theta functions, ten fifth-order functions, and so on.

"The first three pages in which Ramanujan explained what he meant by a 'mock theta function' are very obscure," G. N. Watson, one of the first of those to study them, would allow. But hazy and poorly expressed as they were, they held within them profound mathematical truths that became the subject of Watson's presidential address to the London Mathematical Society sixteen years later and would keep other mathematicians busy for years. Wrote Watson:

> Ramanujan's discovery of the mock-theta functions makes it obvious that his skill and ingenuity did not desert him at the oncoming of his untimely end. As much as any of his earlier work, the mock-theta functions are an achievement sufficient to cause his name to be held in lasting remembrance. To his students such discoveries will be a source of delight and wonder until the time shall come when we too shall make our journey to that Garden of Proserpine where

"In Somewhat Indifferent Health"

Pale, beyond porch and portal
Crowned with calm leaves, she stands
Who gathers all things mortal
With cold, immortal hands.

In 1893, Arthur Conan Doyle, creator of Sherlock Holmes, wrote a story in which the famous detective, grappling with his archfoe Professor Moriarty at Reichenbach Falls in Austria, plunges to his death on the rocks below—or so Holmes's friend, Dr. Watson, is led to believe. Now, four decades later, discussing Ramanujan's surge of mathematical creativity in the months before his death, this other Watson, the mathematician, resurrected the title Conan Doyle had used for Sherlock Holmes's last case. Mock theta functions, he said, were Ramanujan's "Final Problem."

All that year in India Ramanujan worked on mock theta functions, "q-series," and related areas, filling page after page with theorems and computational fragments and infinite series going off in every direction, by one reckoning about 650 formulas in all. When the American mathematician George Andrews began poring through them half a century later, he was stunned by their richness, the surprises they offered, was left mystified over how anyone could think them up. "It is very difficult even for somebody trained in mathematics, but not an expert, to tell them apart," he would say. Among one group of five superficially similar formulas, "the first one took me fifteen minutes to prove, the second an hour. The fourth one followed from the second. The third and fifth took me three months."

Ramanujan's work during the year before he died could be seen to support an old nostrum of the tuberculosis literature—that the tuberculous patient, as he succumbed, was driven to an ever-higher creative pitch; that approaching death inspired a final flurry of creativity impossible during normal times. The idea has been soundly trounced by modern scholars quick to cite artists and other creative workers whose greatest work long preceded their illness and whose deathbed productivity was nil. Still, during Ramanujan's time, the idea had a following. And at least one account of Ramanujan's last days, by P. V. Seshu Iyer, bore its flavor: "There are no papers and researches of his more valued nor more intuitive," he wrote, "than those which he thought out during these fateful days. His physical body was failing no doubt, but his intellectual vision grew proportionately keener and brighter."

But George Andrews, for one, could explain Ramanujan's remarkable productivity in more prosaic, if doubtless sounder, terms. "He had the full power of his collaboration with Hardy behind him and the sort of wild eccentric genius of his youth behind him," too. Ramanujan, in other words, *should* have been at the height of his powers. "And this notebook," said he of the product of that final year, "bears it out."

Sometime during late winter or early spring of 1920, while staying in "Crynant," Ramanujan complained to his mother that the "cry" of that name seemed inauspicious. Komalatammal went to Namberumal Chetty, owner of the bungalows (as well as of a small railroad) and a friend of Sir Francis Spring. Omitting Ramanujan's real reasons, Komalatammal told him her son needed a quieter place. Namberumal obliged, and Ramanujan was moved a little down the road to "Gometra," which means something like "Friend of Cows," a reference to Krishna.

It was a house full of doors. There were banks of doors everywhere, doors leading back outside to an arched portico, doors to an interior court where the cooking was done in a separate outbuilding, literally dozens of doors ranging over the house. When they were all open, the inside of the house, with its floor of large granite slabs, was like a cool, shaded glen open to any stray passing breeze. You'd enter through what from the street seemed a side entrance but which, leading you under one of several arches, actually brought you into the center of the house. There stood a stairway, its banisters fashioned from Burma teak, that led up to the second floor. To your right, toward the front of the house, was a large living room. Ramanujan may briefly have stayed there, but in time he was shifted to a small corner room on the other side of the stairs at the back of the house.

Enhancing its sense of openness was the bungalow's scant furniture. Ramanujan's bed was a mattress and a pillow laid out upon the bare granite floor. He scarcely budged from it. Lying there, he could look up and see a massive wood beam, from which, like a spine, the second floor was supported, spanning the length of the room. It was quiet back here away from the street. Visitors moved silently so as not to disturb him. Tirunarayanan, who fondly recalled his older brother horsing around with him, telling him stories and carrying him on his shoulders, now found access to him jealously guarded, with no one allowed to bother him.

Still, Ramanujan could scarcely have gone unaware of the tensions that gripped the household. The whole family was in a state of jangled nerves. And that he was sick, miserable, angry, and prone to sudden outbursts did nothing to ease matters. Again and again, Janaki and Komalatammal clashed, absorbing energies that might better have gone toward Ramanujan's care. Now, moreover, the contest was not so unequal as before. Janaki was twenty, enjoyed a new and stronger position in the household, and had her brother Srinivasa by her side, too.

Around March, Komalatammal appeared at the door of G. V. Narayanaswamy Iyer, a former student of mathematician P. V. Seshu Iyer, Ramanujan's long-time friend and champion, whose note of introduction she now bore. Narayanaswamy Iyer was a high school teacher in Triplicane, but it was not for his pedagogic abilities that Komalatammal wanted to see him. He was also a noted astrologer.

Narayanaswamy Iyer asked for a copy of the horoscope she wanted read. Komalatammal dictated it from memory. This, he said after studying it for a while, was the chart either of a man of worldwide reputation apt to die at the height of his fame, or one who, if he did live long, would remain obscure. "Who is this gentleman? What is his name?" he asked.

It was, of course, Ramanujan, and Komalatammal, in tears, replied that she had drawn the same inference from his horoscope.

Narayanaswamy backpedaled. *Ramanujan!* "I am sorry I was so hasty," he said. "Please do not carry what I said to any of his relatives."

"I am the mother of Ramanujan," she replied.

Groping for a way out of the uncomfortable scene, Narayanaswamy suggested that maybe his wife's horoscope might mitigate the bad news. "Can you bring her horoscope next time?" he asked. A separate visit was hardly necessary: Komalatammal rattled it off for him.

The astrologer looked for something—*anything*—by which to inspire hope. But after half an hour's study of the two horoscopes, laid side by side, he was reduced to nothing beyond suggesting that perhaps Ramanujan and Janaki might wish to live apart for a while.

That, of course, was just what Komalatammal wanted to hear. "I have been . . . pleading with my son to send her away to her parents' house," she said. "But, alas, he utterly refuses. He has been always an obedient son. But in this matter, his obstinacy is unbreakable."

Did Ramanujan now know he was dying? According to another of those enduring myths that surrounded tuberculosis, the victim was the last to realize his approaching death. "Buoyed up by *spes phthisca,* a delusive hope of recovery," one student of tuberculosis, Nan Marie McMurry, recounts, "the consumption patient was supposed to anticipate a return to health on the deathbed."

But Ramanujan suffered no such delusions and often during his final months told his doctor he had lost the will to live. Earlier, it is true, he had said he would accept a university position offered him once he had recovered his health. And as recently as January, around the time he wrote Hardy about mock theta functions, he expressed interest in subscribing to new math journals. But the dark superstitious side of him saw things more blackly—and, as it happens, more accurately. He had long before read his own horoscope as predicting his death before age thirty-five. And brought to Harrington Road, he punned on the name of the Madras suburb, Chetput, in which it was located; "chat-pat," in Tamil, meant "it will happen soon."

During his last months, Ramanujan drew closer to Janaki, with whom, when he wasn't exploding at her in a fit, he now had a warmer, more relaxed relationship. "He was uniformly kind to me," Janaki recalled. "In his conversation he was full of wit and humor," was forever cracking jokes. As if trying to cheer her up, he plied her with tales of England—of his visits to the British Museum, and the animals he'd seen there, of the time in Cambridge when an English guest eating a South Indian dish he had prepared chomped down on a piece of hot pepper . . .

His life came down a little from the heights of mathematics to small things, human things. He would summon Janaki with a little bell, or would tap with a stick. He told C. S. Rama Rao Sahib, Ramachandra Rao's son-in-law, how much he craved the *rasam* he had enjoyed almost a decade before in those destitute days of 1910 at Victoria Hostel—and loved it when somebody brought him some.

Still, it was not an easy death. Whatever cheerfulness or equanimity he could muster—and some of his friends preferred later to remember him that way—papered over a grim despair. He was sullen and angry much of the time. His mood was volatile, resting on a hair trigger. Janaki felt later—so had Hardy in England, and so did many of his friends in India—that his illness had affected his mind. He was forever raving at one thing or another. There was a tray kept nearby with lentils, spices, and rice, and once, in a fit of anger and pain, Ramanujan pounded it all together with a stick. Narasimha Iyengar, hurt at finding Ramanujan so sullen and cold when he greeted him at the Central Station, visited him now again on Harrington Road. "I found to my great grief that though physically living, he was mentally dead to the world, even to his once dear friends."

Toward the end, "he was only skin and bones," Janaki remembered later. He complained terribly of the pain. It was in his stomach, in his leg. When it got bad, Janaki would heat water in brass vessels and apply hot wet towels to his legs and chest; "fomentation," it was called, standard therapy at the time. But through all the pain and fever, through the endless household squabbles, through his own disturbed equanimity, Ramanujan, lying in bed, his head propped up on pillows, kept working. When he requested it, Janaki would give him his slate; later, she'd gather up the accumulated sheets of mathematics-covered paper to which he had transferred his results and place them in the big leather box which he had brought from England. "He wouldn't talk to anyone who came to the house," said Janaki later. "It was always maths. . . . Four days before he died he was scribbling."

Early on April 26, 1920, he lapsed into unconsciousness. For two hours, Janaki sat with him, feeding him sips of dilute milk. Around midmorning or perhaps a little earlier, he died. With him were his wife, his parents, his two brothers, and a few friends. He was thirty-two years old.

At the funeral later that day, most of his orthodox Brahmin relatives stayed away; Ramanujan had crossed the waters and, too sick on his return to make the trip to Rameswaram for the purification ceremonies his mother had planned, was still tainted in their eyes. Ramachandra Rao arranged the cremation, through his son-in-law and Ramanujan's boyhood friend, Rajagopalachari. At about one in the afternoon, his emaciated body was put to the flames on the cremation ground near Chetput.

The next day, assigning it Registration No. 228, a government clerk officially recorded his death.

4. A SON OF INDIA

The astrophysicist Subrahmanyan Chandrasekhar grew up in Lahore, in what was then northern India, moved to Madras when he was eight, attended Presidency College there, then Cambridge's Trinity College (where he met Hardy), went on to propose the theoretical underpinnings for black holes, and was awarded the Nobel Prize in 1983. In 1920, he was just nine years old, but he well recalls how one day in late April of that year his mother, reading about it in the newspaper, told him of the death of Ramanujan. "Though I had no idea at that time of what kind of a mathematician Ramanujan was or indeed what scientific achievement meant," Chandrasekhar would tell an American audience almost seven decades later,

> I can still recall the gladness I felt at the assurance that one brought up under circumstances similar to my own could have achieved what I could not grasp.
>
> The fact that Ramanujan's early years were spent in a scientifically sterile atmosphere, that his life in India was not without hardships, that under circumstances that appeared to most Indians as nothing short of miraculous, he had gone to Cambridge, supported by eminent mathematicians, and had returned to India with every assurance that he would be considered, in time, as one of the most original mathematicians of the century—these facts were enough, more than enough, for aspiring young Indian students to break their bonds of intellectual confinement and perhaps soar the way that Ramanujan had.

In the India of the 1920s, he would say at another time, "We were proud of Mahatma Gandhi, of Nehru, of [the Nobel Prize–winning poet Rabindranath] Tagore, of Ramanujan. We were proud of the fact that anything we could do would equate to anything else in the world." Within Indian mathematics, of course, Ramanujan's influence extended correspondingly deeper. "I think it is fair to say," Chandrasekhar would observe, "that almost all the mathematicians who reached distinction during the three or four decades following Ramanujan were directly or indirectly inspired by his example." (One brilliant young mathematics student at Madras's Presidency College, T. Vijayaraghavan, deliberately neglected his studies and failed his examinations, the more perfectly to follow in Ramanujan's footsteps.)

Such was Ramanujan's impact on India during the years after his death. But back in the family, preoccupied by the daily cares of life and filled with painful memories of the last difficult years, his death left a wake much smaller, yet just as keenly felt.

In the 1930s, Chandrasekar, the tuberculosis expert who had treated Ramanujan in his final months, was asked by an old Madras friend of Ramanujan whether anything could have been done to save him. Yes, he replied, flying into a rage, Ramanujan could have been saved, *should* have been, and his mother and wife were to blame, at least in part. The day after Ramanujan died, he wrote in his diary:

If he had been allowed to follow my instructions, this double tragedy need not have taken place. The neglect of Ramanujan during his early phase—perhaps partly due to the ignorance of his contemporaries, as well as his relatives' (mother's and wife's) contributory (I almost feel like using the stronger word "criminal") negligence have contributed to this double tragedy—a tragedy which is too deep for tears.

Chandrasekar (who had lost his most illustrious patient, and so may have had his own axe to grind) pictured both wife and mother wallowing in the mire of materialism. But both before Ramanujan's death and now, after it, they were scarcely in a position to ignore the economic facts of their lives. Whatever else he was, Ramanujan was, through his ample fellowships, a breadwinner. Now he was gone, and yes: there was a grappling for what seemed might be the spoils.

On April 29, with Ramanujan's ashes scarcely cold, his eighteen-year-old brother, Lakshmi Narasimhan, wrote Hardy with the news of his death (which Hardy had doubtless already heard). Using language appropriate to a battle dispatch, Lakshmi Narasimhan complained that Ramanujan's books and papers were "entirely under the control of his wife here." He had heard through Ramachandra Rao's son-in-law, he said, that the government would be providing Janaki a small monthly stipend. "I am very sorry that such an arrangement was not made for my family." Janaki, plainly enough, was not part of the family.

It was a pathetic plea, and one that hinted at the pressures Ramanujan had withstood as a young man in order to pursue mathematics rather than establishing himself in a good job. His father had gone blind about when Ramanujan left for England, said Lakshmi Narasimhan. His grandmother was lame. "Besides these," said the young man in whose hands the family's fortunes now rested, "I have got a corpulent mother who resembles my brother in all his physical features." As for himself, he had suspended efforts on his own behalf over the past year, in order to help his brother.

I have no uncles or cousins to protect me. We have no property, as you might have known very well. I have a great desire to study and I wish "to neglect worldly ends, all dedicated to closeness and the bettering of my mind" (Tempest). I have no taste, sorry to say, for Mathematics. I like to read Shakespeare, Wordsworth, Tennyson, and wish to travel in the fairy land "half flying half on foot." I do not know how to feed them. Therefore, I humbly request you to write to the Madras University to give a monthly allowance to us. I have been told that my brother is entitled to get a sum from the Cambridge University.

In short, I am a young man. One day passes with great difficulty. I do not know how to protect them. Unless any arrangement is made to support us, we have to go a-begging from door to door. I entrust the whole case into your hands. It is your bounden duty to protect us.

There is no reason to think Hardy did anything of the kind, and the family drifted back into something like the obscurity that was, save for the accident of Ramanujan's birth, its natural lot.

A few months after Ramanujan's death, his sixty-five-year-old father got sick and was brought to the Pycroft's Road house of Narayana Iyer. There, Komalatammal and Narayana Iyer's wife looked after him until, in November, in the house where Ramanujan and Narayana Iyer had stayed up late doing mathematics, he died.

Komalatammal, meanwhile, never recovered from her son's death. "She lost her optimism and became very often sullen," said one old friend of the family, K. Srinivasa Raghavan, who had become a college professor. After making no effort to contact him for the past seven years, in August 1927 she wrote Hardy, invoking the help of a scribe or family friend to cast her Tamil into lucid English. Her son Lakshmi Narasimhan, she reported, had passed his Intermediate examination (the new name for the F.A. exam Ramanujan had always failed) and now worked for the post office in Triplicane. Tirunarayanan attended Presidency College and had earned his B.A. degree early that year.

Most of the rest of the three-page letter outlined the family's fragile economic footing, detailed the meager pensions it received, and recounted her troubles. "I sent my son to England in 1914 thinking that my family, ever wedded to poverty, would become rich, and that my son would become famous. Like Achilles, he won everlasting fame; like Achilles, he died young." Now, she asked Hardy to intercede on her sons' behalf with the government and with the India Office in London. She wanted for them high positions in the post office department—the better-educated Tirunarayanan as probationary superintendent of post offices, and Lakshmi Narasimhan as inspector of post offices in Madras. Whether or not Hardy interceded, Tirunarayanan did become assistant postmaster. The more flamboyant Lakshmi Narasimhan died while still young.

Long-time family friend Anantharaman, at whose house Ramanujan ate often as a boy, had been in bed recovering from a leg operation when his uncle told him of Ramanujan's death. It came as a great shock. Later, during the late 1930s, Komalatammal would visit him and his family in Triplicane and, he would recall, "console herself by seeing us, and say she was feeling as though she were seeing her son Chinnaswami himself."

Janaki, meanwhile, had gone her own way. Ramanujan had worried about how she'd be treated by the family after his death. She was a twenty-year-old widow, and widows were virtually an oppressed class in India, apt to be ill-treated and despised. She had no education, no skills. Her brother and mother, who had arrived in Chetput two days before Ramanujan's death, figured she could expect no help from Komalatammal and the rest of Ramanujan's family.

So after the cremation she returned with her mother to Rajendram. Then, for the next six years she lived with her brother, who became an income tax officer, in Bombay. When she learned of the twenty-rupee-per-month pension awarded her by the university in return for her rights to Ramanujan's papers, she returned to Madras, briefly staying with her sister in Triplicane. Soon she found her own place—a house on Hanumantharayan Street two doors from where she and Ramanujan had lived before he'd left for England. There she stayed for most of the next half century.

She had learned to embroider and work a sewing machine in Bombay, and so now she scratched out a living making clothes and teaching tailoring to girls. In 1937, S. Chandrasekhar, the astrophysicist, asked by Hardy to try to find a good photograph of Ramanujan next time he was in India, tracked her down in Triplicane. She was, Chandrasekhar reported back, "having a rather difficult life, some of her unscrupulous relatives having swindled her out of such financial resources as R[amanujan] had left her," and unable even to get hold of a copy of her husband's *Collected Papers*, published a few years before.

Around 1948, Janaki began taking care of a small boy, Narayanan, whose mother was in the hospital and who himself suffered from typhus. She visited him in the hospital, nursed him there, brought him books for school. Later, both his parents died, and Narayanan went to live with her. When she couldn't both take care of him and support herself as a seamstress, he briefly left to attend a residential school. But when he was about fifteen, he came back to live with her for good, and she adopted him as her son.

* * *

At the time of Ramanujan's death in April 1920, the editor of the *Journal of the Indian Mathematical Society* had fallen so far behind his publication schedule that the issue bearing the news was dated December 1919. Into copies of that issue, small olive green slips of paper, bordered in black, were inserted:

THE LATE MR. S. RAMANUJAN

We deeply regret to announce the untimely death of Mr. S. Ramanujan, B.A., F.R.S., on Monday, the 26th of April 1920, at his residence at Chetput, Madras. An account of his life and work will appear in a subsequent issue of this journal.

Seven months later, the journal carried two obituary notices. One, by P. V. Seshu Iyer, furnished the facts and dates; it may have been at his request that Lakshmi Narasimhan had assembled the family record of Ramanujan's life. The other, by Ramachandra Rao, reprinted from another magazine, was more lyrical: "And he is no more," it began—

> He whose name shed a lustre on all India, whose career is understood as the severest condemnation of the present exotic [sic] system of education: whose name was always appealed to, if any one forgetful of India's past ventured to doubt her intellectual capabilities.

There was something endlessly appealing about Ramanujan's life to Indian sensibilities. Some might see his early school failures, as Hardy did, as "the worst instance that I know of the damage that can be done by an inefficient and inelastic educational system." To others among the English, his life might seem the stuff of cinema, a rags-to-intellectual-riches story. But to many Indians, it was the idea of the luminous light of his intellect lurking behind the rags that was so compelling, his seeming disdain for the shallow plaudits of the world. "Even when two Continents were publishing his results," wrote Ramachandra Rao,

> he remained the same childish man, with no style in dress or affectation of manner, with the same kind face, with the same simplicity. Pilgrims came to Ramanujan's chambers and wondered if this was he. . . . If I am to sum up Ramanujan in one word, I would say, *Indianality*.

Even before his death, he was being seen as a boon to emerging Indian identity: "By his unique mathematical talents and by the amount of *useful* and *original* work [he has performed]," the *Journal of the Indian*

Mathematical Society said of him, "he has raised India in the estimation of the outside world." S. Chandrasekhar would observe, accurately enough, that "Ramanujan represents so extreme a fluctuation from the norm that his being born an Indian must be considered to a large extent as accidental." And India could take pride in many others among its countrymen who had made it in the West, or upon the great stage of the world generally. Still, at least among its scientists, few were more "Indian" than Ramanujan.

India was a shapeless mass of poverty, of ceaseless struggle for the material necessities of life? Ramanujan, if only reluctantly and only in order to work, shared in that struggle.

India was the Essential East, standing apart from, and independent of, the West? Ramanujan had spent his formative years, and done much of his most individual work, within a day's rail journey of his hometown. He had lived a life almost untouched by the seductive charms of the West, had never had any particular desire to leave India. And when the time came that he did, he bristled at his transformation into an English gentleman.

India was a place of spiritual values triumphant? Ramanujan, at least up until the very end of his life and perhaps even then, never forsook the Indian gods. He invoked the name of the goddess Namagiri. He was steeped in, and accepted, most of the values, beliefs, and ways of life of South India.

The Indian psychologist Ashis Nandy has made a similar point in his book *Alternative Sciences*, contrasting Ramanujan with the Indian physicist and plant physiologist J. C. Bose. Bose, he wrote, was ever troubled by the split within him between East and West, while Ramanujan, the more autonomous of the two, was not. "In his nationalism, as much as in his Westernized modernity, [Bose] was far more deeply bound up with the West, both in admiration and in hatred." Ramanujan, on the other hand, was who he was—a South Indian Brahmin through and through. Less than many other Indians was Ramanujan's head turned by the West and its ways.

Not through anything he professed but in the very model of his life and his achievements, Ramanujan lived on in the soul of India. "The British thought Indians were inferior, and Ramanujan showed otherwise," says P. K. Srinivasan, a Madrasi mathematics teacher who has compiled a book of reminiscences about Ramanujan. "He boosted our morale." In E. H. Neville's view it was the particular playing field on which Ramanujan triumphed—mathematics, in all its austere purity—that magnified his impact. In 1941, Neville wrote out a radio lecture on Ramanujan. His talk was broadcast but, whether due to constraints of time or politics, he did not deliver all he had prepared. Among comments left unaired were these:

> Ramanujan's career, just because he was a mathematician, is of unique importance in the development of relations between India and England. India has produced great scientists, but Bose and Raman were educated outside India, and no one can say how much of their inspiration was derived from the great laboratories in which their formative years were spent and from the famous men who taught them. India has produced great poets and philosophers, but there is a subtle tinge of patronage in all commendation of alien literature. Only in mathematics are the standards unassailable, and therefore of all Indians, Ramanujan was the first whom the English knew to be innately the equal of their greatest men. The mortal blow to the assumption, so prevalent in the western world, that white is intrinsically superior to black, the offensive assumption that has survived countless humanitarian arguments and political appeals and poisoned countless approaches to collaboration between England and India, was struck by the hand of Srinivasa Ramanujan.

In London, at a meeting of the Mathematical Society on June 10, 1920, according to its minutes, "The President referred to the loss that the Society has suffered in the death of Mr. S. Ramanujan and Major MacMahon spoke on the subject of Mr. Ramanujan's mathematical work."

Hardy, after so recently getting the seemingly upbeat letter on mock theta functions, was shocked to learn that Ramanujan had died. He wrote a brief obituary for *Nature*, to which Neville added more information some months later, then a much longer one for the *Proceedings of the London Mathematical Society*.

In America, the *American Mathematical Monthly*, expanding the section it normally reserved for dry accounts of papers appearing in foreign journals, gave ample play to the romantic story of Ramanujan's discovery, even down to his physical appearance: "In conversation he became animated, and gesticulated vividly with his slender fingers."

Ramanujan's papers had somehow wound up with the University of Madras; Janaki later charged Seshu Iyer with spiriting them away during the funeral itself. A few years later, on August 30, 1923, Dewsbury sent all but the original notebooks to Hardy.

In 1921, Hardy brought out the last of Ramanujan's papers, and the same year, he began to take up Ramanujan's work in papers of his own. The early 1920s saw others do so as well. In 1922, Mordell came out with "Note on Certain Modular Relations Considered by Messrs. Ramanujan, Darling and Rogers." The following year, B. M. Wilson's "Proofs of Some Formulae Enunciated by Ramanujan" was published in the *Proceedings of the London Mathematical Society.*

Soon after learning of Ramanujan's death, Hardy had written Dewsbury: "Is it possible that Madras would consider the question of publishing the papers in a collected form? There should be some permanent memorial of so remarkable a genius; and this memorial would certainly be the most appropriate form." Finally, in 1927, after protracted correspondence, Cambridge University Press came out with Ramanujan's *Collected Papers*, 355 pages with almost everything he had ever published. The early Indian work was there. So was the partition function, and highly composite numbers, even questions he had posed readers of the *Journal of the Indian Mathematical Society* and the mathematical parts of his letters to Hardy.

And with its publication, as the wider mathematical world took notice of Ramanujan's work, the floodgates opened. Over the next few years, paper followed paper, dozens and dozens of them, with titles like "Two Assertions Made by Ramanujan," and "Note on a Problem of Ramanujan," and "Note on Ramanujan's Arithmetical Function $\tau(n)$." In 1928, Hardy handed over to G. N. Watson Ramanujan's notebooks along with other manuscripts, letters, and papers, and he and B. M. Wilson set out on the mathematical adventure—editing the notebooks—that was to consume Watson until the eve of World War II and would result in more than two dozen major papers.

So far as the mathematical community was concerned, Ramanujan lived.

"During our generation no more romantic personality than that of Srinivasa Ramanujan has moved across the field of mathematical interest," wrote American number theorist Robert Carmichael in 1932. "Indeed it is true that there have been few individuals in human history and in all fields of intellectual endeavor who draw our interest more surely than Ramanujan or who have excited more fully a certain peculiar admiration for their genius and their achievements under adverse conditions."

Writing in a kindred vein a few years later, Mordell wrote that "few other mathematicians for some generations past have been so full of human interest. The story of his life is that of the rise of an obscure Indian in the face of the greatest difficulties to the position of the most famous mathematician that India has ever produced and of his early death just after he had won the most coveted distinctions."

But if Ramanujan's life exerted a peculiar hold on mathematicians, much more so did his work. In Hungary, in 1931, an eighteen-year-old University of Budapest prodigy, Paul Erdos, had written a paper on prime numbers. His teacher suggested he read a similar proof in Ramanujan's *Collected Papers*, "which I immediately read with great interest." Then, the following year, he saw a Hardy-Ramanujan paper concerned with the number of prime factors in an integer.

Some numbers, recall, are more composite than others, a subject Ramanujan had explored in his longer paper on highly composite numbers. A number like $12 = 2 \times 2 \times 3$ has more prime factors, three, than a number like $14 = 7 \times 2$, which has only two. The number 15 has two, while 16 has four. As you test each integer, the number of its divisors varies considerably. Well, Hardy and Ramanujan had said, we will examine not how this number varies but seek its *average* value—in the same way that you cannot predict the next throw of the dice yet can predict, on average, how often particular dice combinations will appear. Their result, loosely speaking, was that most integers have about $\log \log n$ prime factors: note n, take its logarithm, then take *its* logarithm, and you wind up with a crude estimate that improves as n increases. But as rough as the result was, it beat anything anyone had before, and took twenty-three pages of close-grained mathematical reasoning to prove it.

For almost twenty years, Hardy later told Erdos, their theorem seemed dead in the water, no progress being made in improving it. Then, in 1934, the problem was resurrected, and in 1939 Erdos and Mark Kac were led to a theorem that took it much further. It was only then that mathematicians could look back and pronounce the Hardy-Ramanujan paper of 1917 the founding document of the field that became known as probabilistic number theory.

In Norway, in 1934, a schoolboy named Atle Selberg, who was to become one of the world's most famous number theorists, came upon an article about Ramanujan in a Norwegian mathematical journal. The article termed Ramanujan "a remarkable mathematical genius," but as Selberg told an audience in Madras years later, the Norwegian word most aptly translated as "remarkable" also had connotations of "unusual and somewhat strange."

The article, with some of the results it included, "made a very deep and lasting impression on me and . . . fascinated me very much." Then Selberg's brother, also flirting with mathematics, brought home a copy of Ramanujan's *Collected Papers*. To Selberg, this was "a revelation—a completely new world to me, quite different from any mathematics book I had ever seen—with much more appeal to the imagination." Over the years, it retained its excitement and air of mystery. "It was really what gave the impetus which started my own mathematical work."

Later, his father gave him as a present his own copy of the *Collected Papers* which, as he told the Bombay audience, he carried with him still.

In England, in 1942, Freeman Dyson, a second-year student at Hardy's alma mater, Winchester, won a school mathematics prize and selected a book on the theory of numbers by Hardy and E. M. Wright. "The chapter in Hardy and Wright which I loved the most," he would recall many years later, "was chapter 19 with the title 'Partitions,' " and featuring the congruence properties of the partition function discovered by Ramanujan. Dyson was intrigued, speculated on what the congruence properties might imply, and conceived the notion of "rank." The rank of a partition, as he defined it, was its greatest part minus the number of its parts. Thus, one partition of 9 is

$$6 + 2 + 1 = 9.$$

Its "rank" is 3—the largest part, 6, less the total number of parts, 3. What Ramanujan's congruence properties implied, speculated Dyson, was that certain partitions broke down neatly into equal-sized categories based on their rank.

"That was the wonderful thing about Ramanujan," he would say later. "He discovered so much, and yet he left so much more in his garden for other people to discover. In the forty-four years since that happy day, I have intermittently been coming back to Ramanujan's garden. Every time when I come back, I find fresh flowers blooming."

Two years later, Dyson was working for the Royal Air Force Bomber Command as a statistician, where he saw up close, as all the Allied propaganda could not deny, the staggering losses over Germany. "It was a long, hard, grim winter," he wrote later. In the evenings, he corresponded with another mathematician on ideas first advanced by Ramanujan. "In the cold dark evenings, while I was scribbling these beautiful identities amid the death and destruction of 1944, I felt close to Ramanujan. He had been scribbling even more beautiful identities amid the death and destruction of 1917."

Over the years, then, Ramanujan was never forgotten. A 1940 listing noted 105 papers devoted to his work since his death. In the late 1950s, when Morris Newman delivered a paper at an Institute in the Theory of Numbers conference at the University of Colorado at Boulder, he began: "As with so much in analytic number theory, the study of congruence properties of the partition function originated with Ramanujan," and used Ramanujan's early papers as a jumping-off point. And certainly by the time of the centennial of his birth in 1987, Ramanujan's reputation was secure.

In his *Introduction to the History of Mathematics*, Howard Eves fashioned an outline chronicling the seminal moments in mathematics through the ages. The year 1906, for example, marked the work of Fréchet, the year 1907 that of Brouwer. Then, by Eves's reckoning, came a long dry spell, with nothing, for nine years, that in the sweep of mathematical history ranked sufficiently high to include. In 1916, the dry spell was over with Einstein's general theory of relativity. Then, for 1917, came Eves's next entry:

Hardy and Ramanujan (*analytical number theory*)

Still, through most of the middle years of the century, a sense of tragedy and unfulfilled promise clung to Ramanujan's name, of regret that he had not been greater still. Hardy himself set the tone, pointing to how Ramanujan had inhabited a mathematical desert for so many years. "He would probably have been a greater mathematician if he had been caught and tamed a little in his youth," he wrote; "he would have discovered more that was new, and that, no doubt, of greater importance." Littlewood echoed that sentiment in reviewing the *Collected Papers* in 1927: "How great a mathematician might Ramanujan have been 100 or 150 years ago? What would have happened if he had come into touch with Euler at the right moment?"

But he hadn't, of course; it was a shame, and that was that.

For some years, his work went into eclipse, as new areas of mathematics, wholly distant from those Ramanujan had pursued, became fashionable. The *Collected Papers,* while it had its disciples, was no best-seller, even by scholarly standards. Just 42 books were sold the first year, 209 the second, and an officer of Cambridge University Press predicted in a letter to Hardy late in 1929 that it would be another ten years before the first printing, of 750, was sold out. "When I came to the United States [after World War II]," Freeman Dyson would recall, "I was all by myself as a devotee." To the mathematical avant-garde of the day, Ramanujan's work was no more than a "backwater," a vestige of the nineteenth century.

But that, in time, would change.

5. RAMANUJAN REBORN

One milestone came in the late 1950s. S. Ramaseshan, whose father had known Ramanujan, was visiting friends in Bombay, when he was taken to an inner room of their printing plant. There, he was

> shown a stack of browned old paper with magic squares and beautiful mathematical formulae systematically written in an elegant hand. I could not believe that I was face to face with Ramanujan's notebook, the one I had heard about first from my father in 1937, the famous "frayed notebook" of Ramachandra Rao. I ran the tips of my fingers gently over the old paper—to feel the sheets which Ramanujan himself had filled with a smile on his face when he was without a job and everything else in his life seemed so bleak.

His friends at the Commercial Printing Press had been given "one of the most exciting jobs they had ever undertaken." It was 1957 and the Tata Institute of Fundamental Research, in a publishing venture financed by the Sir Dorabji Tata Trust, was bringing out Ramanujan's notebooks in a facsimile edition of two physically daunting volumes that together weighed in at more than ten pounds.

Until then, only Ramanujan's published papers, along with his letters to Hardy, had been published. But those led Littlewood to think, when he reviewed them in 1929, that "the notebooks would give an even more definite picture of the essential Ramanujan." The published papers mostly showed Ramanujan burnished by Hardy's editing, his work all gussied up and tied in a ribbon. The notebooks, on the other hand, were Ramanujan in the raw.

On October 8, 1962, a group of men met at the three-hundred-year-old Mallikeswarar Temple, at the northern end of Linghi Chetti Street, in Madras's Georgetown district, whose streets Ramanujan had walked half a century before. Here, in the shadow of the temple's ornate gopuram, P. K. Srinivasan, a mathematics teacher at Muthalpiet High School, brought his friends together to launch a project. He had first read about Ramanujan twenty years before. Ever since, he had tried to inspire students with his example. Then, eight years before, a friend had taken him to meet Janaki and Tirunarayanan, Ramanujan's surviving brother. Now, as the seventy-fifth anniversary of Ramanujan's birth approached, he was determined to bring out a memorial book, filled with letters and reminiscences, to honor him.

Recruiting the high school's alumni, or "old boys," to help him, he placed ads in local papers, interviewed people who had known Ramanujan, gathered letters. When he'd get some flicker of interest from an ad or contact, he'd immediately follow up. Often, he found himself just patiently sitting there, while someone rummaged around in an old trunk for some half-remembered letter. Sometimes he'd bring in a stenographer, skilled in both English and Tamil, to record the conversation.

Ramanujan's seventy-fifth birthday was observed across South India. Town High School, in Kumbakonam, named one of its buildings after him. A stamp was issued in his honor; two and a half million copies of his passport photo, reduced to inch-high form, colored sienna, and valued at fifteen new paise, sold out the day they were issued. In Madras, around the time of the anniversary, a birthday celebration was held in his honor, and many of those who had been close to him or his family were in town. Srinivasan exploited the opportunity, stationing old boys at the entrance of the hall to solicit comments, correspondence, and reminiscences.

At another point, he visited Ramanujan's old house in Kumbakonam; Tirunarayanan had given him permission to look through the almirah, a sort of wardrobe, kept in a separate locked room of the house. In the presence of the tenant, Srinivasan unlocked the room. When he opened the almirah, which was covered with dust and cobwebs, cockroaches swarmed out. But in it, despite Tirunarayanan's assurance that any such find was unlikely, he found a letter Ramanujan had written his father *from* England.

P. K. Srinivasan's compilation of letters and reminiscences came out in 1968. Brief biographies of Ramanujan appeared, in English, in 1967, 1972, and 1988; in Tamil in 1980 and 1986; and in Hindi, Kannada, Malayalam, among other Indian languages.

Then, in 1974, Deligne proved the tau conjecture.

Almost sixty years before, in 1916, Ramanujan had published a paper with the unprepossessing title, "On Certain Arithmetical Functions." An arithmetical function is one that originates in trying to learn certain properties of numbers; pi (n), the number of prime numbers, and $p(n)$, the number of partitions, both problems on which Ramanujan worked, were two of them. The tau function, τ (n), was another.

"Let $\sigma_s(n)$ denote the sum of the sth powers of the divisors of n," Ramanujan had begun. If $n = 6$, for example, its divisors are 6, 3, 2, and 1. So that if, say, $s = 3$, "the sum of the sth powers of the divisors," σ_3 (6), is just $6^3 + 3^3 + 2^3 + 1^3 = 252$. But how calculate $\sigma_s(n)$ generally? That question led Ramanujan, after fifteen pages, to the tau function, whose properties, Hardy would write twenty years later, "are very remarkable and still very imperfectly understood." As in so much of analytical number theory, stepping through the open door of a simple-seeming problem had led into a mathematical labyrinth of formidable complexity.

Ramanujan's hypothesis, as Hardy called it, or the tau conjecture as it became more generally known, did not offer an explicit formula for τ (n). Rather, it merely stated that τ (n) was "on the order of . . ." something. In other words, like the prime number theorem, it was a kind of approximation. "There is reason for supposing," Ramanujan wrote, "that τ (n) is of the form 0 ($n^{11/2+\epsilon}$)." The notation was a way of saying that its value, whatever it actually was, was always less than something. What was "something"? Leaving out constants, it was, at most, n to some power only slightly greater than $^{11}/_2$ (which means the square root of n to the eleventh power).

Ramanujan's conjecture, in the words of S. Raghavan of the Tata Institute of Fundamental Research in Bombay, "kept at bay a whole galaxy of distinguished mathematicians for nearly six decades," remaining one of the major open problems in number theory. An account by Hardy in 1940 hinted at both the interest it generated and its resistance to solution. Ramanujan himself, he reported, had proved that τ (n) = 0(n^7); but 7, the power of n here, was *much* more than $^{11}/_2$, and so still far short of proving the tau conjecture. Two years later, Hardy himself cut a little closer, to $0(n^6)$. Kloosterman got closer in 1927, Davenport and Salie closer still in 1933. The Scottish mathematician Robert Rankin, a student of Hardy's, proved in 1939 that $\tau(n) = 0(n^{29/5})$.

All this was a little like showing, first, that the murderer lived at a house on Union Street numbered less than 2170; then, that the house number was less than 2160; then, less than 2158 . . . And yet *still* you couldn't prove he lived at 2155, which is where your prime suspect lived.

It was only in 1974 that the Belgian mathematician Pierre Deligne, in what would be described as "one of the celebrated events of 20th century mathematics," proved the conjecture using powerful new tools supplied by the field known as algebraic geometry. Deligne was subsequently awarded the Fields Medal, the mathematical community's counterpart to the Nobel Prize, for his triumph.

And when he did it, it helped solidify Ramanujan's reputation all the more.

In 1988, looking back upon the fall and rise of Ramanujan's reputation over the years, Bruce Berndt compared him to Johann Sebastian Bach, who remained largely unknown for years after his death in 1750. For Bach, the big turnaround came on March 11, 1829 with Felix Mendelssohn's performance of the *St. Matthew Passion*. For Ramanujan, suggested Berndt, the roughly analogous event was George Andrews's discovery of the Lost Notebook in 1976.

In late April of that year, Andrews was a young University of Wisconsin visiting professor bound for a one-week conference in France. The airline fare structure made it cheaper to go for three weeks than one, so he looked for something to keep him, his wife, and his two daughters occupied during the extra time. A side trip to Cambridge? A colleague there had suggested he might be interested in rummaging through some papers left behind by G. N. Watson at his death in 1965. So Cambridge it was.

Watson, who had worked on Ramanujan's papers for many years before World War II, was a major classical analyst from the prewar period and a Fellow of the Royal Society. When he died, the society asked J. M. Whittaker, son of one of Watson's collaborators, to write his obituary. Whittaker wrote Watson's widow: could he come see his papers? She invited him to lunch, then took him upstairs to the study. There, Whittaker recalled, papers

> covered the floor of a fair sized room to a depth of about a foot, all jumbled together, and were to be incinerated in a few days. One could only make lucky dips [into the rubble] and, as Watson never threw away anything, the result might be a sheet of mathematics but more probably a receipted bill or a draft of his income tax return for 1923. By an extraordinary stroke of luck one of my dips brought up the Ramanujan material.

This "material," all 140 pages of it, was part of a batch of papers Dewsbury had sent Hardy in 1923 that had later gone to Watson. After his "lucky dip," Whittaker passed them on to Robert Rankin, who in 1968 handed them to Trinity College.

Whittaker and Rankin, both professional mathematicians of high standing, but whose backgrounds ill-suited them to distinguish this material from what had already appeared in the published notebooks, had failed to see in it what Andrews saw now. Within a few minutes, he realized that some of it bore on mock theta functions, the subject of his own Ph.D. thesis, and related subjects. Which meant these were papers Ramanujan could have generated only during *the last year of his life in India.*

He was thrilled, "extremely excited that I had my hands on something spectacular." But what now? Get the papers photocopied, of course.

The interior of the Trinity College library is a visually arresting place that owes its beauty to the great English architect Christopher Wren. After its completion in 1695, someone said of the Wren Library that it "touches the very soul of any one who first sees it." Everywhere are delicate woodcarving, stained-glass windows, and statues filling apses on the wall. Busts of Cambridge greats are arrayed along it length, and Thorwaldsen's statue of Lord Byron stands at the far end. It is at a long table behind the Byron statue that library staff seat scholars studying medieval manuscripts, or the papers of Newton, or the poems of Milton. And in 1976, George Andrews was one of them. To him, the Wren was "a shrine, with busts of Newton glowering down on you. The idea of going up to the desk and having something xeroxed terrified me."

He mustered up his courage and told them what he wanted. "That will be about seven pounds" for airmail postage to the United States, they warned him. "Will that be all right?"

It would be fine, he assured them. "I was ready to take a second mortgage on my house to get it."

In a paper appearing soon after he had unearthed it, Andrews styled his find "the Lost Notebook." Its discovery, mathematician Emma Lehmer was moved to say, was "comparable to the discovery of a complete sketch of the tenth symphony of Beethoven."

But the appellation ruffled feathers in Britain. Robert Rankin pointed out, for example, that it was not a "notebook" but loose papers; and that, pristinely secure within the walls of the Wren Library, it had never been lost. Still, as Andrews observed later, "the manuscript and its marvelous results disappeared from any mention or account by the mathematical community for more than 55 years." In that sense, they had indeed been "lost." His contribution lay "not in saving them from oblivion as [Whittaker and Rankin] did," but in recognizing them for what they were.

Sometime before, in January 1974, Bruce Berndt had been on sabbatical at the Institute for Advanced Study and came upon papers bearing upon his own work of two years before that proved some formulas in Ramanujan's notebooks. The institute library had no copy of the facsimile edition, but neighboring Princeton's did. In it, he found a slew of related formulas that seemed similar to the others but that, try as he might, he couldn't prove. "And that," he recalled, "bothered me." He set an informal task for himself—to prove all the results in chapter 14. That took him a year. Later, Andrews visited Illinois and mentioned that Watson and Wilson had spent years trying to prove the theorems in the notebooks, and that their own notes were still around. Ever since, beginning in May 1977, Berndt has worked on editing Ramanujan's notebooks, "and I haven't done anything else since."

For the decade ending in 1988, a computer search of the literature revealed, some three hundred papers referred to Ramanujan in their titles or their abstracts. Ties and cross-links to other areas of mathematics, what Freeman Dyson calls "connections to deep structure and more general abstract notions," were showing up everywhere. Over the past twenty years or so, Dyson says, "It has become respectable again to take Ramanujan seriously. So much that he conjectured was not just pretty formulas but had substance and depth." The great tree that was Ramanujan's work sent its roots down deep and far.

He had planted it for himself, not to better the material condition of India or of the world. And yet, its underground tendrils ranged into fields far distant from pure mathematics—and into applications which, Hardy might have cringed to learn, were by no means "useless."

6. BETTER BLAST FURNACES?

"If I am asked to explain how, and why, the solution of the problems which occupy the best energies of my life is of importance to the general life of the community, I must decline the unequal contest," wrote Hardy, just after the guns of World War I had stilled.

> I have not the effrontery to develop a thesis so palpably untrue. I must leave it to the engineers and the chemists to expound, with justly prophetic fervour, the benefits conferred on civilization by gas-engines, oil, and explosives. If I could attain every scientific ambition of my life, the frontiers of the Empire would not be advanced, not even a black man would be blown to pieces, no one's fortunes would be made, and least of all my own. A pure mathematician must leave to happier colleagues the great task of alleviating the sufferings of humanity.

Whether and how the engineers and chemists might indeed apply Ramanujan's work to the common purposes of life strikes a sensitive chord in India, beset as it is by practical problems of great urgency and less naturally inclined to trust in research whose rewards may accrue only decades or centuries later. "Several theorems of Ramanujan are now being widely used in subjects like particle physics, statistical mechanics, computer science, cryptology and space travel in the United States—subjects unheard of during Ramanujan's time," *The Hindu* assured its readers in its December 19, 1987 issue. But efforts to justify Ramanujan's work on utilitarian grounds go back almost to his lifetime.

At the Third Conference of the Indian Mathematical Society, held in Lahore soon after his death, Ramanujan's life and work were on every speaker's lips. "Mention of Ramanujan's name," declared the society's president, Balak Ram, "suggests the question of the organization and endowment of scientific research in India," in particular the balance between applied and pure. The engineer might be rewarded for his inventiveness, become rich the way Edison did. But what of learning for learning's sake? Ought society reserve a place for it? Yes, he insisted.

> Every settled community is sub-consciously or consciously convinced that attempts at progress are apt to be wasteful unless guided by thinkers and teachers who have leisure to think and teach clearly; therefore, we find these communities giving special protection and advice to learned men and thus providing them with leisure and the psychological incentives which replace the missing pecuniary reward.

If India was to be assured of "progress," he was saying, it had to allow for its Ramanujans. In the long run, it would pay off.

When the Ramanujan stamp came out in 1962, the Indian postal service took pains to point out the potential applications of his work: "His work and the work of other mathematicians on Riemann's zeta function, done in another context, has now been geared to the technological mill. It has been applied to the theory of pyrometry, the investigation of furnaces aimed at building better blast furnaces." And his work on mock theta functions, modular equations, and in other realms was being studied for its possible application to atomic research.

Down through the years, Ramanujan's mathematics has indeed been brought to bear on practical problems, if sometimes tangentially. For example, crystallographer S. Ramaseshan has shown how Ramanujan's work on partitions sheds light on plastics. Plastics, of course, are polymers, repeating molecular units that combine in various ways; conceivably, you might have one that's a million units long, another of 8251, another of 201,090, and so on. Ramanujan's work in partitions—on how smaller numbers combine to form larger ones—plainly bears on the process. As it does, for example, in splicing telephone cables, where shorter subunits, of varying lengths, again add up to make a whole.

Blast furnaces? Plastics? Telephone cable?

Cancer?

At a meeting of the southeastern section of the American Physical Society in Raleigh, North Carolina, in November 1988, three University of Delhi researchers presented a paper entitled "A Study of Soliton Switching in Malignancy and Proliferation of Oncogenes Using Ramanujan's Mock-Theta Functions." They were using Ramanujan's mathematics to help understand cancer, if only as a small, tangential contribution to a vast and complex subject. When the *Hindu* noted the paper, however, it assigned the headline "Ramanujan's Maths Help Fight Cancer."

In other fields, Ramanujan's mathematics has played a more decisive role—as in, for example, string theory, which imagines the universe as populated by infinitesimally short stringlike packets whose movement produces particles. Grounded in the real world or not—the jury is still out—the mathematics required to describe these strings demands twenty-six dimensions, twenty-three more than the three on which we manage in everyday life. Partition theory and Ramanujan's work in the area known as modular forms have proved essential in the analysis.

An important problem in statistical mechanics has also proved vulnerable to Ramanujan's mathematics—a theoretical model that explains, for example, how liquid helium disperses through a crystal lattice of carbon. As it happens, the sites helium molecules may occupy in a sheet of graphite, say, can never lie adjacent to one another. Since each potential site is surrounded by six neighbors in a hexagonal array, once it is filled, the six around it define an unbreachable hexagonal wall. Reviewing in 1987 the work for which he received the prestigious Boltzmann Medal seven years earlier, R. J. Baxter of the Australian National University in Canberra set out the thinking behind his "hard hexagon model." Mathematically, he showed, it was built on a particular set of infinite series. And "these series," he observed, "are *precisely* those that occur in the famous Rogers-Ramanujan identities" (though he hadn't realized it back in 1979 when he'd found them on his own). Based on them, he found a way to determine the probability that any particular site harbored a helium molecule; predictions borne of the model agreed closely with experiment. "The Rogers-Ramanujan identities," Baxter concluded, "are perhaps not so remote from 'ordinary human activity' as Hardy would have liked!"

Computers, scarcely the dream of which existed in 1920, have also drawn from Ramanujan's work. "The rise of computer algebra makes it interesting to study somebody who seems like he had a computer algebra package in his head," George Andrews once told an interviewer, referring to software that permits ready algebraic manipulation. Sometimes in studying Ramanujan's work, he said at another time, "I have wondered how much Ramanujan could have done if he had had MACSYMA or SCRATCHPAD or some other symbolic algebra package. More often I get the feeling that he was such a brilliant, clever and intuitive computer himself that he really didn't need them." Then, too, a modular equation in Ramanujan's notebooks led to computer algorithms for evaluating pi that are the fastest in use today.

"Hype," Freeman Dyson calls some of what he deems undue fanfare for practical applications of Ramanujan's work. To apply his mathematics to string theory is "stretching the point," he says. "You don't have to read Ramanujan to do string theory." It was all true enough, and completely valid as far as it went—but peripheral, if not irrelevant, to most mathematicians.

What makes Ramanujan's work so seductive is not the prospect of its use in the solution of real-world problems, but its richness, beauty, and mystery—its sheer mathematical loveliness. Hardy was enraptured by it. And Dyson was, and Selberg, and Erdos, and many, many others. "The best seem to appreciate Ramanujan early," says Richard Askey, a University of Wisconsin mathematician deeply involved in Ramanujan's work. "The rest of us have to need some of his work before really appreciating it." George Andrews once told an interviewer how, as a young man, he was beguiled by the Ramanujan-Hardy partition formula. "I was stunned the first time I saw this formula. I could not believe it. And the experience of seeing it explained, and understanding how it took shape . . . convinced me that this was the area of mathematics I wanted to pursue."

Because it lies on a cool, ethereal plane beyond the everyday passions of human life, and because it can be fully grasped only through a language in which most people are unschooled, Ramanujan's work grants direct pleasure to only a few—a few hundred mathematicians and physicists around the world, perhaps a few thousand. The rest of us must either sit on the sidelines and, on the authority of the cognoscenti, cheer, or else rely on vague, metaphoric, and necessarily imprecise glimpses of his work.

Some Ramanujan experts have resorted to the language of other, more accessible realms of knowledge to suggest his hold on the mathematical imagination.

Emma Lehmer, recall, likened *The Lost Notebook* to a Beethoven symphony.

Watson concluded his presidential address to the London Mathematical Society in 1937 by saying that one Ramanujan formula gave him "a thrill which is indistinguishable from the thrill which I feel when I enter the Sagrestia Nuova of the Capelle Medicee and see before me the austere beauty of the four statues representing, 'Day,' 'Night,' 'Evening,' and 'Dawn' which Michelangelo has set over the tombs" of the Medicis.

Berndt compared Ramanujan to Bach and, alluding to Ramanujan's devotion to mathematics, quoted from Shelley's "Hymn to Intellectual Beauty":

I vowed that I would dedicate my powers
To thee and thine; have I not kept the vow?

For the layman, to be sure, this is an ultimately unsatisfying way to confront Ramanujan's mathematics, for it keeps us at several removes from *what he did,* leaves us having to take others' word for it, looking at his mathematical achievements through a blurry film of metaphor, poetry, and, yes, ignorance. True, the composition of a sonata may be equally mysterious; but the *result* more intimately involves the five senses.

What Ramanujan did will live forever. It will not, to be sure, live in the hearts of the masses of men, like the work of Gandhi, Shakespeare, or Bach. Still, his ideas and discoveries, percolating through those few minds tuned to them, will mingle with the intellectual energy of the cosmos, and thence into the deep, broad pool of human knowledge. "What we do may be small, but it has a certain character of permanence," wrote Hardy of the work of pure mathematicians, "and to have produced anything of the slightest permanent interest, whether it be a copy of verses or a geometrical theorem, is to have done something utterly beyond the powers of the vast majority of men."

By the time of the centennial of his birth in 1987, Ramanujan's reputation shone with a new luster. In India, he was compared to Nehru and Nobel Prize–winning physicist C. V. Raman, both of whose centennials were being celebrated at about the same time. Three Indian films were made about his life. A Ramanujan Mathematical Society, started in 1986, published the first volume of its journal.

Celebrations were held all across South India. Andrews, Askey, and Berndt, the three American mathematicians who had most contributed to the restoration of Ramanujan's name, were kept busy shuttling all over the country, giving lectures at Annamalainagar, and Bombay, and Pune, and Gorakhpur and Madras.

In Madras, when the Narosa Publishing House issued *The Lost Notebook*, Prime Minister Rajiv Gandhi was there to sign the first copy and present it to Janaki.

In Kumbakonam, the framed and garlanded poster-sized portrait of Ramanujan, fat again and wearing the mortarboard he wore when he received his Cambridge degree, was borne through the streets atop a gaily decorated elephant, to the accompaniment of traditional street musicians. The National Cadet Corps was there, and girl scouts in pink blouses and rose-colored skirts, and traditional dancers, all of them showered with flower blossoms as they paraded down the street in front of Ramanujan's house. Loudspeakers shouted praise of Kumbakonam's favorite son to the assembled throngs in the street.

At Anna University, which named its computer center for Ramanujan,

Andrews, his voice choked with emotion, presented Janaki with a shawl. It was she who deserved the credit for the Lost Notebooks, he said, since she had kept his papers together while he lay dying.

After World War II, the university had noticed that Janaki's 20-rupee-per-month pension no longer went very far and raised it to 125. Janaki financed her son Narayanan's way through college. He later got a job with the State Bank of India, married, and had three children. Recently, he retired from the bank job he held for twenty-five years and began to spend more time caring for his mother.

After a hard life and years of anonymity, Janaki herself began to garner attention as her husband was rediscovered. Along the way, she grew more outspoken. At one point, presented a pension by the University of Madras, she remarked that while the cash was fine, it would have done her more good sixty years earlier. A little earlier, in 1981, she had told an Indian newspaper reporter, "They said years ago a statue would be erected in honor of my husband. Where is the statue?"

Dick Askey learned of Janaki's lament. "If she wants a bust of her husband," he thought, "we owe that much to Ramanujan, and to her." *We* meant the mathematicians of the world. Askey knew the Minnesota sculptor Paul Granlund, from whom he had bought some works, and contacted him. Granlund agreed to do it, provided at least three busts would materialize from the project. (Ultimately, ten did.) Askey and his wife would buy one. So would Chandrasekhar, the astrophysicist, to whom Askey had written about the idea. But where would the $3000 come for Janaki's? Askey got some of it from institutions, including Trinity College. But most came in individual $25 contributions from mathematicians around the world. Today, Granlund's bronze bust of Ramanujan, based on the passport photograph, stands on a pedestal in Janaki's house in Madras.

With the approach of the centennial, the house became a pilgrimage site. Mathematicians passing through Madras paid her homage. She appeared in a British television special about Ramanujan, *Letters from an Indian Clerk,* later shown in America as well. In August of the centennial year, a foundation presented her with a purse of twenty thousand rupees and a monthly pension of one thousand rupees; she asked that a Srinivasa Ramanujan Trust be created for awards and scholarships to bright young mathematics students. Early the following year, Trinity College made its own gesture, its council in February 1988 agreeing to "a grant of £2,000 a year until further notice."

As the centennial approached, T. V. Rangaswami, a Tamil language journalist in Madras, set out on what became a thirty-one-part series of articles on Ramanujan. He spent months collecting letters, documents, and photographs, and interviewing Janaki, who lived near him in Triplicane. Each afternoon they would meet and talk. One afternoon, he showed her a photograph of "Gometra," the house off Harrington Road where she had nursed Ramanujan and in which he had died. The wrinkled old woman, removed by more than sixty years from the events that took place there, broke down and cried.

7. *SVAYAMBHU*

A bittersweet tang sometimes slipped into the encomiums India lavished on Ramanujan over the years, sad reminders of the poverty, bureaucracy, and institutional rigidity that almost crushed him in 1905. Ramanujan was an inspiration to India, yes—but also a rebuke. How could India let him come so close to being lost to the world? Why hadn't he gotten more encouragement? Why was it left to foreigners to make him famous?

J. B. S. Haldane, the distinguished English biologist who lived in India toward the end of his life, in the early 1960s complained that

> today in India Ramanujan could not get even a lectureship in a rural college because he had no degree. Much less could he get a post through the Union Public Service Commission. This fact is a disgrace to India. I am aware that he was offered a chair in India *after* becoming a Fellow of the Royal Society. But it is scandalous that India's great men should have to wait for foreign recognition. If Ramanujan's work had been recognized in India as early as it was in England, he might never have emigrated and might be alive today. We can cast the blame for Ramanujan's non-recognition on the British Raj. We cannot do so when similar cases occur today. . . .

On his birthday in 1974, a Professor Srinivasa Ramanujan International Memorial Committee published a commemoration volume littered with advertisements expressing "respectful homage" and like sentiments by every little South Indian company that wanted to briefly bask in Ramanujan's reflected glory, from Madras Aluminum Co. in Coimbatore to Smart Dresses on Ranganathan Street in Madras. Prime Minister Indira Gandhi wrote from New Delhi that Ramanujan's "untutored genius" dazzled science and that "his achievement will inspire successive generations of Indian youth." But editor S. Ramakrishnan added: "Let not Free India lose sight of her living Ramanujans, languishing in obscurity."

Languishing in obscurity. There was another side to Ramanujan's service as symbol of India and, in 1946, Nehru himself had referred to it in his *Discovery of India*:

> Ramanujan's brief life and death are symbolic of conditions in India. Of our millions how few get any education at all; how many live on the verge of starvation . . . If life opened its gates to them and offered them food and healthy conditions of living and education and opportunities of growth, how many among these millions would be eminent scientists, educationists, technicians, industrialists, writers, and artists, helping to build a new India and a new world?

R. Viswanathan, a later headmaster of Ramanujan's alma mater, Town High School in Kumbakonam, would insist that, given the resources, he could turn out many Ramanujans. Taken literally, he was quite wrong; all the wealth of the British Empire, all the rich intellectual tradition of Europe, all the freedom and opportunity of America, have made for but a handful of Ramanujans through the centuries. Still, Viswanathan had expressed a larger truth—that India held vast stores of talent and ability denied the means to develop fully. Ramanujan represented his country's intellectual and spiritual strengths—but also its untapped potential.

But didn't Ramanujan's story prove, quite to the contrary, that genius in the end overcomes? If Ramanujan, with all his disadvantages, could command the attention of the world and leave so indelible a mark on it, couldn't anyone endowed with special gifts do it, too?

Not so. So long did Ramanujan languish, so many times did his future hang on a knife edge, so close did he come to dying unknown—and so plainly was his full promise never realized—that his life's lesson bears as much on the stumbling blocks he faced as on his success, such as it was, in overcoming them.

Ramanujan had much more going for him than millions of others in India did. His family was poor but hardly destitute. He was a Brahmin, part of a culture that encouraged learning. His mother was tolerant of his whims and forceful in advancing his interests. His innocent charm won over those his eccentricities might otherwise have put off. And he enjoyed a peculiarly stubborn faith in himself and his powers.

What, one must ask, if he had had none of these? What if he had been every inch the genius he was, with just as much to give the world, but his mother had been a little less supportive? Or he had been the barest bit less likeable or less sure of his abilities? Doubtless he would have wound up like his brothers, an anonymous government bureaucrat, or otherwise consigned to obscurity. For those who have biographies written about them, the System by definition works; the measure of its failure lies in those who never bask in the warm glow of the world's acclaim. Those you never hear about.

Ramanujan, then, was an embarrassment to India as well as an inspiration, a reminder of the gauntlet India's other Ramanujans must run in order to achieve anything. As symbol of South Indian genius, he was a delight to contemplate. In the sometimes harsh reality of his life, much less so.

When in November 1968, on receiving the Srinivasa Ramanujan Medal of the Indian National Science Academy, S. Chandrasekhar revealed that Ramanujan had tried to commit suicide, his comments sparked a furor. "I was shocked and surprised," he wrote later, "that I was accused by several [including his uncle, C. V. Raman] of defaming Ramanujan's name." Someone charged him with trying to enhance his own reputation at Ramanujan's expense. Chandrasekhar had sinned: he had torn down Ramanujan as icon, replaced him with Ramanujan the man.

Today, as in Ramanujan's day and in all the years in between, many of India's best minds leave to go overseas, where they are nourished by Western ideas and sidestep the sort of "inefficient and inelastic" educational system that stifled Ramanujan. India, a poor country, can lavish praise on the long-dead Ramanujan more easily than it can lavish resources on finding and nurturing new Ramanujans.

In 1951, a wealthy merchant and patron of higher education, Alagappa Chettiar, founded the Ramanujan Institute. After his death, the institute faced severe financial problems and almost shut down. In 1957, it was absorbed by the University of Madras, where it became the Centre for Advanced Studies in Mathematics. In 1972, it moved into a low-slung modern building just across a little concrete bridge spanning the Buckingham Canal from the main university campus. The institute, with nineteen teachers, lecturers, and research students, remains there today. But it does not specialize in areas of mathematics Ramanujan pursued, nor does its name bear anything like the luster of its namesake.

Even the centennial festivities were marred by reminders of the limited opportunities India offered its Ramanujans and the obstacles it placed in their way. Why, some wondered, were Ramanujan's research reports from 1914 lost while in Indian care? Why was it left to American, more than Indian, mathematicians to restore Ramanujan's reputation? "I want you all to sit back and think about this," one speaker, S. Ramaseshan, asked his listeners at a centennial event in Kumbakonam:

> How many registrars in this country today, or for that matter how many vice chancellors of today, 100 years after Ramanujan was born, would give a failed pre-university student a research scholarship of what is now equivalent of Rs. 2000/- or Rs. 2500/- today? This is after 40 years of independence, when we can no longer blame a colonial power for not encouraging Indian talent.

A writer for the *Illustrated Weekly of India*, referring to Ramanujan, suggested that "perhaps the luminescence of his mind is too harsh for Indians in this age of their intellectual bleakness. . . ."

Ramanujan as inspiration. Ramanujan as rebuke. Ramanujan as . . . ? Ramanujan's life can be made to serve as parable for almost any lesson you want to draw from it. At an early Indian Mathematical Society conference, his name was invoked to point up the society's meager resources: "When the famous mathematician, Mr. Ramanujan, F.R.S., sought some modest help from our Society to enable him to devote his attention to mathematical studies and research, our bankruptcy was made manifest. . . ." In the 1950s, a library director recounted Ramanujan's story for readers of *Wilson Library Bulletin*, emphasizing the impact on him of Carr's *Synopsis*. Ramanujan's life, he was saying, was a testimonial to the books that had nurtured him. All across the years and up to the present, it was like that—Ramanujan invoked as model, inspiration, warning, or instructive case history. Ramachandra Rao had done it. Hardy had. Nehru had.

Cut cruelly short, Ramanujan's life bore something of the frustration that a checked swing does in baseball; it lacked follow-through, roundedness, completion. It never had a second half to give it shape. So we continue to give it shape now, years after his death.

His life was truncated, like a cone sliced off short of its vertex. Or like an economic graph that stops with the present, leaving forecasters to fill in a vast and uncertain future. Ramanujan's life, littered with what-might-have-beens, was like that: it was so easy to see in it what you wanted to see. Its bare facts fairly cried out for interpretation.

Was his failure in school testimony to India's failure to nurture its own? Or was his rescue by Ramachandra Rao and Narayana Iyer proof that, in the end, India recognized and appreciated him?

Was he an example of the oppressiveness of the raj, a case of native genius nearly quashed—then making its contribution, when it did emerge, to "English mathematics"? Or was his discovery a testament to British beneficence?

Would he have achieved more had he found mentors early on? Would he have become the next Gauss or Newton? Or did working on his own, under less than ideal conditions, make him more mathematically resourceful, even contribute to his stunning originality?

Was his genius the product of sheer intellectual power, different only in degree from other brilliant mathematicians? Or was it steeped in something of the mystical or the supernatural?

Was Ramanujan's life a tragedy of unfulfilled promise? Or did his five years in Cambridge redeem it?

In each case, the evidence left ample room to see it either way. In this sense, Ramanujan's life was like the Bible, or Shakespeare—a rich fund of data, lush with ambiguity, that holds up a mirror to ourselves or our age.

There were no mathematicians in Ramanujan's family, no strain of unusual mathematical aptitude. So *where did he come from?*

Legion are those who have taken credit for discovering Ramanujan or for otherwise figuring largely in his life. Ramaswami Iyer would later say he was proud of two things—starting the Indian Mathematical Society and discovering Ramanujan. Narayana Iyer's family today points with pride to its patriarch's role in rescuing Ramanujan from oblivion, and numberless people later claimed credit for convincing him to go to England. Then, too, many saw Ramanujan as the creation of his mother; the two of them, plainly, were made from the same cloth, and her heart was bound up in his success.

Among the English, Hardy was not alone in savoring his role in Ramanujan's life; Neville, too, would sweetly recall his days with him in Madras a quarter century before, and point out that had he "failed to win the confidence of Ramanujan and his friends," Ramanujan might never have reached England. But Hardy remains the prime candidate. Paul Erdos has recorded that when Hardy was asked about his greatest contribution to mathematics, he unhesitatingly replied, "The discovery of Ramanujan." At another time, indeed, Hardy went on record as calling him "my discovery."

In a sense, of course, he was. But a more satisfying way to explain Ramanujan's emergence as a figure on the world mathematical stage was given in an article appearing in India during the centennial, by a writer for *The Illustrated Weekly of India* known only as "RGK." Ramanujan, he suggested, was *svayambhu*—"self-born." He had sprouted up out of the soil of India of his own accord. He had created himself.

You cannot say much about Ramanujan without resorting to the word *self*. He was self-willed, self-directed, self-made. Indeed, some might conceivably label him "selfish" for his preoccupation with doing the mathematics he loved without any great concern for the betterment of his family or his country.

Ramanujan did what he wished to do, went his own way. It was only later, *after* he had indulged in an orgy of mathematical creation, that he might wake up and realize how far he had strayed from the common run of human intercourse. Only then might he begin to care, sometimes painfully much, how others thought of him.

When he was a teenager, he lent support to the local crazy man, though the whole rest of Kumbakonam dismissed him as a crank. He gave himself over to mathematics, throwing aside everything else, even the degree-as-meal-ticket his mother so much wanted for him. He knocked on doors all over South India, introducing himself to one mathematician after another. Then, when he had exhausted the mathematical resources of India, he turned to England. He wrote Baker. He wrote Hobson. He wrote Hardy.

Hardy discovered Ramanujan? Not at all: a glance at the facts of 1912 and 1913 shows that Ramanujan discovered Hardy.

And what of the dream of Namakkal, when the goddess Namagiri presumably gave her blessing for his trip to England? What fierce drive to live out his other, truer "dream" did Ramanujan need to contrive, subconsciously, that it turn out as it did?

Ramanujan was a man for whom, as Littlewood put it, "the clear-cut idea of what is *meant* by proof . . . he perhaps did not possess at all"; *once* he had become satisfied of a theorem's truth, he had scant interest in proving it to others. The word *proof*, here, applies in its mathematical sense. And yet, construed more loosely, Ramanujan truly had *nothing to prove*.

He was his own man. He made himself.

"I did not invent him," Hardy once said of Ramanujan. "Like other great men he invented himself." He was *svayambhu*.

Just what did Ramanujan want?

He wanted nothing—and everything.

He sought no wealth, certainly none beyond what he needed to carry out his work, and to give to his family what he felt was expected of him.

He did crave respect, understanding, perhaps even a favorable judgment from history.

But what Ramanujan wanted more, more than anything, was simply the freedom to do as he wished, to be left alone to think, to dream, to create, to lose himself in a world of his own making.

That, of course, is no modest wish at all. He wanted "leisure." And he got it.

In South India today, everyone has heard of Ramanujan. College professors and bicycle rickshaw drivers alike know his story, at least in sketchy outline, just as everyone in the West knows of Einstein. Few can say much about his work, and yet something in the story of his struggle for the chance to pursue his work on his own terms compels the imagination, leaving Ramanujan a symbol for genius, for the obstacles it faces, for the burdens it bears, for the pleasure it takes in its own existence.

Epilogue

By the time he learned of Ramanujan's death, Hardy had already left Trinity.

In December 1919, about when Ramanujan, at his doctor's insistence, prepared to leave Kumbakonam for Madras, Hardy was writing J. J. Thomson, master of Trinity College, with the news that he had accepted the Savilian Professorship at Oxford. "The post carries with it a Fellowship at New College, the acceptance of which will vacate my Fellowship here automatically," he wrote. One reason for the move was the increasing load of administrative responsibility he bore at Cambridge. "If I wish to preserve full opportunities for the researches which are the principal permanent happiness of my life," he had decided, he would need a position offering "more leisure and less responsibility." At Oxford, he'd been assured, he would get that.

Unmentioned in the letter but probably weighing more on him than administrative chores were the hard feelings left over from the war, the infighting that surrounded the Russell affair, and the departure of Ramanujan. "If it had not been for the Ramanujan collaboration, the 1914–1918 war would have been darker for Hardy," wrote Snow. "It was the work of Ramanujan which was Hardy's solace during the bitter college quarrels." Now Ramanujan was gone. Trinity, his home for thirty years, had grown ugly to him. He scarcely spoke with some of his colleagues. Earlier, he had urged W. H. Young, an older mathematician who had spent much of his professional life abroad, to apply for the Savilian chair, only to ask him to withdraw his name, which Young did. "Hardy," recalled Young's son, Laurence, also a mathematician, "felt he must get away."

Oxford was the other great English university, less than a hundred miles away. In *Camford Observed,* Jasper Rose and John Ziman tried to bring it and Cambridge within the compass of a single account: "Oxford is a city of wide and noble thoroughfares, the High, the Broad, St. Giles; in Cambridge all the streets straggle. The great buildings of Oxford face the streets, tall and imposing, and form a series of breathtaking vistas. The great buildings of Cambridge are more isolated, less emphatic, more secretive, giving on to college courts and gardens. Oxford is more coherent; Cambridge more diffuse. Oxford overwhelms—Cambridge beguiles." Academically, Cambridge tipped slightly more toward the sciences, Oxford toward the classics.

New College was one of Oxford's two dozen or so distinct colleges. The place was like a walled city, with medieval battlements, pierced by tall, narrow slots through which archers could fire their bows, still enclosing two corners of it and forming a backdrop to the shrubs, trees, and bushes of the College Garden. In moving to New College, Hardy was coming full circle. The college was founded by William of Wykeham in 1379, eight years before he founded Winchester as a feeder to it. It had been the destination of many of Hardy's abler Winchester classmates twenty-five years before and would probably have been his as well had he not, his head turned by that St. Aubyn book, opted for Trinity instead.

Hardy had been in Oxford just a few months when he received the news from Madras:

> By direction of the [University] Syndicate, I write to communicate to you, with feelings of deep regret, the sad news of the death of Mr. S. Ramanujan, F.R.S., which took place on the morning of the 26th April.

"It was a great shock and surprise to me to hear of Mr. Ramanujan's death," Hardy replied in a letter to Dewsbury. But was there, in what he wrote next, the barest breath of defensiveness?

> When he left England the general opinion was that, while still very ill, he had turned the corner towards recovery; he had even gained over a stone in weight (at one period he had wasted away almost to nothing). And the last letter I had from him (about two months ago) was quite cheerful and full of mathematics.

Ramanujan *had*, after all, been entrusted to Hardy's care. Was Hardy—in a not uncommon sort of response to a loved one's death—now trying to assure himself that Ramanujan's final decline had come only *after* he had been placed safely aboard the ship to India?

About the impact of Ramanujan's death on Hardy there can be no doubt:

> For my part, it is difficult for me to say what I owe to Ramanujan—his originality has been a constant source of suggestion to me ever since I knew him, and his death is one of the worst blows I have ever had.

At Oxford, the specter of the Great War still hung heavily over Hardy, as it did all across Europe. Feeling against Germany ran deep. "Let us trust," one English scientist had written *Nature* in the closing months of the war, "that for the next twenty years at least all Germans will be relegated to the category of persons with whom honest men will decline to have any dealings." Mathematicians were not immune to the bitterness; many in England and France felt that Central European mathematicians should be banned from international mathematical congresses.

Hardy had been revolted by the war's stupid savagery, hated the whole idea of old men sending boys off to die, and felt cruelly cut off from his mathematical friends on the Continent. Now, the war over, he tried to heal the wounds. He wrote the London *Times* protesting some of the vengeful imbecilities being bruited about. He cooperated in the peacemaking efforts of Gösta Mittag-Leffler, long-time editor of *Acta Mathematica*, a Swedish mathematical journal founded in 1882 midst similar tensions among mathematicians following the Franco-Prussian War. He wrote with his views of the war to the great German mathematician Edmund Landau; his own views, Landau wrote back with a mathematician's touch, had been the same—except "with trivial changes of sign."

While visiting Germany in 1921, Hardy wrote Mittag-Leffler: "For my part, I have in no respect modified my former views, and am in no circumstances prepared to take part in, subscribe to, or assist in any manner directly or indirectly, any Congress from which, for good reasons or for bad, mathematicians of particular countries are excluded." He had boycotted one such congress in Strasbourg in 1920 from which Germans, Austrians, and Hungarians had been kept out, and he would boycott another in Toronto four years later.

The armistice, the departure of Ramanujan, his own move to Oxford, and Ramanujan's death had all come within eighteen months. But by all accounts, Hardy fell in easily at New College, felt at home there in a way he never had in Cambridge. It was, Snow tells us, "the happiest time of his life." He was accepted. His new Oxford friends made a fuss over him. His conversational flamboyance found new and appreciative ears. Sometimes, it seemed, everyone in the Common Room—the Oxford term for what back at Cambridge was the Combination Room—waited to hear what Hardy was going to talk about.

Meanwhile, his collaboration with Littlewood, conducted largely by mail, continued. He was at the height of his mathematical powers, the zenith of his fame. Mary Cartwright would recall how her bare mention of "Professor Hardy's class" to a college porter drew a response revealing "a far greater respect for Hardy than the customary deference of those days of any college porter to any don."

For the academic year 1928–1929, Hardy exchanged places with Princeton's Oswald Veblen and spent the year in the United States, mostly at Princeton University. While in America, he kept up a busy lecture schedule; he spoke at Lehigh University, for example, on January eleventh, at Ohio State on the eighteenth, at the University of Chicago on the twenty-first. During February and March, he was in residence at California Institute of Technology. At the end of the year, Princeton's president asked him to stay a bit longer. Hardy wrote back that though he'd had "a delightful time," his duties at Oxford demanded his return.

On this or one of his other trips to America, he developed a taste for baseball. Babe Ruth, it was said of him, "became a name as familiar in his mouth as that of the cricketer Hobbs." One time, he was sent a book, inscribed by "Iron Man" Coombs, stuffed with problems in baseball tactics. "It is a *wonderful* book," he wrote a postcard saying. "I try to solve one problem a day (e.g. 1 out, runners at 1st and 2nd, batsman [the cricket term] hits a moderate paced grounder rather wide to 2nd baseman's left hand—he being right handed. Should he try for a double play 1st to second or 2nd to first? I *think* the former.)"

His love of cricket and tennis, of course, continued unabated. In tennis, he steadily improved his game. In one snapshot taken while Hardy was at Oxford, it is a bright, sunny day in late spring or early fall and Hardy, looking absolutely smashing in his white tennis gear, stands midst a group of about a dozen other players. They are Beautiful People, ca. 1925 or so, and Hardy, clutching his racket, wearing long-legged tennis togs and a heavy shawl sweater under a jacket, is one of them.

Though happy at Oxford, by 1931 Hardy was back in Cambridge, as Sadleirian Professor, following the death of E. W. Hobson; Cambridge was, after all, still the center, far more than Oxford, of English mathematics, and he was now being offered its senior mathematical chair. Another reason, according to Snow, was that the two universities had different rules about retirement; whereas Oxford would turn him out of his rooms at sixty-five, at Trinity he could occupy them until he died.

For a time, Hardy would periodically return to Oxford for a few weeks at a time to captain the New College Senior Common Room cricket team. And, of course, he was always there at Lord's for the annual match between Cambridge and Oxford. "There he was at his most sparkling, year after year," wrote Snow. "Surrounded by friends, men and women, he was quite released from shyness; he was the centre of all our attention, which he by no means disliked; and one could often hear the party's laughter from a quarter of the way round the ground."

Hardy's formerly unpopular antiwar views were now, when not forgotten, actually applauded. The young Cambridge mathematicians, Snow records, "were delighted to have him back: he was a *real* mathematician, they said, not like those Diracs and Bohrs the physicists were always talking about: he was the purest of the pure." It was, as Laurence Young portrayed it later, a golden age of Cambridge mathematics. "Spiritually and intellectually, Cambridge was suddenly at least the equal of Paris, Copenhagen, Princeton, Harvard, and of Warsaw, Leningrad, Moscow." A sprinkle of foreign visitors to Cambridge had now, as Jews and others sought escape from Hitler's Germany, become a torrent.

Beginning around 1933, Hardy, in cooperation with the Society for the Protection of Science and Learning, used his influence to get Jews and others driven from their jobs to England and other safe havens. Mathematicians of the stature of Riesz, Bohr, and Landau were among those who got out. "Hardy, in many ways, was other-worldly," A. V. Hill wrote, "but in his deep solicitude for the dangers and difficulties of his colleagues he showed not only a broad humanity but a fine and resolute loyalty to the universal integrity and brotherhood of learning."

Hardy resigned from at least one German organization of which he had been a member—not because it was German, but because of what it *did*. "My attitude towards German connections of this kind," he wrote Mordell in the early Nazi period, "is that I do nothing unless I am positively forced to; but if anti-Semitism becomes an ostensible part of the programme of any periodical or institution, then I cannot remain in it."

In 1934, Hardy wrote to *Nature* responding to a University of Berlin professor who purported to show the influence of blood and race upon creative style in mathematics. There were, it seems, "J-type" and "S-type" mathematicians, the former of good Aryan stock, the latter Frenchmen and Jews. Hardy icily surveyed Professor Bieberbach's assertions, made a show of seeking ground on which to excuse them, finally found himself "driven to the more uncharitable conclusion that he really believes them true."

Hardy's sympathies lay invariably with the underdog, and his political views were decidedly left-wing. Until about 1927, he was active in the National Union of Scientific Workers, even made recruiting speeches on its behalf. In one, as J. B. S. Haldane paraphrased it later, he said to his audience of scientists "that although our jobs were very different from a coalminer's, we were much closer to coalminers than capitalists. At least we and the miners were both skilled workers, not exploiters of other people's work, and if there was going to be a line-up he was with the miners." Visitors to Hardy's rooms often noted that on his mantelpiece stood photographs of Einstein, the cricketer Jack Hobbs—and Lenin.

But within the mathematical community, he, Littlewood, and those in their camp stood squarely in the Establishment. English mathematicians, Hardy wrote in 1934, no longer labored under "the superstition that it is impossible to be 'rigorous' without being dull, and that there is some mysterious terror to exact thought." The revolution he had helped usher in a quarter century before had won the day. Indeed, some would grumble later that it had actually impeded progress in such fields as algebra, topology, functional analysis, and other topics within pure mathematics. By the 1930s, in any case, Hardy was seen as part of the older generation.

During these years, the honors, large and small, rolled in. On March 6, 1929, the one-hundredth anniversary of the death of the great Norwegian mathematician Abel, Hardy, in the presence of the king of Norway, received an honorary degree from the University of Oslo.

On December 27, 1932, he got the Chauvenet Prize, awarded every three years for a mathematical paper published in English, for his "An Introduction to the Theory of Numbers."

On February 29, 1934, he received a letter, on the hammer-and-sickle embossed stationery of the Soviet Union, from J. Maisky, the Soviet ambassador to Britain, congratulating him on his election as an honorary member of the Academy of Sciences in Leningrad.

The universities of Athens, Harvard, Manchester, Sofia, Birmingham, and Edinburgh awarded him degrees. He received the Royal Medal of the Royal Society in 1920, its Sylvester Medal in 1940. He was made an honorary member of many of the leading foreign scientific academies. Without a doubt, he was the most distinguished mathematician in Britain.

To this period, his prime, much Hardy lore is owed. One year, Hardy's New Year's resolutions were to:

1. Prove the Riemann hypothesis.
2. Make 211 no out in the fourth innings of the last test match at the Oval [which was something like hitting a grand slam home run while behind by three runs in the ninth inning of the World Series' final game].
3. Find an argument for the nonexistence of God which shall convince the general public.
4. Be the first man at the top of Mt. Everest.
5. Be proclaimed the first president of the U.S.S.R. of Great Britain and Germany.
6. Murder Mussolini.

Another story neatly combined his love of cricket, his pleasure in the sun, his warfare with God, and his madcap bent. One of his collaborators, Marcel Riesz, was staying at the place Hardy shared with his sister in London. Hardy ordered him to step outside, open umbrella clearly in view, and yell up to God, "I am Hardy, and I am going to the British Museum." This, of course, would draw a lovely day from God, who had nothing better to do than thwart Hardy. Hardy would then scurry off for an afternoon's cricket, fine weather presumably assured.

In long talks with Hardy beginning in 1931 and extending over the next fifteen or so years, C. P. Snow came away steeped in Hardy's "old brandy" sensibilities. By old brandy Hardy meant any "taste that was eccentric, esoteric, but just within the confines of reason." For example, he once wrote Snow that "the half-mile from St. George's Square to the Oval [in London] is my old brandy nomination for the most distinguished walk in the world." Old brandy was a sort of studied eccentricity—youthful foolishness transformed into a "mature" form, made a little self-conscious, ossified . . .

And that, indeed, is what had happened to Hardy. Somehow, he had become an old man. Even by the fall of 1931, when he was fifty-four, you could see signs of it. Back in Cambridge for the year, Norbert Wiener noticed that "by now, Hardy had become an aged and shriveled replica of the young man whom I had met in Russell's rooms" twenty years before. Hardy knew it, too. On his return to Cambridge, he was distressed by all the new, young faces he saw among the mathematicians.

"There is," he wrote, "something very intimidating to an older man in such youthful quickness and power."

One day in 1939, while dusting his bookcase, Hardy had his first heart attack. He was sixty-two at the time. In its wake, he could no longer play tennis, or squash, or cricket. His creativity waned. One listing of his most important papers (it included every one of those on which he had worked with Ramanujan) included none beyond 1935. Now, his output declined by even a crude quantitative yardstick—from half a dozen or so per year in the late 1930s to one or two per year.

His waning mathematical powers depressed him. So did the new war with Germany. But around 1941, when young Freeman Dyson came up to Cambridge from Hardy's old school, Winchester, and for two years attended his lectures, Dyson couldn't see it. To him Hardy was still a god. He and three other advanced students all sat around a table in a small room in the old Arts School, listened, and watched Hardy from a few feet away:

> He lectured like Wanda Landowska playing Bach, precise and totally lucid, but displaying his passionate pleasure to all who could see beneath the surface. . . . Each lecture was carefully prepared, like a work of art, with the intellectual denouement appearing as if spontaneously in the last five minutes of the hour. For me these lectures were an intoxicating joy, and I used to feel sometimes an impulse to hug that little old man in the white cricket-sweater two feet away from me, to show him somehow how desperately grateful we were for his willingness to go on talking.

Hardy retired from the Sadleirian Chair in 1942.

The year before, a photographer for the British magazine *Picture Post* snapped his picture at a rugby match. There he was on a chilly winter day, cigarette in hand, all rolled up in flannels, watching Cambridge defeat Oxford, 9 to 6. The photograph later appeared in one of the volumes of his collected papers. His sister didn't like it. "It makes him look so old," she said.

But he *was* old.

By 1946, he was virtually an invalid. Snow pictured him as "physically failing, short of breath after a few yards' walk." His sister came to nurse him (though the Trinity rules were so strict that she had to leave his rooms at night).

In early 1947, he tried to kill himself by swallowing barbiturates. But he took so many that he vomited them up, hit his head on the lavatory basin, and wound up with an ugly black eye for his trouble.

Later that year, the Royal Society notified him that he was to receive its highest honor, the Copley Medal. "Now I know that I must be pretty near the end," he told Snow. "When people hurry up to give you honorific things there is exactly one conclusion to be drawn."

On November 24, Snow wrote his brother Philip: "Hardy is now dying (how long it will take no one knows, but he hopes it will be soon) and I have to spend most of my spare time at his bedside."

It *was* soon. Hardy died on December 1, 1947, the day he was to be presented the Copley Medal. He left his substantial savings and the royalties of his books, once having provided for his sister, to the London Mathematical Society. "His loss," wrote Norbert Wiener, "brought us the sense of the passing of a great age."

The 1939 heart attack began the long physical and emotional slide that led to his suicide attempt. And it was in its wake, about a month after France fell to the Nazis, that he put the finishing touches to *A Mathematician's Apology*, his paean to mathematics. Snow saw the *Apology* "as a book of haunting sadness," the work of a man long past his creative prime—and knowing it. "It is a melancholy experience for a professional mathematician to find himself writing about mathematics," wrote Hardy. Painters despised art critics? Well, the same went for any creative worker, a mathematician included. But writing *about* mathematics, rather than doing it, was all that was left him.

And yet, the sadness is at the prospect of a rich, full life nearing its end, not bitterness at a life ill-spent. Pride runs through the *Apology*, too, and pleasure, and deep satisfaction.

> I still say to myself when I am depressed, and find myself forced to listen to pompous and tiresome people, "Well, I have done one thing *you* could never have done, and that is to have collaborated with both Littlewood and Ramanujan on something like equal terms."

Ramanujan. All these twenty years later, Ramanujan remained part of him, a bright beacon, luminous in his memory.

"Hardy," said Mary Cartwright, his student during the 1920s and whom Hardy would describe as the best woman mathematician in England, "practically never spoke of things about which he felt strongly." Yet at one remove from his listener, on the printed page, he became a little freer. And there he revealed Ramanujan's hold on him: "I owe more to him," he wrote, "than to any one else in the world with one exception [Littlewood?] and my association with him is the one romantic incident in my life."

In the years after his death, Hardy began rummaging through Ramanujan's papers and notebooks. That, as many other mathematicians were to learn, could be tough going. After arriving in Oxford, Hardy wrote Mittag-Leffler that he had prepared a short paper from Ramanujan's manuscripts, "but it was hardly substantial enough for the *Acta* [the journal Mittag-Leffler edited]. I am now trying to make a more important one. But it is not possible to do it very rapidly, as all of Ramanujan's work requires most careful editing."

By 1921, he had culled from Ramanujan's papers enough to prepare a sequel to Ramanujan's work on congruence properties of partitions. The manuscript from which he was working, Hardy wrote in a note appended to the paper, which appeared in *Mathematische Zeitschrift*, "is very incomplete, and will require very careful editing before it can be published in full. I have taken from it the three simplest and most striking results, as a short but characteristic example of the work of a man who was beyond question one of the most remarkable mathematicians of his time."

Hardy's own papers over the years were fairly littered with Ramanujan's name: "Note on Ramanujan's Trigonometrical Function $c_q(n)$ and Certain Series of Arithmetical Functions" appeared in 1921; "A Chapter From Ramanujan's Notebook" in 1923; "Some Formulae of Ramanujan" in 1924. Then more in the mid-1930s: "A Formula of Ramanujan in the Theory of Primes"; "A Further Note on Ramanujan's Arithmetical function $\tau(n)$." Hardy appreciated the debt he owed Ramanujan and Littlewood: "All my best work," he wrote, "has been bound up with theirs, and it is obvious that my association with them was the decisive event of my life."

Hardy was thirty-seven when he met Ramanujan, living out his boyhood dream as a Fellow of Trinity, already an F.R.S. But his collaboration with Littlewood had only just begun, and he would come to view his early contribution to mathematics, though formidable by standards other than his own, as unspectacular.

Then, abruptly, Ramanujan entered his life.

Ramanujan was, if nothing else, a living, breathing reproach to the Tripos system Hardy despised. Sheer intuitive brilliance coupled to long, hard hours on his slate made up for most of his educational lacks. And he was so devoted to mathematics that he couldn't bother to study the other subjects he needed to earn a college degree. This "poor and solitary Hindu pitting his brains against the accumulated wisdom of Europe," as Hardy called him, had rediscovered a century of mathematics and made new discoveries that would captivate mathematicians for the next century. (And all without a Tripos coach.)

Is it any wonder Hardy was beguiled?

From then on, over the next thirty-five years, Hardy did all he could to champion Ramanujan and advance his mathematical legacy. He encouraged Ramanujan. He acknowledged his genius. He brought him to England. He trained him in modern analysis. *And*, during Ramanujan's life and afterward, he placed his formidable literary skills at his service.

"Hardy wrote exquisite English," the *Manchester Guardian* would say of him, citing especially his obituary notice of Ramanujan as "among the most remarkable in the literature about mathematics." To mathematician W. N. Bailey, it was "one of the most fascinating obituary notices that I have ever read." And it was Hardy's book on Ramanujan, more than anything he knew about him otherwise, that convinced Ashis Nandy to make Ramanujan a prime subject of his own book. Hardy's pen fired the imagination, shaping Ramanujan's reception by the mathematical world.

It began in 1916, when Hardy reported to the university authorities in Madras on Ramanujan's work in England; one look at it and they asked that it be prepared for publication. In it, Hardy wrote of the "curious and interesting formulae" Ramanujan had in his possession, of how Ramanujan possessed "powers as remarkable in their way as those of any living mathematician," that his gifts were "so unlike those of a European mathematician trained in the orthodox school," that his work displayed "astonishing individuality and power," that in Ramanujan "India now possesses a pure mathematician of the first order." This was scarcely the sort of language apt to pass unnoticed among mathematicians, Indian or British, accustomed to the sort of flat, gray prose normally appearing in their journals. Someone once said of Hardy that "conceivably he could have been an advertising genius or a public relations officer." Here was the evidence for it.

Hardy's long obituary of Ramanujan appeared first in the *Proceedings of the London Mathematical Society* in 1921, a little later in the *Proceedings of the Royal Society*, then again in Ramanujan's *Collected Papers* in 1927. In it, he told Ramanujan's *story*. He invested it with feeling. His language lingered in memory. "One gift [Ramanujan's work] has which no one can deny," he concluded—"profound and invincible originality."

> He would probably have been a greater mathematician if he had been caught and tamed a little in his youth; he would have discovered more that was new, and that, no doubt, of greater importance. On the other hand he would have been less of a Ramanujan, and more of a European professor, and the loss might have been greater than the gain.

Snow, who first met Hardy in 1931, revealed that for all Hardy's shyness, "about his discovery of Ramanujan, he showed no secrecy at all." Mary Cartwright recalled that "Hardy was terribly proud, and rightly, of having discovered Ramanujan." Ramanujan had enriched his life. He didn't *want* to forget Ramanujan, and he didn't.

On February 19, 1936, Hardy wrote S. Chandrasekhar from Cambridge: "I am going to give some lectures (here and at Harvard) on Ramanujan during the summer." They would become the basis for *Ramanujan: Twelve Lectures on Subjects Suggested By His Life and Work*. "A labor of love," one reviewer called it.

The Harvard lecture was part of the great university's celebration of the three-hundredth anniversary of its founding. The bash culminated in three grand Tercentenary Days, from September 16 to 18. Harvard Yard, now a great outdoor theater with seventeen thousand seats, was awash with silk hats, crimson bunting, and colorful academic costumes. On the second evening, at 9:00 P.M., upward of half a million people lined both banks of the Charles River for two hours of fireworks.

The following morning, under a steady drizzle and dark, brooding clouds (the leading edge of a hurricane moving up the Atlantic Coast), sixty-two of the world's most distinguished biologists, chemists, anthropologists, and other scholars received honorary degrees. They marched in a procession from Widener Library and took their places on stands erected in front of the pillars of Memorial Church at the yard's north end. Psychoanalyst Carl Jung was among them. So was Jean Piaget, the pioneer student of child development. So was English astrophysicist Sir Arthur Eddington. So was Hardy. The citation honoring him, slipped within the red leather presentation book stamped with Harvard's *Veritas* seal, called him "a British mathematician who has led the advance to heights deemed inaccessible by previous generations."

During his stay at Harvard, Hardy was put up at the house of a prominent lawyer, who later became a United States senator. Both he and his host, according to one account, were worried: whatever would they talk about? "The lawyer was no better prepared to discuss Zeta functions than the mathematician to comment upon the rule in Shelley's case." So they seized on their common enthusiasm—baseball. The Red Sox were in town, and Hardy was at Fenway often to watch them.

In the weeks prior to the grand finale, a Tercentenary Conference of Arts and Sciences brought to Harvard more than twenty-five hundred scholars for lectures under broad rubrics of knowledge like "The Place and Functions of Authority," and "The Application of Physical Chemistry to Biology." Einstein's wife was ill, so he sent word that he couldn't come. Nor could Werner Heisenberg, author of the uncertainty principle, who was advised at the last minute that he was needed for eight weeks' service in Hitler's army. Their absence notwithstanding, it was an august group, including no fewer than eleven Nobel Prize winners. "Highbrows at Harvard," *Time* headed its account. The *New York Times* covered some of the public lectures, including Hardy's.

At about nine in the evening of the conference's first day, Hardy—wizened, gray, and almost sixty now—rose before his audience in New Lecture Hall. "I have set myself a task in these lectures which is genuinely difficult," he told them, in the measured cadences that were the mark of his speech and of his prose.

> and which, if I were determined to begin by making every excuse for failure I might represent as almost impossible. I have to form myself, as I have never really formed before, and to try to help you to form, some sort of reasoned estimate of the most romantic figure in the recent history of mathematics; a man whose career seems full of paradoxes and contradictions, who defies almost all the canons by which we are accustomed to judge one another, and about whom all of us will probably agree in one judgment only, that he was in some sense a very great mathematician.

And then Hardy, the memory still fresh of the day a quarter century before when an envelope stuffed with formulas arrived in the mail from India, began to tell about his friend, Ramanujan.

Notes

I have supplied notes for most quotations and for most statements of fact that the inquiring reader might be moved to question.

I have not usually supplied citations for amply documented historical events, such as World War I. Information about South Indian towns and regions, population, geography, climate, temples, agriculture, literacy statistics, and the like is largely drawn from standard gazetteers, published both in Ramanujan's time and in the years since, including the *Imperial Gazetteer*, Madras Presidency, 1908, and various district gazetteers.

For many of the book's supporting characters, I have not usually cited biographical information from the *Dictionary of National Biography*, from the series of published biographical memoirs of Fellows of the Royal Society, or from similar standard biographical references.

In the case of unpublished letters, I have tried to note the college or other institution in whose files they may be found, or the person who supplied them to me.

References to Hardy's *Collected Papers* are to volume 7 unless indicated otherwise; those to P. K. Srinivasan are to volume 1 of *Ramanujan Memorial Number*, unless indicated otherwise. Those to Ragami, the pen name of T. V. Rangaswami, carry no page number but are based on informal translations from the Tamil of his book *Ramanujan, the Mathematical Genius*, by V. Anantharaman and his wife Malini, of Baltimore. The "Reading Manuscript" refers to the original version of E. H. Neville's broadcast talk on Ramanujan, now stored at the University of Reading, in England.

Much of the information contained in the Family Record is ambiguous, and I have referred to it only when, in full context, the evidence seems clear.

Information drawn from written materials and personal interviews has been supplemented by personal observation and informal conversation in England and India.

PROLOGUE

p. 1. "*your wonderful countryman Ramanujan.*" S. R. Ranganathan, 80–81.

p. 7. "*was near to religion.*" *Oxford Magazine*, 2 January 1948.

"*a thought of God.*" S. R. Ranganathan, 88.

CHAPTER ONE

p. 11. *September 1887*. Family Record.
And so widely observed. Slater, *Southern India*, 121.
"*wet skull.*" *Coimbatore District Gazetteer*, 1966.
Teppukulam Street. Interview, T. V. Rangaswami.

p. 12. Smallpox incident. Family Record; Ragami; P. K. Srinivasan, Margosa is also known as *neem*.
case study in the damning statistics. Births and deaths in Family Record.

p. 13. *case of itching and boils*. Family Record.
he scarcely spoke. Ragami.
Atshara Abishekam. Ragami.
bristled at attending. Family Record.
Ramanujan was fond of asking. Seshu Iyer, 81.

p. 14. *crush him to pieces*. Family Record.
shuffled between schools. Family Record.
arms folded in front of him. Ragami.
loan dispute. Interview, T. V. Rangaswami.
bounced back to his maternal grandparents. Family Record.
the family enlisted a local constable. Family Record.
Back in Kumbakonam by mid-1895. Family Record.

p. 15. *a small courtyard*. Padfield, 14; Hemingway, *Tanjore District Gazetteer*.
Mahammakham festival. T. R. Rajagopalan, 34; Hemingway, *Tanjore District Gazetteer*, supplies 1897 as a festival year.
three-quarters of a million pilgrims. Slater, *Southern India*, 117.
its water level was said to rise several inches. Urwick, 68.

p. 16. *seventy-two-bed hospital*. *Imperial Gazetteer*, Madras Presidency, 1908.
Rice growing. Interview, J. M. Victor. See also Hemingway, *Tanjore District Gazetteer*.
little room in which to graze. Hemingway, *Tanjore District Gazetteer*.
that same number of villages. *Imperial Gazetteer*, Madras Presidency.

p. 17. Kumbakonam saris and silk. Interviews in Kumbakonam, especially with R. Viswanathan, headmaster, Town School, and with cloth merchants.
two thousand small looms. *Imperial Gazetteer*, Madras, 1908.
could cost as much as a hundred rupees. Hemingway.
a year's income to many poor families. For example, Compton, 164–165, pictures five annas per day as a typical wage for unskilled labor (16 annas = 1 rupee). Fuller, 63, says field laborers could be hired for two or three pence a day, the English pence being equal to an anna. A more recent South Indian district gazetteer, looking back, gives the average agricultural wage between 1901 and 1912 as three rupees, eight annas per month, or again something like three annas per day. *Imperial Gazetteer*, 1908, also gives the going rate for unskilled labor at three annas per day (with skilled labor worth seven or eight annas per day). Thus, if we figure three hundred working days a year at, say, four annas per day, we get about seventy-five rupees as a representative year's income.
normally the husbands. Slater, *Southern India*, 119.

p. 17. Clerk's life. Based on interviews in Kumbakonam.
Srinivasa was good at appraising fabrics. K. R. Rajagopalan, 4.

p. 18. Indian father's role. See Carstairs, 67–69.
"Very quiet." Bharathi, 51.
"weightless." Nandy, 102.
reminders to keep up the house. P. K. Srinivasan, 170.
Ramanujan wrote his mother. Ibid., 168.
Goats and Tigers. Interview, M. Vinnanasan, Kumbakonam. Janaki, in P. K. Srinivasan, 171, says Ramanujan and his mother played the "15 points game," another name for Goats and Tigers. A recent Salem District gazetteer refers to a game, called *Pulikatlam,* which may be the same thing.
"*a shrewd and cultured lady.*" Seshu Iyer, 81. Other information about Ramanujan's mother and her family is derived from biographical material in Port Trust File and interviews with S. Sankara Narayanan, K. Bhanumurthi, Janaki, and others.

p. 19. *managed a choultry.* Interview, T. V. Rangaswami.
Komalatammal fed him his yogurt. This picture drawn largely from interview with K. Bhanumurthi.
into the principal's office. P. K. Srinivasan, 85.

p. 20. "*An exceptionally gifted lady.*" Ibid., 114.
Caste system. See Thurston; Mayo; Bhattacharaya; Fuller; Padfield; Compton; M. N. Srinivas.

p. 21. *to secure divine grace.* See, for example, Chopra et al.

pp. 21–22. *Brahmin families . . . would pull over to the side of the road.* Fuller, 147. Interview, A. Saranathan.

p. 22. Hindu eating practices. Fuller, 147. Numerous interviews.

p. 23. *that gave food a reddish cast reminiscent of blood.* Interview, A. Saranathan, Kumbakonam. "If I see meat, I begin to vomit."
"*As the child learned* . . ." Carstairs, 67.
a fastidiousness about Hindu life. See Carstairs, 80.
"*Asceticism and mysticism* . . ." In Singer, 8.

p. 24. "*Simple living and high thinking.*" Interview, T. S. Bhanumurthy.
of 650 graduates of the University of Madras. Cited in M. N. Srinivas, 102.

p. 25. "*a language made by lawyers and grammarians.*" Quoted in Slater, *Southern India*, 132.
distinct from . . . Hindi. G. Ramakrishna et al., 459.
boasted a verse form reminiscent of ancient Greek. Slater, *Southern India*, 136.
almost twenty million people. Thurston, 122.
11 percent of Tamil Brahmins literate. Cited by M. N. Srinivas, 179.
Town High School. Interview, R. Viswanathan, the current headmaster. See also "History of Our School," in *Centenary Celebration Souvenir.*

p. 26. *partial to impromptu strolls between classes.* "My Reminiscences," R. Kandaswamy Moopanar, in *Centenary Celebration Souvenir.*
coming to him for help. S. R. Ranganathan, 61.
"*But is zero divided by zero also one?*" S. R. Ranganathan, 105.
Boarders. P. K. Srinivasan, 84.

p. 27. Loney's *Trigonometry.* S. R. Ranganathan, 105. Most likely, according to Richard Askey, this was only part 1 of the two-volume text.
learned from an older boy. Madras Port Trust.
understand trigonometric functions. Seshu Iyer, 82.
to any number of decimal places. Ibid.
finish in half the allotted time. Ibid.
solve them at a glance. P. K. Srinivasan, 84.
Ramanujan did Ganapathi Subbier's job. P. K. Srinivasan, 104. The current headmaster, R. Viswanathan, gives the number of students in the school at about one thousand. N. Govindaraja Iyengar, quoted in P. K. Srinivasan, puts the figure at fifteen hundred.
deserved higher than the maximum possible marks. P. K. Srinivasan, 121.

p. 28. Sarangapani Temple. See T. R. Rajagopalan, 36; G. Ramakrishna et al., 258; Das, 135; Balasubrahmanyan, 196.

p. 29. *temple built by Nayak kings.* Das, 137.

p. 30. *fall asleep in the middle of the day.* Ragami, "Ramanujan, 'A Gift From Heaven,' " *Indian Express,* 19 December 1987.

p. 31. *Once a year during the years he was growing up.* Seshu Iyer, 82. Interviews, at the temple, with S. Govindaraja Battachariar and P. Vasunathan. For more on Uppiliapan Koil see Das, 143.
Sacred thread ceremony. Padfield, 63.
Moonlight walk to Nachiarkovil. S. R. Ranganathan, 66.
God, zero, and infinity. Ibid., 84.

p. 32. "*Immensely devout.*" Ibid., 73.
"*A true mystic.*" Ibid., 88.

p. 33. *deity of the anthill.* Whitehead, 15.

p. 34. "*Brahminical Hinduism is here a living reality.*" Hemingway.
"*as Westminster Abbey and St. Paul's are to the other churches of London.*" Urwick, 72.
Rameswaram. See G. Ramakrishna et al., 383.
Hindu deities. See Das; Friedhelm Hardy, "Ideology and Cultural Contexts of the Srivaishnava Temple," in Stein, 119; Fuller.

p. 35. *One contemporary English account.* Fuller, 160.
"*fusion of village deities . . .*" Chopra et al.
direct them toward something higher and finer. V. Subramanyam, a member of Narayana Iyer's family in Madras, offers a delightful metaphor: Consider an architect's drawing, whose arcane visual language might represent a masterpiece. One ignorant of that language may, as he examines the drawing, be unable to see the genius it embodies. And yet, he *can* appreciate the care and reverence with which the drawing is unrolled, handled, and preserved, perhaps be inspired to search out its hidden meaning. Likewise, a stone deity.

p. 36. *to enter a trance.* S. R. Ranganathan, 13.
a bizarre murder plot. P. K. Srinivasan, 98.
speak through her daughter's son. S. R. Ranganathan, 13.

CHAPTER TWO

p. 39. *It first came into his hands.* Seshu Iyer and Ramachandra Rao, xii. But S. R. Ranganathan, 19, says it was an "elderly friend."
"not in any sense a great one." Hardy, *Ramanujan*, 2.
Tripos. See chapter 4.

p. 40. G. S. Carr. Hardy, *Ramanujan*, 3. Carr, iv. Cambridge University records.
from his desk in Hadley. Carr, x.
first statement on the first page. Ibid., 33.

p. 44. *"I have, in many cases . . ."* Ibid., iv.

p. 45. *"his* methods." Littlewood, *Miscellany*, 87.
"Through the new world thus opened to him . . ." Seshu Iyer and Ramachandra Rao, xii.
Government College. Hemingway, *Tanjore District Gazetteer. Imperial Gazetteer*, Madras Presidency, 1908. Government College *Calendar* for 1975–76, part B.
bridge today spanning the river. T. V. Rangaswami, in an interview, claims that during Ramanujan's time there was a feeble footbridge across the river, perhaps constructed from palmyra palms. But Government College *Calendar*, 5, records the building of the bridge only in 1944, replacing a ferry service.

p. 46. *a hostel for seventy-two students.* Hemingway, *Tanjore District Gazetteer.*
"College regulations could secure his bodily presence." Neville, "Ramanujan" (*Nature* 149), 292.
"He was quite unmindful." P. K. Srinivasan, 122. See also S. R. Ranganathan, 20.
foreign-language math texts. P. K. Srinivasan, 106.

p. 47. *left him to do as he pleased.* Ibid., 122.
"ingenious and original." Seshu Iyer, 82.
Professor demanded book's return. P. K. Srinivasan, 122.
Seshu Iyer "indifferent." Ibid., 99.
"To the college authorities . . ." Neville, "Ramanujan" (*Nature* 149), 292.
went to see the principal. P. K. Srinivasan, 85.
Tuition was thirty-two rupees per term. Hemingway, *Tanjore District Gazetteer.*

p. 48. *In early August 1905.* Family Record.
two-thirds the load, at two-thirds the speed. Hemingway, *Tanjore District Gazetteer.*
could still take three weeks. Kameshwar C. Wali, *Chandra* (University of Chicago Press, 1990), 42.

p. 49. *"When you get to the third-class railway carriage . . ."* Compton, 27.
Vizagapatnam. Thurston, 40; Urwick, 83; *Imperial Gazetteer*, Madras Presidency, 1908.
Fragmentary accounts. The scanty information available about Ramanujan's flight to Vizagapatnam is drawn from Family Record; Suresh Ram, 10; Seshu Iyer, 83; S. R. Ranganathan, 46, 94; K. R. Rajagopalan, 11. Also, interviews with P. K. Srinivasan and T. V. Rangaswami.
probably by September. Family Record.

p. 50. Primary exam humiliation. P. K. Srinivasan, 106. Ragami.
he secreted the papers . . . in the roof of his house. Seshu Iyer and Ramachandra Rao, xii.
"too sorry for his failure." P. K. Srinivasan, 27.
"An obligatory aspect of shame . . ." Wurmser, 52.

p. 51. *"wince at ourselves in the mirror . . ."* Ibid., 17.
"Hiding is intrinsic . . ." Ibid., 29.
a "mental aberration." Quoted in Nandy, 108
"a temporary unsoundness of mind." Madras Port Trust. See also Seshu Iyer, who says that "being too sensitive to ask his parents for help" after his failure at Government College, he left for Vizagapatnam.
he breathlessly inquired. S. R. Ranganathan, 77.

p. 52. Pachaiyappa Mudaliar. Muthiah, 186.
modeled on the Temple of Theseus. Ibid., 185.
arrived at Egmore Station. Interview, T. V. Rangaswami, who says he heard it from Janaki.
"To appear and succeed . . ." Fuller, 175.
half failed. Fuller, 176.

p. 53. *lived a few blocks . . . off the fruit bazaar.* S. R. Ranganathan, 73.
bout of dysentery. Family Record.
at Pachaiyappa's College. Ibid., 66, 73.
show how to solve it. S. R. Ranganathan, 65–67.
Singaravelu Mudaliar. Ibid., 65.
Ramanujan's gifts. "Of the problems given in our textbooks in geometry, algebra, and trigonometry," recalled a Pachaiyappa's classmate, Ramayana Ratnakara T. Srinivasa Raghavacharya, "he used to remark, 'These are all mental sums.' " S. R. Ranganathan, 75.

p. 54. *"At the upper left-hand part of the stomach . . ."* Foster and Shore, 17.
"Procure a rabbit." Ibid., 13.
a big, anesthetized frog. S. R. Ranganathan, 68.
". . . the Digestion chapter." Ibid., 69.
He'd take the three-hour math exam. P. K. Srinivasan, 119.
December 1906. There is some disagreement here. For example, according to K. R. Rajagopalan, Ramanujan did not take the F.A. examination in December 1906, but only the following year. And the Family Record has him taking an "Intermediate Exam" in 1908. None of which much affects the story: Ramanujan took exams, always failed them, and wound up credentialless.

p. 55. *"fall far short of the multitude."* Compton, 171.
Ramanujan's cap. P. K. Srinivasan, 117.
Agricultural workers in surrounding villages. See note for chapter 1, page 17.
an old woman . . . would invite him in for a midday meal. S. R. Ranganathan, 22.
Ramanujan fed *dosai*. P. K. Srinivasan, 96.

p. 56. Ramanujan tutors Viswanatha Sastri. Ibid., 89.
Tutor to Govindaraja Iyengar. S. R. Ranganathan, 62.

p. 57. *"In proving one formula, he discovered many others . . ."* Neville, "Ramanujan" (*Nature* 149) (1942): 292.
a succession of handwritten accounts. This is what I have called the Family Record.
"a peculiar green ink." G. N. Watson, "Ramanujan's Note Books," *Journal of the London Mathematical Society* 6 (1931): 139.

p. 58. *"Two monkeys having robbed an orchard . . ."* Quoted by A. C. L. Wilkinson, "Presidential Address," *Journal of the Indian Mathematical Society*, February 1919, 24.

p. 63. "*Every rational integer.*" Hardy, in Ramanujan, *Collected Papers*, xxxv.
"*from which most of us would shrink.*" Lecture notes, B. M. Wilson. Trinity College.

p. 64. William Thackeray. *Madras Civil Servants* (London: Longman, Orme, Brown, and Co., 1839).
"*very proper that in England.*" Quoted in Chopra et al., vol. 3.

p. 65. "*into the icy water.*" Bell, 330.
to well-worn paths. For a kindred instance, consider Weierstrass: "The creative ideas with which he fertilized mathematics were for the most part thought out while he was an obscure schoolteacher in dismal villages where advanced books were unobtainable. . . .Being unable to afford postage, Weierstrass was barred from scientific correspondence. Perhaps it is well that he was; his originality developed unhampered by the fashionable ideas of the time. The independence of outlook thus acquired characterized his work in later years." Bell, 416.
"*the carefree days.*" Neville, "Ramanujan" (*Nature* 149): 293.
Ramanujan warned the parents of a sick child. S. R. Ranganathan, 85.

p. 66. "*the primordial God and several divinities.*" P. K. Srinivasan, 92.
Satyapriya Rao. S. R. Ranganathan, 85; P. K. Srinivasan, 90. "He was a sturdy strong man and he taught the students native exercises—'Dandal,' 'Baski,' etc." *Centenary Celebration Souvenir.*
Theory of Zero and Infinity. S. R. Ranganathan, 82.
"*spoke with such enthusiasm.*" Ibid., 83.

p. 67. *he'd drop by the college.* Ibid., 61.
legs pulled into his body. Ibid., 62.
"*the Indian character has seldom been wanting.*" *Hindu*, 11 February 1889.

p. 68. "*a blank and vacant look.*" Bharathi, 51.
"*that time-tested Indian psychotherapy.*" Nandy, 109.

CHAPTER THREE

p. 69. Ramanujan's marriage. Ragami; P. K. Srinivasan, K. R. Rajagopalan, 17–18; Nandy; Suresh Ram; S. R. Ranganathan. Interviews with Janaki, Mr. and Mrs. T. U. Bhanumurthy.
The family had once been better off. Interview, Janaki.

p. 70. *would not so much as glimpse his face.* In an interview, Janaki confirms this standard practice.
Father's response to word of son's marriage. Interviews, Janaki, Mr. and Mrs. T. S. Bhanumurthy.
a family in Kanchipuram. Family Record.
ever sensitive to local custom. Compton, 117.
Mysore ban. Ibid., 119.
six months' income. Fuller, 143.
local usurers. Marvin, 102.
a double wedding. Interview, Janaki. Family Record.

p. 71. *for three years.* Family Record.

p. 72. Hydrocele. Interview, Jacek Mostwin.

p. 72. *Dr. Kuppuswami volunteered to do the surgery for free.* P. K. Srinivasan, 100; K. R. Rajagopalan, 19; Family Record.
As chloroform administered. Seshu Iyer, 82.
the wound began to bleed. P. K. Srinivasan, 100.

p. 73. *he would often have to depend.* Interview, T. V. Rangaswami.
more than a week's pay. Compton, 27, puts the third-class fare (in 1904) at about a farthing per mile. A farthing was one-quarter of a penny, and a penny equaled an anna. Thus, a trip to Madras, about a four-hundred-mile round-trip from Kumbakonam, would have been about a hundred annas, or six rupees. Ramanujan's father made twenty rupees per month. Plainly, getting around South India during this period must have represented a real financial burden to Ramanujan and his family.
at the house of a friend. S. R. Ranganathan, 87.
he stayed with Viswanatha Sastri. P. K. Srinivasan, 90.

p. 74. *on the sufferance of two old Kumbakonam friends.* S. R. Ranganathan, 69.
Narasimha squeaked by. P. K. Srinivasan, 108.
One day probably soon after this. S. R. Ranganathan, 74.

p. 75. "*incorrigible idler.*" Neville, "Ramanujan" (*Nature* 149): 292.
had made him marry. S. R. Ranganathan, 24.
"*Like regiments we have to carry our drums.*" Chaudhuri, 101.

p. 76. *never amount to anything.* P. K. Srinivasan, 123.
"*so friendly and gregarious.*" Ramaseshan, 4.
"*He would open his notebooks.*" P. K. Srinivasan, 123.
"*such a simple soul.*" S. R. Ranganathan, 61.
puppet shows. P. K. Srinivasan, 97.

p. 77. "*knew nothing of mathematics.*" S. R. Ranganathan, 22.
Sometime late in 1910 . . . S. R. Ranganathan, 23; P. K. Srinivasan, 99; Family Record.
Everyone called him "Professor." K. G. Ramanathan, 20.
"*I was struck by the extraordinary mathematical results.*" P. K. Srinivasan, 129.
Not seen one another since 1906. Seshu Iyer, 83.

p. 78. "*I was not big enough.*" P. K. Srinivasan, 102.
Ramachandra Rao. Bhargava, *Who's Who in India; Who's Who in Madras,* 1934.
Ramachandra Rao meeting. P. K. Srinivasan, 86; Ramachandra Rao; S. R. Ranganathan, 24, 74; Family Record; Seshu Iyer, 83. Neville, "The Late" (*Nature* 106): 66.

p. 79. *he probably didn't "work it out."* Richard Askey supplied this insight.

p. 81. Fermat's leisure. Bell, 59.

p. 82. *fifth-largest city in the British Empire.* After London, Calcutta, Bombay, and Liverpool. Lewandowski, 51.
Origins of name. Thurston, 2; Krishnaswami Nayadu, 1.
History of Madras. See Slater, *Southern India;* Urwick; Steevens; Lewandowski; Srinivasachari; Lanchester.

p. 83. *more than fifty feet above sea level.* Lanchester, 90.
"*hutments.*" Srinivasachari, 292.
the city had expanded horizontally. Lewandowski, 46.
large rural tracts. Singer, 142.

p. 83. "*Madras is more lost in green.*" Steevens, 299.
p. 84. *a Brahmin-run restaurant.* Bharathi, 50.
spin occult stories. S. R. Ranganathan, 16.
pp. 84–85. *a refreshing* Gangasnanam. Ibid., 67.
p. 85. Public works projects in Madras. Srinivasachari, 298.
a letter from Ramaswami Iyer. Journal of the Indian Mathematical Society 11 (April 1919): 42–44.
Indian mathematics. See, for example, Eves, 161–183; M. S. Rangachari, "The Indian Tradition in Mathematics," *Journal of Indian Institute of Science* (Ramanujan Special Issue 1987): 3–9.
p. 86. "*a mixture of pearl shells and sour dates.*" Eves, 171.
p. 87. "*Ramanujan's first love.*" Letter, Bruce Berndt to author.
p. 90. *Ramanujan had stumbled on Bernoulli numbers.* Madras Port Trust Archives, in Askey.
"*Some Properties of Bernoulli's Numbers.*" Ramanujan, *Collected Papers,* 1–14.
p. 91. *By one reckoning.* K. Srinivasa Rao, "Srinivasa Ramanujan: His Life and Work," in Nagarajan and Soundararajan, 4.
"*Mr. Ramanujan's methods were so terse and novel.*" Seshu Iyer, 83.
"*by no means light.*" P. K. Srinivasan, 131.
p. 92. "*numerical evidence as sufficient.*" Mordell, *Nature* (1941); 645.
"*they call you a genius.*" S. R. Ranganathan, 25.
p. 93. "*a piece of thin string.*" Bharathi, 50.
on paper already written upon. S. R. Ranganathan, 76.
prevailed on him to copy it over. S. R. Ranganathan, 58. Here, "during this period" means between early 1911, when Ramanujan first came under Ramachandra Rao's influence, and early 1913, when he left for England.
"*Paper, The Great Immortalizer.*" Bharathi, v.
For about a year. Neville, "Ramanujan" (*Nature* 149), 292.
grew to bother him. S. R. Ranganathan, 26. "It was Ramanujan, unwilling to be longer a burden, who brought to an end his dependence" by taking the Port Trust job: Neville, Reading Manuscript.
a temporary job in the Madras Accountant General's Office. Family Record. Srinivasan, 176, puts the dates of the job as January 12 to February 21, 1912, but there is some reason to doubt this. Conceivably, it was the previous year.
"*a clerkship vacant.*" P. K. Srinivasan, 31.
p. 94. "*quite exceptional capacity in Mathematics.*" Ibid., 49.
The port of Madras. See Thurston; Urwick; Srinivasachari; Lanchester; *The Port of Madras: Past, Present and Future;* gazetteers. Interview, V. Meenakshisundaram.
p. 95. *his own motorcar.* Muthiah, 183.
Sir Francis Spring. See *The Port of Madras: Past, Present and Future,* 8.
Narayana Iyer. Interviews with his family in Madras.
Sir Francis relied on him heavily. "Whatever Narayana Iyer said," says S. Sankara Narayanan, "he agreed."
"*the most convenient dress.*" *Hindu,* 24 April 1896.
p. 96. *late in 1912.* P. K. Srinivasan, 172. Family Record.
a third of the city's population. Lewandowski, 46.

p. 97. *three rupees per month.* Interview, Janaki.
"sinecure post." Ramachandra Rao, 87. Bharathi, 47.
"running to his office." Bharathi, 50.
Ramanujan's Port Trust routine. Interview, Janaki. "Since a year [after taking his job at the Port Trust] Ramanujan still found his clerkship a bondage from which he craved release, he must have been applying himself conscientiously to duties which his patron did not for a moment intend him to take seriously": Neville, Reading Manuscript.
establishing cash balances. Interview, T. V. Rangaswami.
"pilotage fund clerk." Internal memo, 11 February 1913, Madras Port Trust. Ramanujan's regular job was as a "bill clerk."

p. 98. *prowling for packing paper.* S. R. Ranganathan, 76.
sternly regarding his aide. Ibid., 27.
slates propped on their knees. Bharathi, 48.
"You must descend to my level." Ibid.
Ramanujan as diamond. P. K. Srinivasan, 172.

p. 99. The British Raj and the Indian Civil Service. See Worswick and Embree; Fuller; Compton.
"treason to our trust." Quoted in Worswick and Embree, 140.
two-thirds those of the British. Fuller, 276.

p. 100. *"Plato's ideal rulers."* Worswick and Embree, 142.
"awakened by responsibility." Fuller, 276.
"regarded with awe, not affection." Urwick, 55.
"the English official is an incomprehensible being." Ibid.
An Englishman typically had his own washerman. Compton, 190.
brush and fold his clothes. Ibid., 246.

p. 101. *steward stooped to serve him tea.* Ibid., 207.
"no assimilation between black and white." Ibid., 247.
Narasimha had introduced him to E. B. Ross. Family Record.
E. W. Middlemast recommendation. P. K. Srinivasan, 49.
"cunning and contentious in argument." Compton, 40.

p. 102. *"you can polish the Hindu intellect."* Ibid., 171.
"constantly disconcerted." Fuller, 179.
"hardworking, docile and enduring." Urwick, 50.

p. 103. *at Ramachandra Rao's behest.* Family Record refers to a letter from Ramachandra Rao to Griffith on 3 November 1912. See also Neville, Reading Manuscript.
"You have in your office." P. K. Srinivasan, 50.
"If his genius is so elusive." Ibid., 51.

pp. 103–104. *"He gives me the impression of having brains."* Ibid., 52.

p. 104. *"I think I was right in writing to Prof. Hill."* Letter, Griffith to Spring, 28 November 1912. Madras Port Trust.
Ramanujan had fallen into some pitfalls. P. K. Srinivasan, 53.

p. 105. *"evidently a man with a taste for mathematics."* Letter, Griffith to Hill, 7 December 1912. Madras Port Trust.

p. 106. Many had advised Ramanujan to write to England. S. R. Ranganathan, 70; P. K. Srinivasan, 91.
as long as two weeks. Compton, 178.

p. 106. *letters drafted with the help.* Hardy ("Obituary," 494) reports that Ramanujan told him the personal introduction had been written by "a friend." And S. R. Ranganathan, 32, argues that the letter was "largely worded for him by the seniors who had helped him," Ramanujan presumably being too shy and humble to write anything so boastful. Plainly, Ramanujan got help. But as I argue later, I don't think he was as humble as some think.

In Ramanujan's letter to Hardy of January 22, 1914, he claims that "all letters written to you, except this one [and one other], did not contain my language. Those were written by the superior officer mentioned before, though the mathematical results and handwriting were my own." But there is reason to believe (see chapter 5) that Ramanujan was dissembling. Besides, Ramanujan was always scrupulous about acknowledging the help of others and may have deemed *any* help from his friends to mean that the language was not his "own." Finally, even *this* letter, the letter he acknowledges as his own, was not so dramatically different from the rest. In short, I believe there was invariably a large chunk of Ramanujan in any letter that bore his signature.

pp. 106–107. Baker and Hobson said no. Snow, *Apology* foreword, 33–34: "I mentioned that there were two persons who do not come out of the story with credit. Out of chivalry, Hardy concealed this in all that he said or wrote about Ramanujan. The two people concerned have now been dead, however, for many years, and it is time to tell the truth. It is simple. Hardy was not the first eminent mathematician to be sent the Ramanujan manuscripts. There had been two before him, both English, both of the highest professional standard." But Snow went no further. In *Alternative Sciences*, 146–147, Nandy reports that Littlewood identified them as "Baker and Popson." In "Ramanujan—A Glimpse," 78, Bollobás corrects the error. In a letter to Bruce Berndt, Indian mathematician K. Venkatachaliengar dismisses the assertion that Baker was among those who ignored Ramanujan. But he seems to do so solely on the basis that anyone who would, as Baker at one point did, recommend not an Englishman but an Indian, C. Hanumantha Rao, to a position at an Indian university could scarcely have spurned Ramanujan. Baker, writes Venkatachaliengar, was "a very sincere man of upright conduct. . . . [He] would have acknowledged his error if it were so to [Ramanujan] direct or through Hanumantha Rao." This assertion, however, is not enough to undercut our confidence in what comes down to us directly through Littlewood who, according to Bollobás ("Ramanujan—A Glimpse," 78), "often chuckled over the embarrassment of his colleagues who failed to recognize a genius."

CHAPTER FOUR

p. 109. "*He . . . looks a babe of three.*" Michael Holroyd, *Strachey: A Critical Biography*, vol. 1, *The Unknown Years* (London: Heinemann, 1967), 516, footnote.
Hardy was sometimes refused beer. Bollobás, *Littlewood's Miscellany*, 120.
His college rooms had no mirrors. Snow, *Apology* foreword, 16.
See also Alexanderson, 65.
"*Red Indian bronze.*" Snow, *Apology* foreword, 9.

p. 110. *Hardy took six pages to attack the plan.* Letter, Hardy to Jackson, 9 November 1910. Trinity College.
enthusiasms, peeves, and idiosyncrasies. Titchmarsh, 452.
"*There's a match due to begin.*" Collins, 27.
leaving the benighted. M. H. A. Newman, in BBC obituary. Trinity College.
Sister read to him about cricket when he died. Snow, *Apology* foreword, 58.
Chapel incident. Ibid., 20.
"*happiest hours of my life.*" Snow, *Apology* foreword, 21.
would not shake hands. Interview, Mary Cartwright.

p. 111. "*strange and charming of men.*" Woolf, *Sowing*, 123.
"*the most extravagant and fanatical kind.*" Hardy, *Collected Papers,* 532.
"*the one great permanent happiness of my life.*" Letter, Hardy to Thomson, December 1919. Cambridge University.
He admitted the pro-God position was stronger. Interview, Mary Cartwright.
"*what his real opinions were.*" Titchmarsh, 450.

pp. 111–112. "*more delicate, less padded, finer-nerved.*" Snow, *Apology* foreword, 16.

p. 112. "*a slightly startled fawn.*" Woolf, *Sowing,* 124.
"*To sit that way.*" Alexanderson, 64.
"*personal powers and capabilities.*" A. Ostrowski, in *Experentia* 5 (1949): 131–132.

p. 113. Middle-class schools. *Cranleigh School Magazine,* July 1887, 133, 198. See also Megahey.
"*the labourer's son.*" Megahey, 12.
55 were the sons of tradesmen. Minutes, Cranleigh School, 29 January 1880.
Isaac Hardy. *Surrey Advertiser and County Times,* 7 September 1901; *The Cranleighan,* 1901, 267–271; *Surrey County School Register,* 1872, 1; other Cranleigh School records.
an extra fifty pounds per year. Minutes, Surrey County School, 23 October 1874.
rich industrialists. Brandon, 97.

p. 114. Genealogy of Hardy's parents. Information furnished by Robert A. Rankin, Glasgow.
"*with more than a touch of the White Knight.*" Snow, *Variety of Men,* 204.
singing lessons. Surrey County School Register, Christmas 1872.
earnestly and sincerely mourned. Surrey Advertiser and County Times, 7 September 1901.
"*rare and precious souls.*" *The Cranleighan,* 1901, 270.
Sophia Hall Hardy. Information furnished by Robert A. Rankin, Glasgow, through D. H. J. Zebedee, *Lincoln Diocesan Training College, 1862–1962.*
to church two or three times. Interview, Mary Cartwright.

p. 115. *taught piano.* Minutes, Cranleigh School, 29 January 1890.
Handel concert. *Cranleigh School Magazine,* July 1878.
Gertrude Hardy. References to her are scattered through *St. Catherine's School Magazine* over more than thirty years.
"*There is a girl I can't abide.*" Gertrude Hardy, "Lines Written Under Provocation," *St. Catherine's School Magazine* (October 1933), 575.

p. 116. "*a little obsessive.*" Snow, *Apology* foreword, 14.
Only "good" books. Titchmarsh, 447.
writing down numbers into the millions. Ibid.
Eustace Thomas Clarke. *Cranleighan,* June 1905, 18. *Cranleighan,* 1937, 120.
give wrong answers. Snow, *Apology* foreword, 14.
"*He seems to have been born with three skins too few.*" Ibid., 15.

pp. 116–117. Gertrude "*shy and diffident.*" *St. Catherine's School Magazine,* obituary, August 1963, 6. "A teacher of outstanding ability, she combined keen interest in her brighter pupils and, as I have good reason to remember, boundless patience with the slow. Her reproofs were astringent, but never undeserved. Her scrupulous fairness gave that feeling of security which is so valuable to the young."

p. 117. "*I was very shy.*" *St. Catherine's School Magazine,* October 1938, 229.
The clergyman and the kite. Stanislaw Ulam, *Adventures of a Mathematician* (New York: Scribners, 1976), 60.
Gertrude the "Mohammedan." Letter, Marjorie Dibden to Robert A. Rankin, 24 November 1983. Robert Rankin.
lost her eye as a child. Interview, Mary Cartwright.
Running it would be Mr. and Mrs. Hardy. School Minutes, 29 January 1880.
At least around 1881. The census records, obtained at the Guildford Library, are RG 792 97b schedule 153. Hardy is listed as "Godfred." The records clearly indicate the Hardy house as "Mt. Pleasant," cottages that still stand. However, school records refer to the twenty-four boys as "at Mr. Hardy's house," which could only mean the much larger House. It is conceivable, then, that Mr. and Mrs. Hardy actually lived, along with their children, at the House, while maintaining official residence in the Mt. Pleasant cottage.

p. 118. *ubiquitous in Surrey since the seventeenth century.* Nairn and Pevsner, 14. See also Jekyll, 4.
low-ceilinged sitting room. Tour of the house. My thanks to its current resident, whose name I failed to record.
who helped Mrs. Hardy at the House. Sketch of the History of the Surrey County School, 17.
to rent out sleeping space. It is unlikely that the servants mentioned in the census records served the Hardy family. Megahey, 16, reports that an upper-middle-class income of five hundred pounds per year was enough to hire three servants. Isaac Hardy made less than one-fifth of that.
a thousand pounds a year. Megahey, 16.
second master made only a hundred. Ibid., 17.
"*a typical Victorian nursery.*" Titchmarsh, 447.

p. 119. *little more than a mud track.* Megahey, 20.
"*Cranleigh boys may wander where they please.*" Ibid., 27.
"*enlightened, cultivated, highly literate.*" Snow, *Apology* foreword, 14.
a sixth-former who was twenty. Megahey, 31.
reached the sixth form at twelve. He was placed in the third form at ten. Megahey, 52.

pp. 119–120. Ward's assessment of young Hardy. Cranleigh School Register, Report to the Council of the Surrey County School, 26 July 1889.

p. 120. *137 boys competed.* Sabben-Clare, 54.
Hardy placed first. Megahey, 52.
beginning to shed. Ibid., 44.
"*Connel.*" Register of Winchester School, courtesy Robert Rankin.
originally endowed. Dunning and Sheard, 48.
"*gentlemanly rebels and intellectual reformers.*" Bishop, 18.

p. 121. *architectural details. Winchester College: A Guide,* school publication, rev. and repr. May 1984.
"*notions.*" See, for example, Sabben-Clare, 144; Bishop, 26; *Winchester College Notions,* "by Three Beetleites" (Winchester: P. & G. Wells, 1904).
thirty strokes, with a ground-ash stick. Sabben-Clare, 44.
"*Morning Hills.*" Ibid., 151.
circled through the stone-arched walkways. Ibid., 152.

p. 122. "*Mrs. Dick.*" Ibid., 45.
Cricket. John and Emma Leigh, of Cranleigh, assure me that a foreigner possesses the essential knowledge of cricket when he fully understands the following:

> You have two sides. One out in the field and one in.
> Each man that's in the side that's in goes out and when he's out he comes in and the next man goes in until he's out.
> When they are all out the side that's out comes in and the side that's been in goes out and tries to get those coming in out.
> Sometimes you get men still in and not out.
> When both sides have been in and out including the not outs
> That's the end of the game.

(with acknowledgments to the Marylebone Cricket Club)
Hardy saw Richardson and Abel. Snow, "The Mathematician," 67.
practices were sacrosanct. Megahey, 27.
"*the village did much better.*" *Cranleigh School Magazine,* October 1888, 251.

p. 123. "*a thing of personal art and skill.*" Neville Cardus, *Play Resumed with Cardus* (London Souvenir Press, 1979), 14.
"*Canvassing, Clarendon type, and professional cricket.*" Records of Magpie & Stump, 966th meeting, 3 June 1910. Cambridge University.
with walking stick and tennis ball. Woolf, *Sowing,* 124.
"*one long game of cricket.*" *Wykehamist,* 20 June 1893.
"*small, taut, and wiry.*" C. J. Hamson, text of a talk given in the Senior Combination Room at Trinity, 24 November 1985, and published in *Trinity Review,* 1986, 23.
Accounts of Hardy as athlete. *Wykehamist,* December 1895.
frustrating . . . a brilliant cricket career. Snow, *Apology* foreword, 48.

p. 124. *crushed any artistic ability.* Titchmarsh, 447.
so sick he almost died. Snow, *Apology* foreword, 17.
Hardy and mutton. Titchmarsh, 452. Bollobás, 120.
never returned to visit. Wykehamist, 18 February 1948.
Of twenty-six class hours. Bishop, 32. The figures are from 1900.
Richardson. Sabben-Clare, 62.
Duncan Prize. Winchester School records.

p. 124. *physics on his own. The Wykehamist,* 18 February 1948.
a weakness for detective stories. Interview, Mary Cartwright. Hardy told her he had once read thirty-six of them in a week. (His sister shared the weakness: *St. Catherine's School Magazine,* August 1963, 6.)
One day when he was about fifteen. Hardy, *Apology,* 145.
"*Herbert was back again at Trinity.*" *A Fellow of Trinity,* Alan St. Aubyn (London: Chatts and Windus, 1892), 261.

p. 125. "*a decent enough fellow.*" Hardy, *Apology,* 146.
Wykehamists to Cambridge, Oxford. *Wykehamist,* December 1896.
"*Congratulations are due.*" *Wykehamist,* December 1895.

p. 126. *the school's annual Speech Day. Surrey Advertiser,* 1 August 1896.
74 percent to Oxford or Cambridge. Ellis, 144.

p. 128. *assigned a room in Whewell's Court.* Trinity College records.
Tripos. See "Old Tripos Days at Cambridge," by A. R. Forsyth, *Mathematical Gazette* 19 (1935): 162–179; "Old Cambridge Days," by Leonard Roth, *American Mathematical Monthly* 78 (1971): 223–226 (both of these are reprinted in Campbell and Higgins, vol. 1, 81–103); Rouse Ball; Littlewood, *Miscellany,* and "Mathematical Life and Teaching"; Hardy, *Collected Papers,* 527–553.

p. 129. "*the most difficult mathematical test.*" Roth, in Campbell and Higgins, 97.
in earliest form, to 1730. Rouse Ball, 11.
"*wish them Joy of their Honour.*" Quoted in *Oxford English Dictionary,* entry for "Wrangler."

p. 130. Wooden Spoon. Cambridge Folk Museum display.
"*All-American, a Rhodes scholar, and Bachelor of the Year.*" Campbell and Higgins, vol. 1, 81.

p. 131. "*You,* sir." Roth, in Campbell and Higgins, 98.
1881 Tripos results. Littlewood, "Mathematical Life and Teaching," 21.
Typical Tripos problems. See, for example, Forsyth, in Campbell and Higgins, 86–89; Roth, in Campbell and Higgins, 97–98; Thomson, 56–60; Littlewood, *Miscellany,* 72–75.
Must be a trick. The future Senior Wrangler was Littlewood. He tells the story in his *Miscellany,* 74.
excellent training—for the bar. Thomson, 58.
like a racehorse. Indeed, practically all who have written about the coaching system employ this metaphor.

p. 132. *two dozen of them.* Forsyth, in Campbell and Higgins, 86.
"*beyond the pale of accessible criticism.*" Forsyth, in Campbell and Higgins, 84.
handed over to R. R. Webb. Titchmarsh, 448.
"*altogether too much like a crossword puzzle.*" Quoted in R. Clark, *Life of Bertrand Russell,* 43.

p. 133. "*without serious prejudice to his career.*" Hardy, *Collected Papers,* 537.
"*the verdict of my elders.*" Hardy, *Apology,* 144.
Hardy's boyhood history of England. Titchmarsh, 447.
Hardy's essay impressed his examiners. Littlewood, "Reminiscences," 12.
Headmaster Fearon's treatment. Titchmarsh, 448.

p. 134. Love's suggestion. Hardy, *Apology,* 147.
"*at last in [the] presence of the real thing.*" Hardy, *Collected Papers,* 722.

p. 134. *into two regions.* See Ibid., 723.
p. 135. "*that remarkable work.*" Hardy, *Apology*, 147.
"*even the rebel Hardy.*" Littlewood, quoted in Young, 270.
"*to take so much trouble and to learn no more.*" Hardy, *Collected Papers*, 530.
"*It is a great triumph.*" Letter, G. M. Trevelyan to Hardy, 14 June 1898. Trinity College.
"*he ought to have won it.*" Snow, *Apology* foreword, 24.
p. 136. Shakespeare Society meeting. Minutes, 23 May 1901.
Hardy in Cranleigh. *Surrey Advertiser and County Times*, 7 September 1901.
p. 137. "*final and irrevocable.*" Trinity College, records of clubs and societies.
The Apostles. See Deacon; Levy; Woolf, *Sowing*.
"*a very nice scientist.*" Deacon, 33.
Hardy inducted. Levy, 195.
"*tenacity of a bulldog.*" Woolf, *Sowing*, 148.
"*his peculiar passion for truth.*" Woolf, *Beginning Again*, 24.
p. 138. "*Does Youth Approve of Age?*" Levy, 196.
"*absolute freedom of speculation.*" Quoted in Deacon, 70.
"*breathe the magic air.*" Quoted in Deacon, 59.
G. L. Dickinson. See Furbank, 59.
who once advised Hardy. Snow, *Apology* foreword, 31.
"*Walt Whitmanesque feelings of comradeship.*" Levy, 227.
"*The Higher Sodomy.*" Deacon, 55.
p. 139. "*the womanisers pretend to be sods.*" Deacon, 62.
R. K. Gaye. Woolf, *Sowing*, 124; Cambridge University records.
"*they were never seen apart.*" Ibid., 124.
"*rumor of a young man.*" Interview, Mary Cartwright.
"*just another English intellectual homosexual atheist.*" Hodges, 117.
"*his beloved John Lomas.*" Snow, "The Mathematician," 72. Deacon, 73, reports that on one previous occasion Hardy tried to commit suicide, "the first time following the death of a close male friend."
"*a non-practicing homosexual.*" Interview, Béla Bollobás.
p. 140. Vote on "self-abuse." Levy, 207.
"*school-boy affections.*" Ellis and Symonds, 39.
"*relentlessly Victorian while it could.*" Hynes, 185.
p. 141. "*frenetic sexual affairs.*" Himmelfarb, 42.
fathered a child. See chapter 5.
p. 142. "*his knowledge of higher mathematics.*" Minutes, Shakespeare Society, 16 February 1903.
26 percent never married. Ellis, 152. Writes Ellis: "A passionate devotion to intellectual pursuits seems often to be associated with a lack of passion in the ordinary relationships of life, while excessive shyness really betrays also a feebleness of the emotional impulse."
Cranleigh teachers were men. Megahey, 38–39.
Winchester was the same way. Writes Bishop, 28: "The English boarding school was not the only system in history to treat women and the family as potentially subversive elements: it has even been argued that virtually every training establishment which has exalted direct service to the community has

guarded itself against the feminine threat. Like the ancient Spartans, the Turkish *janissary* corps and the Chinese commune, the British public school has tacitly recognized that women 'are incapable of putting the interests of any outside body above the interests of those they love.' " The quote is by John Pringle, "The British Commune," *Encounter*, London, February 1961.

p. 142. "*society of bachelors.*" Quoted in Fowler and Fowler, 233.
"*Get you to Girton.*" Photo, Cambridge Folk Museum.

p. 143. "*string little sentences together.*" Barham, 17.
selected for their plainness. Ibid., 1.
"*perfectly happy marriage.*" Quoted in Deacon, 58.
"*the ascetic ideal is adopted.*" Ellis and Symonds, 40.

p. 144. "*scattered through his life.*" Snow, *Apology* foreword, 26.
"*many forms of contact with life very painful.*" Buxton and Williams, 117.

p. 145. *hoping to lure Americans. Biographical Memoirs of Fellows of the Royal Society* 19 (December 1983): fn, 495.
six hours a week of lectures. Titchmarsh, 448.
"*but very little of importance.*" Hardy, *Apology*, 147.
Hardy F.R.S. Royal Society records.

p. 146. The article objecting to Mendel. *Proceedings of the Royal Society of Medicine.* 1:165.
Hardy's reply. Hardy, *Collected Papers*, 477. Hardy-Weinberg Law. See entry for Hardy, by Victor Cassidy, in *Thinkers of the Twentieth Century.*

p. 147. *a "manifesto of liberation."* Himmelfarb, 35.
"*never done anything 'useful.'*" Hardy, *Apology*, 150.
"*Hardyism.*" Davis and Hersh, 87.
"*The 'real' mathematics.*" Hardy, *Apology*, 119.

p. 148. "*they do not fit the facts.*" Ibid., 135.
"*a long garden party on a golden afternoon.*" Hynes, 4.
"*the greatest disaster.*" Campbell and Higgins, vol. 1, 96.

p. 149. "*The Great Sulk.*" Ibid.
"*why it should do so.*" R. Clark, 43.

p. 150. "*Corky.*" Hardy, *Collected Papers*, 829. Editor's Note.
"*instead of getting on with the real job.*" Littlewood, "Mathematical Life and Teaching," 15.
"*marooned in its limitations.*" Campbell and Higgins, vol. 1, 85.
"*Oh, we never read anything.*" Bell, 433.
"*with the splendour of a revelation.*" E. H. Neville, obituary of Andrew Russell Forsyth. *Journal of the London Mathematical Society* 17 (1942), 245.

p. 151. "*not very good at delta and epsilon.*" Quoted in Campbell and Higgins, vol. 1, 100.
"*nothing else in the world.*" Titchmarsh, 451.
"*never heard the equal.*" Wiener, *Ex-Prodigy*, 190.
"*but you will regret it.*" Barnes, 35.
Hardy sometimes diverted to mathematics those ill-equipped for it. See, for example, *Manchester Guardian*, 2 December 1947.
"*the only profession.*" Hardy, *Apology*, 150.
"*some of the most perfect English of his time.*" Snow, "The Mathematician," 68.

p. 152. Hardy's early style "vulgar." Bollobás, *Littlewood's Miscellany*, 118.
"*light with grace, order, a sense of style*." Snow, *Apology* foreword, 49.
Hardy wrote about Russell. "The New New Realism," *Cambridge Magazine*, 11 and 18 May 1912.
Hardy's obituaries. See the *Collected Papers*.
wrote up the joint paper. Cited in Bateman and Diamond, 31. But in an interview, Mary Cartwright reports that Littlewood wrote up their very last one.
"*devices to stimulate the imagination*." Hardy, *Collected Papers*, 598.
"*which is to my mind unequalled*." Quoted in H. T. H. Piaggio, "Three Sadleirian Professors," *The Mathematical Gazette* 15 (1931): 464.
Thought impossible without words. Hardy, *Collected Papers*, 837.
Act of writing gave him pleasure. Bollobás, *Littlewood's Miscellany*, 118.

p. 153. Hardy on Johnson's book. Hardy, *Collected Papers*, 819.
"*of far greater intrinsic difficulty*." Hardy, *Pure Mathematics*, vi.

p. 154. "*look and behave very much like a line*." Ibid., 4.
"*little better than nonsense*." Arthur Berry, *The Mathematical Gazette* 5 (July 1910): 304.

p. 155. "*to reform the Tripos*." Hardy, *Collected Papers*, 537.
"*so long as its standard is low*." Ibid., 528.
"*into a marginal first*." Ibid., 537.
"*Our English way*." H. B. Heywood, "The Reform of University Mathematics," *The Mathematical Gazette* 12 (1925): 323.

p. 156. "*high-souled frustration*." Littlewood, *Miscellany*, 71.
"*an acrobat*." C. J. Hamson, [untitled], *Trinity Review* (1986): 23.
"*he was so kind to the weak ones*." A variant of the story appears in M. L. Cartwright, "Some Hardy-Littlewood Manuscripts," *Bulletin of the London Mathematical Society* 13 (1981), 294.
"*in the same generous terms*." Titchmarsh, 454.

p. 157. Hardy would discuss Bergson. R. Clark, 169.
an undergraduate. Wiener, *Ex-Prodigy*, 183.
"*truths and values*." Woolf, *Beginning Again*, 20.
Hardy's routine. Snow, *Apology* foreword, 31.

CHAPTER FIVE

p. 159. *covered with Indian stamps*. Snow, *Apology* foreword, 30. Unless otherwise noted, most of the account describing the receipt of Ramanujan's letter is drawn from Snow, 30–38.
"*I beg to introduce myself*." Ramanujan, *Collected Papers*, xxiii.

p. 160. "*enclosed papers*." Ramanujan's letters are in the Cambridge University Library. Most of them appear in Ramanujan, *Collected Papers*, xxiii–xxx and 349–355.

p. 161. *late January 1913*. Snow, *Apology* foreword, 31, gives the time as simply "January." Ramanujan wrote him on January 16. Time for mail to reach England was usually less than two weeks. A day between January 26 and January 31 seems most likely.

p. 162. *a practical joke*. Snow, 30–31, speaks of Hardy getting manuscripts from "cranks," viewing Ramanujan's as possibly a "fraud." Neville, in the Reading Manuscript, expands on this theme considerably.

p. 163. "*some curious specialization of a constant.*" Hardy, "Obituary," 494.
"*prodigy.*" Littlewood, *Miscellany*, 66.
The books Littlewood studied. Ibid.

p. 164. "*a rough-hewn earthy person.*" Bateman and Diamond, 33.
"*unfortunately very saintly.*" Littlewood, "Reminiscences."
long-term relationship with a married woman. The relationship was one of many decades' standing. The woman was the wife of Dr. Streatfeild, Bertrand Russell's physician. Littlewood's daughter is Ann Johannsen. Bollobás, *Littlewood's Miscellany*, 18. Interviews with Béla Bollobás, Charles Burkill.
"*storm and smash a really deep and formidable problem.*" Hardy, quoted in Burkill, 61.
it went to Hardy as referee. Littlewood, *Miscellany*, 78.
turned out to be flawed. Bateman and Diamond, 29.

p. 165. "*only three really great English mathematicians.*" This has become a commonplace in mathematical circles.
recently moved into rooms on D Staircase. Trinity College records.
probably in Littlewood's rooms. Snow, 33, says it was Hardy's rooms. But Bollobás, in "Ramanujan—A Glimpse," 76, says it was Littlewood's. Mary Cartwright, who was Littlewood's student and collaborator, pictures Littlewood as invariably expecting people to come to see him. She deems it unlikely that they met in Hardy's rooms.
"*seemed scarcely possible to believe.*" Hardy, "Obituary," 494.
"*I should like you to begin.*" Hardy's response to Ramanujan's letter over the next few pages is largely drawn from Hardy, *Ramanujan*, 7–10.

p. 169. "*curious and entertaining.*" Hardy, in Ramanujan, *Collected Papers*, xxv.
"*theorems sent without demonstration.*" Neville, "Ramanujan" (*Nature* 149), 293.
"*a man of altogether exceptional originality and power.*" Hardy, *Nature* (1920): 494–495.
before midnight. This is the canonical version, according to Snow. But in an interview, Béla Bollobás emphasizes that Ramanujan's formulas were complicated, his results astounding; that even Hardy and Littlewood may have needed more than an evening to evaluate them; that, while true to the spirit of the events, Snow may conceivably have collapsed several evenings' work into one.
"*really an easy matter.*" Mordell, "Ramanujan," 643.

p. 170. "*another of those monstrosities.*" Bell, 314.
"*a distinguished and conspicuous figure.*" Hardy, *Collected Papers*, 751.
suggesting great imagination or flair. H. T. H. Piaggio, "Three Sadleirian Professors," *The Mathematical Gazette* 15 (1931): 463.
"*An old stick-in-the-mud.*" Interview, Mary Cartwright.

p. 171. *new and original—but not better.* I owe to inventor Jacob Rabinow this insight into the natural and normal human resistance to the new.
"*Indian Bazaar.*" *Cranleigh School Magazine*, May 1887, 112.
"*I was a depressed class.*" Interview, Mary Cartwright.

p. 172. "*the large bottomed.*" Snow, *Apology* foreword, 42.

p. 172. "*for the downfall of their opposites.*" Snow, "The Mathematician," 70.
"*forget the sensation.*" Neville, "Ramanujan" (*Nature* 149), 293.
Hardy had sprung into action. Snow, *Apology* foreword, 34, says it was the next day. In any case, Hardy got in touch with the India Office before he even wrote Ramanujan.

pp. 172–173. "*I was exceedingly interested by your letter.*" Letter, Hardy to Ramanujan, 8 February 1913.

p. 174. *Later in the month, Davies met with Ramanujan.* Letter, Ramanujan to Hardy, 22 January 1914.
On February 25. Memo, Sir Francis Spring. S. R. Ranganathan, 29.

p. 175. "*anxiety as to his livelihood.*" P. K. Srinivasan, 55.
Hardy's imprimatur. That Walker's assessment was formed largely by Hardy's letter is not the way the story usually comes down to us. Hardy (*Collected Papers*, 703) says Walker "was far too good a mathematician not to recognize [the quality of Ramanujan's work], little as it had in common with his own. He brought Ramanujan's case to the notice of the Government." But the contrary conclusion, that Walker's recommendation came only in the wake of Hardy's letter, seems unavoidable. For one thing, in Spring's memo (Suresh Ram, 27–28), he notes that Walker "disclaimed ability to judge Mr. Ramanujan's work and said that Mr. Hardy of Trinity College, Cambridge, was in his opinion the most competent person to arrive at a judgement of the true value of the work. Mr. Ramanujan had already been in correspondence with Mr. Hardy, a letter of whose dated 8 February 1913—just 17 days before Dr. Walker's visit here—is in this file." Lining up Walker behind Ramanujan smacks of an orchestration of Madras opinion by Spring and Narayana Iyer.

p. 176. "*who views my labours sympathetically.*" Letter, Ramanujan to Hardy, 27 February 1913. Cambridge University Library.
fortified by Hardy's letter. Hardy's letter to the Secretary for Indian Students had also reached Madras. See S. R. Ranganathan, 34.
"*invincible originality.*" Hardy, *Collected Papers*, 720.

p. 177. Ramanujan's second letter. 27 February 1913.
"*perfect in manners, simple in manner.*" Neville, "The Late" (*Nature* 106), 662.

p. 178. "*his name would live for one hundred years.*" *Letters from an Indian Clerk.*
"*not done by 'humble' men.*" Hardy, *Apology*, 66.
"*what we can do for S. Ramanujan.*" Suresh Ram, 29.
syndicate's action. See Suresh Ram, 28–31; S. R. Ranganathan, 30–31; letter, K. Venkatachaliengar to Bruce Berndt, supplied by Berndt.
"*with all their vehement speeches.*" Letter, K. Venkatachaliengar to Bruce Berndt.

p. 179. *By April 12, Ramanujan had learned the good news.* Letter, Ramanujan to Registrar, University of Madras, 12 April 1913. Madras Port Trust.
"*a vice in Kumbakonam.*" Neville, "Ramanujan" (*Nature* 149), 293.
a corruption of Tiru Alli Keni. Krishnaswami Nayadu, 30. The description of the Triplicane tank and adjacent temple is drawn from personal observation and Das, 245.
Pallava king's gift of land. Krishnaswami Nayadu, 30.

p. 179. *moved back to Triplicane.* Family Record.

p. 180. *threshold for paying full dues. Journal of the Indian Mathematical Society* (February 1917), 19.
work with Narayana Iyer. S. R. Ranganathan, 31.
math books from K. B. Madhava. P. K. Srinivasan, 132.
Ramanujan at the Connemara. S. R. Ranganathan, 31.
Serve food in his hand. Ibid., 90.
his mother made it for him. Interview, Janaki.

p. 181. *a little science experiment.* Interview, Janaki. Ashis Nandy, 131, speaks of Ramanujan "teaching her the elements of science," but places this after Ramanujan's return from England, not before.
"*the one you were working on before eating.*" Interview, Janaki. Nandy pictures Ramanujan as "using her as a secretary," again referring to the post-England period.
She didn't ask him to, and he didn't volunteer. Interview, Janaki.
"*How maddening his letter is...*" Letter, Littlewood to Hardy, in Ramanujan, *The Lost Notebook*, 383.
"*your obvious mathematical gifts.*" P. K. Srinivasan, 56.

p. 182. "*what little I have.*" Letter, Ramanujan to Hardy, 17 April 1913. Cambridge University Library.
August get-together. S. R. Ranganathan, 12.
submitted some theorems. Journal of the Indian Mathematical Society 5, no. 5: 185.
perhaps out to win over . . . Littlehailes. Letter, K. Venkatachaliengar to Bruce Berndt.
he looked forward to studying Ramanujan's results. Memo, Narayana Iyer, 31 October 1913. Madras Port Trust.
"*Does Ramanujan know Polish?!?*" S. R. Ranganathan, 12.
Quarterly Reports. See Berndt, *Ramanujan's Notebooks*, part I, 295; and Berndt, "Quarterly Reports."

p. 183. "*to discover new pathways through the forest.*" Berndt, "Quarterly Reports," 516.

p. 184. " '*new and interesting.*' " Berndt, *Ramanujan's Notebooks*, part I, 297.
"*the result is wrong.*" P. K. Srinivasan, 57.
"*no human agency.*" Snow, *Apology* foreword, 34.
Ramanujan's version. Letter, Ramanujan to Hardy, 22 January 1914. Cambridge University Library.

p. 185. "*a dead man in their estimation.*" T. Ramakrishna.
"*disregard the orders of the caste?*" Gandhi, 37.

p. 186. *with men from the villages bringing their wives and children.* J. Chartres Molony, *A Book of South India* (London: Methuen & Co., 1926). Cited in Lewandowski, 53.
the area around Triplicane's Parthasarathy Temple. A detailed map in Lanchester, unpaged, shows residential patterns in Madras.
"*not a man like Littlewood.*" Letter, Hardy to Mittag-Leffler, about 1920. Robert Rankin, Glasgow.

p. 187. Senate House. Muthiah, 106; personal observation.
"*a man at once diffident and eager.*" Neville, Reading Manuscript.
"*opposition . . . withdrawn.*" Neville, "Ramanujan" (*Nature* 149), 293.

p. 188. Influences on Ramanujan to go to England. S. R. Ranganathan, 36.
"weight of my influence." Ramachandra Rao, 88.
"the fulfillment of his life's purpose." Neville, "Ramanujan" (*Nature* 149), 293.

p. 189. *in late December of 1913.* Family Record.
Ramanujan wrote back home. Family Record..
three nights. P. K. Srinivasan, 114; Bharathi, 48.

p. 190. *"I locked myself up in the attic."* Quoted by C. T. Rajagopal, *Mathematics Teacher (India)* 11A (1975), 119.
"within a very few months." Letter, Ramanujan to Hardy, 22 January 1914.
he had come to say good-bye. P. K. Srinivasan, 100.

p. 191. *Ramanujan's other doubts.* Letter, Ramanujan to Hardy, 22 January 1914; and Neville, "Ramanujan" (*Nature* 149), 293.
Father-in-law's concerns. K. R. Rajagopalan, 33.
Mother's concerns. Ibid., 32–33.
actually shook his hand. K. R. Rajagopalan, 34.
"the glory that belonged to Madras." Neville, Reading Manuscript.
"where they longed to see him." Neville, Reading Manuscript.
Neville wrote Hardy. Neville, "Ramanujan" (*Nature* 149), 294.
a worried reply. Letter, Mallet to Hardy, 11 February 1914. The Library, University of Reading.

p. 192. *"I'm writing in a hurry."* Letter, Hardy to Neville. The Library, University of Reading.
"unknown geniuses." Neville, "Ramanujan" (*Nature* 149), 294.
"the most interesting event." P. K. Srinivasan, 59.

p. 193. *"hidden under a bushel in Madras."* Ibid., 61. If Venkatachaliengar's account of the syndicate meeting is accurate, Littlehailes had by now changed his tune.
"if not transcendental, order of genius." Ibid., 64.
"by no means the worst brain." Quoted in Slater, *Southern India*, 262.
"for the full development of personality." Ibid., 263.
"His Excellency cordially sympathizes." S. R. Ranganathan, 35.

p. 194. *On March 11.* S. R. Ranganathan, 37.
On March 14. Interview, Janaki. "Three days before he left . . ."
Now, he wept. Interview, Janaki.
while her mother-in-law was at the temple. Kalyanalakshmi Bhanumurthy, "The Man Behind the Mathematics," *The Hindu*, 20 December 1987.
she was so young and pretty. K. R. Rajagopalan, 35.
"as if . . . obeying a call." Ramachandra Rao, 88.
driving him around town on his motorcycle. S. R. Ranganathan, 36.
"nothing but vegetable food." Ramachandra Rao, 88.

p. 195. *How, he pleaded, was he to wear them?* P. K. Srinivasan, 134; S. R. Ranganathan, 62.
she wouldn't recognize him. K. R. Rajagopalan, 35.
his friends stayed up with him. S. R. Ranganathan, 71.
a special wharf. The Port of Madras: Past, Present and Future, 8. Ramanujan's friend K. Narasimha Iyengar reports, in S. R. Ranganathan, 70, that Spring prevailed on Ramanujan and Neville to leave from Madras, not the more usual Bombay.

p. 195. The *Nevasa*. Details drawn from (British) National Maritime Museum Information File; *BI Centenary*, 1856–1956, by George Blake (London: Collins, 1956); "The Notable 'Nevasa,' " by J. H. Isherwood, in *Sea Breezes*, new series 21 (1956): 28–31; "End of the 'Nevasa,' " *Sea Breezes*, new series 5 (1948): 148–149. Ramanujan's send-off. P. K. Srinivasan, 91–92.

p. 196. "*he may get an inspiration.*" Bharathi, 49.

CHAPTER SIX

p. 197. *seasick and unable to eat*. P. K. Srinivasan, 3.

On March 19. The picture of Ramanujan's voyage to England is drawn from Family Record; details about the *Nevasa* (see chapter 5); Ramanujan's letters in P. K. Srinivasan; S. R. Ranganathan, 38; K. R. Rajagopalan, 36; *P & O Pocket Book* (London: Adam and Charles Black, 1908); *Lloyds Weekly Shipping Index* for the period of the voyage.

withdrew to his . . . cabin. K. R. Rajagopalan, 36.

"*his own history or psychology*." Hardy, *Ramanujan*, 11.

p. 198. *four-hundred-rupee fare*. The second-class fare between London and Bombay was 32 pounds in 1914, or about 480 rupees. *British Passenger Liners of the Five Oceans*, by C. R. Vernon Gibbs (London: Putnam, 1963), 63.

orthodox Brahmins among them. S. R. Ranganathan, 36.

p. 199. *POSH. A Hundred Year History of the P & O, 1837–1937*, by Ernest Andrew Ewart (London: I. Nicholson and Watson, 1937), 112—113.

he posted at least four letters back to India. Family Record.

stamps showing the pyramids. P. K. Srinivasan, 92.

"*warmest thanks to your uncle*." Ibid., 3.

Reception by Neville. Letter, Raymond Neville to Robert A. Rankin, 15 May 1982.

London. See, for example, Geoffrey Marcus, 47.

p. 200. "*a tempo clearly faster*." Woolf, *Beginning Again*, 16.

Cromwell Road. *Report of the [Lytton] Committee on Indian Students*.

met A. S. Ramalingam. Rankin, "Ramanujan as a Patient," 84.

Chestertown Road. Information on the street, and specifically the house at 113, is drawn from an interview with Constance Willis, the current owner, and tour of her house; photos from the Cambridgeshire Collection; and records furnished through J. D. Webb, Building Control Officer, Cambridge City Council.

for two months. In his dedication to Ramanujan of *The Farey Series of Order 1025* (Cambridge: Cambridge University Press, 1950), xxvii, Neville remembers it as three months. But this conflicts with Ramanujan's letter to Krishna Rao of 11 June 1914, in P. K. Srinivasan, 7. There, he says he's already moved into the college (after less than two months in Cambridge).

p. 201. *Hardy and Neville took care of most of it*. P. K. Srinivasan, 7.

dipped his pen in black ink. Terms Book, 1899–1915. Trinity College.

graced by a wondrous spring. Marcus, 112.

King George visited Cambridge. *Cambridge Independent Press*, 1 May 1914.

Littlewood . . . saw him about once a week. P. K. Srinivasan, 154.

p. 201. "*very unassuming, kind and obliging.*" Ibid., 7.
Berry stood at the blackboard. P. K. Srinivasan, 145.
p. 202. "*even this was pronounced wrongly.*" Quoted in K. G. Ramanathan, "Ramanujan: A Life Sketch," *Science Age,* December 1987, 22.
he "waddled" across Trinity's Great Court. Lecture notes, B. M. Wilson. Trinity College. Wilson later crossed out "waddled," perhaps because he thought the word undignified, but it is clearly visible in the original manuscript.
"*inconvenient for the professors.*" P. K. Srinivasan, 7.
sad to leave. Interview, Mary Cartwright.
consummate hosts. T. A. A. Broadbent, "Eric Harold Neville," *Journal of the London Mathematical Society* 37 (1962): 482.
p. 203. *nearly every day.* Hardy, *Ramanujan,* 11.
perhaps a third. Ibid., 10.
p. 204. "*a mass of unpublished material.*" Hardy, *Collected Papers,* 718.
"*a really searching analysis.*" Hardy, "A Chapter from Ramanujan's Notebook," *Proceedings of the Cambridge Philosophical Society* 21 (1923): 503.
Polya. Berndt, *Ramanujan's Notebooks,* part I, 14.
"*not a light one.*" G. N. Watson, "Ramanujan's Notebooks," *Journal of the London Mathematical Society* 6 (1931): 140.
a month to prove. Ibid., 150.
p. 205. "*at least a Jacobi.*" Letter, Littlewood to Hardy, probably early March, 1913. Trinity College.
"*Euler or Jacobi.*" Hardy, *Collected Papers,* 720.
p. 206. "*greatest formalist of his time.*" Hardy, *Ramanujan,* 14.
p. 207. "*delighted surprise.*" Littlewood, *Miscellany,* 87.
"*never heard of most of it.*" Snow, *Apology* foreword, 36.
"*without a rival.*" Hardy, *Collected Papers,* 720.
"*what a treasure I had found.*" Hardy, *Ramanujan,* 1.
"*I edited them very carefully.*" Hardy, *Collected Papers,* 717.
pp. 207–208. *the 2:15 P.M. train out of Cambridge.* Hardy, *Collected Papers,* 751.
p. 208. London Mathematical Society meeting. Minutes.
within the radius of a hydrogen atom. Jonathan M. Borwein and Peter B. Borwein, *Scientific American,* February 1988, 112.
p. 209. "*having no other business at the time.*" Ibid., 114.
p. 210. "*new results.*" Hardy, *Collected Papers,* 494.
p. 211. "*an exciting foxhunt.*" *Cambridge Daily News,* 18 April 1914.
Caesar died. *Cambridge Daily News,* 19 April 1914.
"*But he was a happy man.*" Neville, "Ramanujan" (*Nature* 149), 294.
p. 212. "*no war in this country.*" P. K. Srinivasan, 168.
muffled the echoes. First Eastern General Hospital Gazette, 6 July 1915, 135.
p. 213. *strung from the ceiling.* Ibid., 22 June 1915, 100.
ambulance with a great red cross. Ibid., 97.
would see them on his way to Hall. Parry, 200.
"*We have a new Cambridge.*" Ibid., 95.
inches-deep mud on unasphalted roads. Keynes, 182.
On September 20. Butler, 200.
"*The depravation of Germany.*" Parry, 96.

p. 213. "*driven in front of the enemy.*" Butler, 200.
"*Germans set fire.*" P. K. Srinivasan, 168.
p. 214. "*Now here's a problem for you.*" S. R. Ranganathan, 81. Mahalanobis didn't cite the December 1914 issue, but the content of the problem, and the issue in which it appears, square perfectly with his recollection.
p. 215. "*The limitations of his knowledge.*" Hardy, *Collected Papers*, 714.
p. 216. "*found a function which exactly represents.*" Ramanujan, *Collected Papers*, 349.
"*The stuff about primes is wrong.*" Letter, Littlewood to Hardy, probably early March 1913, in Ramanujan, *The Lost Notebook*, 380.
p. 217. "*Euclid's theorem assures us.*" Hardy, *Apology*, 99.
p. 218. De la Vallée-Poussin. A professor at the University of Louvain; many of his works were lost when the library was set to the torch in the first month of the war. *Journal of the Indian Mathematical Society* 10: 27.
p. 219. "*Proofs will be supplied later.*" *Journal of the Indian Mathematical Society* 5 (1913): 50–51.
came to see where he had stumbled. See Hardy, *Ramanujan*, chapters 1 and 2.
"*no complex zeroes.*" Hardy, *Collected Papers*, 706.
Riemann hypothesis. For a nice introduction, accessible to the layperson, see Campbell and Higgins, vol. 2, 149–153.
p. 220. *Has the Riemann hypothesis been proved?* This is one of many bits of Riemann hypothesis lore.
"*the Analytic Theory of Numbers is not one of them.*" Hardy, *Collected Papers*, 706.
"*his achievement . . . is most extraordinary.*" Littlewood, *Miscellany*, 87.
"*more wonderful than any of his triumphs.*" Hardy, *Collected Papers*, 706.
"*His instincts misled him.*" Hardy, *Ramanujan*, 38.
p. 221. "*as the climax of a conventional pattern of propositions.*" Ibid., 16.
Take the sequence of integers. These illustrative examples are drawn from Ivars Peterson, "A Shortage of Small Numbers," *Science News*, 9 January 1988.
One Hardy liked to cite. Hardy, *Ramanujan*, 16.
"*the largest number which has ever served any definite purpose in mathematics.*" Hardy, *Ramanujan*, 17.
p. 222. The 2 = 1 "proof." A high school classic.
p. 223. *slide into trouble in numerous ways.* Examples due to mathematicians Clare Friedman, Richard Askey, and Bruce Berndt.
"*He disregarded entirely all the difficulties.*" Hardy, *Collected Papers*, 706.
p. 224. "*of what is meant by a proof.*" Littlewood, *Miscellany*, 88.
"*very much mitigated.*" R. Clark, 176.
"*hardened to some extent.*" Hardy, *Ramanujan*, 10.
"*I have changed my plan.*" P. K. Srinivasan, 15.
"*My notebook is sleeping in a corner.*" Ibid., 21.
p. 225. "*no symptom of abatement.*" Hardy, *Collected Papers*, 715.
"*a very curious function.*" P. K. Srinivasan, 23.
"*he sets to work to manufacture a proof.*" Hardy, *Ramanujan*, 16.
p. 226. "*first-rate importance.*" Littlewood, *Miscellany*, 88.
"*Mathematics has been advanced.*" Klein, in Mordell, "Ramanujan," 647.
an informal scale. Berndt, *Ramanujan's Notebooks* I, 14.
"*break the spell of his inspiration.*" Hardy, *Collected Papers*, 715.

p. 227. *"excerpts from a more interesting lecture."* Young, 282.
"At Cambridge we are in darkness." Parry, 98.
"In France and Flanders we make no progress." Ibid.
Cambridge schoolgirl. Keynes, 62.
"the front was like a first-rate club." Parry, 96.

p. 228. *"cheerful indifference."* Williams, 15.
"Even Littlewood." Hardy, *Apology*, 140.
"did not fill all of Littlewood's working hours." Burkill, 63.
Hardy deemed "unfit" to serve. On what basis I have not learned.
at least one obituary. Norbert Wiener, "Godfrey Harold Hardy," *Bulletin of the American Mathematical Society* 55: 72–77. The record was set straight in a subsequent letter to the *Bulletin* from Littlewood and others.
"I don't like conscientious objectors as a class." Letter, Hardy to Jenkinson, 15 June 1918. Cambridge University.
Hardy "wrote passionately." Littlewood, "Reminiscences," 13.
Its first public meeting. Hardy, *Russell*, 12.
the authorities moved to block it. Ibid., 19.

p. 229. *risen 32 percent.* Marwick, 125.
three or four times a month. Family Record, 55.
a parcel full of books. P. K. Srinivasan, 162.
preoccupied . . . with seeing his work in print. The letters are all in P. K. Srinivasan, 7, 11, 13, and 32, respectively.

p. 230. *he went straight to Hardy.* S. R. Ranganathan, 77.
the following year. P. K. Srinivasan, 28.
"no help nor references in Madras." Ibid., 29.
students sometimes taunted him. S. R. Ranganathan, 71.
"a thrill to me to discover." Ibid., 78.
Ramanujan at tea parties. Ibid., 76.
reserved in large groups. Ibid., 83.
Ramanujan during vacations. Ibid., 76.

p. 231. *Charley's Aunt.* Suresh Ram, 43.
By mid-October. Trinity College records.
Highly composite numbers paper. See Ramanujan, *Collected Papers*, 78–128.
"I gave up the struggle earlier." W. N. Bailey, quoted in George E. Andrews, *Q-Series.*

p. 232. *"of an elementary but highly ingenious character."* Hardy, *Collected Papers*, 497.
"extraordinary insight and ingenuity." Ibid., 499.

p. 233. *perhaps the most brilliant.* Barnes, 33.
"entirely justifying the hopes." P. K. Srinivasan, 66.
"the most remarkable mathematician I have ever known." Seshu Iyer and Ramachandra Rao, xvii.
Spring wrote Dewsbury. This and other correspondence concerning Ramanujan's possible return to India is in Madras Port Trust.
about 75 pounds per year. Marwick, 23.
threshold for paying income taxes. Ibid., 21.

p. 234. *"almost ludicrously simple tastes."* Hardy, *Collected Papers*, 714.

p. 234. Ramanujan, B.A. Rankin, "Ramanujan's Manuscripts and Notebooks, II," 364.
Ramanujan didn't return to India. But there is a third reason, beyond the two given. See chapter 7.
"*most unfortunate.*" Hardy, *Collected Papers,* 491.
p. 235. *four thousand casualties per day.* Marwick, 133.

CHAPTER SEVEN

p. 237. Chatterji dinner. Suresh Ram, 40–42. Details about Chatterji and the other guests come from *Who's Who in India, 1937–38.* Chatterji was apparently still in India, at Punjab University, until late 1915, and he was married in 1916, leading to my guess on the date of the dinner.
Kasturirangar's appraisal. S. R. Ranganathan, 90.
p. 239. "*instinctive perfection of manners.*" Neville, "Ramanujan" (*Nature* 149), 294.
the tablecloths of peacetime were gone now. Told to me at High Table.
"*one does not want to* work *at conversation.*" Littlewood, "Reminiscences."
p. 240. *little book kept over the years. Trinity College Kitchens: Suggestions and Complaints, 1909–1955.* Trinity College.
unusually strict in his orthodoxy. S. Chandrasekhar has estimated that even back in India, Ramanujan's family would have been in the top quarter of some imaginary Orthodoxy Scale. Those traveling to England were typically far less orthodox.
Traveler's story. Interview, T. M. Srinivasan.
p. 241. *Fried in lard.* Suresh Ram, 42. See also S. R. Ranganathan, 78.
a little gas stove. Ibid., 76.
stirring vegetables over the fire. See also Snow, *Apology* foreword, 35: "Hardy used to find him ritually changed in his pyjamas, cooking vegetables rather miserably in a frying pan in his own room."
"*think of my home and country.*" Gandhi, 40.
about twenty Indians per year. These figures are derived from the "government studies" cited. Fuller, 186, gives a figure of about seventeen hundred for the whole United Kingdom.
p. 242. "*The initial difficulty.*" Amar Kumar Singh, *Indian Students in England.*
A still later student. Interview, Rajiv Krishnan, Christ's College, Cambridge.
dream of dying. Ibid.
"*the reticence of his English friends.*" Neville, Reading Manuscript.
pp. 242–243. "*detachment almost amounting to indifference.*" Quoted in Barbara Tuchman, *The Proud Tower* (New York: Macmillan, 1966), 66.
p. 243. "*without the slightest suggestion of discourtesy.*" Chaudhuri, 90.
"*Oh, were we introduced?*" Interview, S. Sankara Narayanan. A similar story is told by Sud in *How to Become a Barrister and Take a Degree at Oxford or Cambridge.*
"*an air of indifference.*" Satthianadhan, 28.
"*that they were not greatly interested.*" Barnes, 35.
"*his face dropped.*" Young, 271.
"*no one will blink an eyelid.*" Bollobás, *Littlewood's Miscellany,* 149.

p. 244. *ritually pure dhoti.* Interview, T. V. Rangaswami.
could not long forget his foreignness. See also S. R. Ranganathan, 47.
"*you bathe only* once *a day?*" David E. Fisher, *Fire & Ice* (New York: Harper & Row, 1990), 46. Told to Fisher, he reports, by a Polish mathematician, at Cornell, in the 1960s.

p. 245. "*very rarely seen.*" B. M. Wilson. Unpublished notes. Trinity College.
in the cool of the night. Suresh Ram, 40.
"*a serious psychological effect.*" *Colonial Students in Britain,* 94.
just 17 percent did while in England. Amar Kumar Singh, *Indian Students in Britain.*
Trinity Sunday Essay Society. Trinity College.
Ramanujan not in Majlis. Suresh Ram, 44.
"*their nationalism and radicalism.*" Nandy, 125.

p. 246. "*wrapped in a medieval gloom.*" W. E. Heitland, *After Many Years,* quoted in *Cambridge Commemorated,* ed. Laurence and Helen Fowler (Cambridge University Press, 1984), 283–284.
"*greatest power and skill to resolve.*" Norbert Wiener, *Bulletin of American Mathematical Society* 55 (1949): 77.

p. 248. "*A moment's consideration.*" Hardy, *Lectures by Godfrey H. Hardy on the Mathematical Work of Ramanujan* [Notes by Marshall Hall] (Ann Arbor, Michigan, 1937), 25.
"*unable to discover.*" Ramanujan, *Collected Papers,* 277.

p. 249. "*as Ramanujan imagined.*" Hardy, *Ramanujan,* 9.
"*an extremely obvious one.*" Hardy, *Quatrième Congrès des Mathématiciens Scandinaves,* 46.

p. 250. "*hardly the type to be chosen by Central Casting.*" Gian-Carlo Rota, Introduction to *Collected Papers of P. A. MacMahon,* ed. George Andrews, xiii.
and regularly thrash him. Ibid., xiv.

p. 251. "*Then the problem is completely solved.*" Bollobás, *Littlewood's Miscellany,* 98.
"*a very great step.*" Littlewood, *Miscellany,* 89.
"*or even bounded.*" Ramanujan, *Collected Papers,* 283.
"*we proceeded to test this hypothesis.*" Ibid.

p. 252. Selberg's argument. Atle Selberg, "Ramanujan and I," *Science Age,* February 1988, 39.
two men. Hindu, 22 December 1987.

p. 253. "*not the only mathematician.*" *Letters from an Indian Clerk.*
"*a singularly happy collaboration.*" Littlewood, *Miscellany,* 90. See also A. Ostrowski, in *Experentia* 5 (1949): 131–132: "They were very different scientific personalities . . . On the one side, a man who had completely given himself up to calculation, who guessed intuitively and worked out the most complicated and hidden formulae, who overflowed with such discoveries, only to become entangled again and again in the net of analytical difficulties. On the other side, a man who was a supreme master of all the finesses and most subtle arguments of modern analysis, and who was continually enriching them by new inspirations, devices, and subtleties."
"*any man I have ever known.*" Snow, *Variety of Men,* 194.

p. 254. "*His sense of excellence was absolute.*" Lionel Charles Robbins, in Buxton and Williams, 117. See also Philip Snow, C. P. Snow's brother: "Hardy was a stern critic of the irrelevant, the ignorant, the brash." Philip Snow, 44.

p. 254. *"on a high plane."* Interview, Charles Burkill.
Polya's zoo story. George Polya, "Some Mathematicians I Have Known," *American Mathematical Monthly* 76 (1969): 746–753.
"a morning's work gone west." Letter, Hardy to Mordell, undated. St. John's College.

p. 255. *a personality of expectations.* The expectations applied fully to himself: "In Hardy's letters [to Littlewood], 'Please check,' 'Please check very carefully,' seems like a theme song, and when he was wrong he swore, which confirms my impression that he really minded very much when he himself made bad mistakes." Mary Cartwright, "Some Hardy-Littlewood Manuscripts," *Bulletin of the London Mathematical Society* 13 (1981): 273–300.
"some splendid problems to work at." Ramanujan, *The Lost Notebook,* 389.
Oliver Lodge. P. K. Srinivasan, 143.
"the bath rooms are nice and warm." Bollobás, "Ramanujan—A Glimpse," 79.

p. 256. *For thirty hours at a stretch.* Unpublished notes, B. M. Wilson. Trinity College.
"Only cabbages have no nerves." Bell, 329–330.
"Meals were ignored or forgotten." Bell, 109.

p. 257. *"their mental development."* Satthianadhan, 38–39.
"study-worn, consumptive-looking." Ibid.
toward the end of their second year. Lytton Committee, testimony of S. M. Burrows.

p. 258. *"inclined to develop tuberculosis."* Lytton Committee, testimony of S. S. Singara.
"a regime unsuited to . . . conditions." Lytton Committee, testimony of Sir R. W. Philip.
"to cook one or two things myself." P. K. Srinivasan, 5.
powdered rice in tin-lined boxes. Bharathi, 48.

pp. 258–259. Ramanujan's eating habits. Interview, Janaki.

p. 259. *weird hours in the early morning.* S. R. Ranganathan, 39.
"a little salt and lemon juice." P. K. Srinivasan, 27.

p. 260. *" 'infants, Indians and invalids.' "* Basil Willey, *Spots of Time: A Retrospect of the Years 1897–1920* (London: Chatto and Windus, 1965), 118.
on a cricket ground. *Historical Register of the University of Cambridge,* Supplement, 1911–1920 (Cambridge: Cambridge University Press, 1922), 203.
Marching boots. *Cambridge Magazine,* 5 May 1917.
"our pride and sad privilege." Larmor address, 2 November 1916. *Proceedings of the London Mathematical Society,* 2d series, 16 (1917): 6.

p. 261. *"for the most part wasted."* Barnes, p. 35.
Robert Collins. Seymour & Warrington, caption, fig. 41. The death of Graham is recorded in *Madras Times,* 6 April 1919.
"memorial services." Wootton, *In a World I Never Made,* quoted in Fowler and Fowler, 284.
"Things here are sad and sorrowful." Parry, 101.

p. 262. *two-fifths of her meat.* Marwick, 18.
Kipling poem. Cited in Marcus, 41.
one sure way to tell rich from poor. Marwick, 24: "Poor dieting gave many members of the lower orders yet another characteristic which distinguished them from the rest of society: their small stature."

p. 262. "*rosy, well-covered men.*" Russell, 51.
had climbed 65 percent. Marwick, 125.

p. 263. "*At Wrexham.*" Quoted in Marwick, 192.
"*portions strangely dwarfed.*" Parry, 107.
"*good milk and fruits.*" P. K. Srinivasan, 5.
"*overwork, overplay, overworry.*" R. C. Wingfield, *Modern Methods in the Diagnosis and Treatment of Pulmonary Tuberculosis*, 1924. Quoted in Bryder, 166.

p. 264. "*We travelled in convoy.*" Russell, 54–55.
Nursing home on Thompson's Lane. Rankin, "Ramanujan's Manuscripts and Notebooks," 95.
by special dispatch. P. K. Srinivasan, 73.
Hardy . . . nursed him. K. R. Rajagopalan, 41.
"*to take proper care of himself.*" P. K. Srinivasan, 69.
"*Ramanujan was not exemplary.*" S. R. Ranganathan, 74.
"*A difficult patient.*" Seshu Iyer and Ramachandra Rao, xviii.
"*Whenever he started with a doctor.*" Béla Bollobás, in *Nova* 1508, 14. The transcript attributes this statement to S. Chandrasekhar, but is in error.

p. 265. Few doctors could abide Ramanujan. Letter, A. S. Ramalingam to Hardy, in Rankin, "Ramanujan as a Patient," 87.
Probably around October. Ibid., 81.
Chowry-Muthu. I am guided in these assertions by Rankin, Ibid. See also P. K. Srinivasan, 110.
exploratory surgery [for gastric ulcer] *considered.* Rankin, "Ramanujan as a Patient," 87, 92.
Not hydrocele, but cancer. Ibid., 91.
Blood poisoning. P. K. Srinivasan, 76.
Tuberculosis—disease, treatment, and history. See, for example, Bryder, Smith, Dubos, Keers.
one in eight deaths in Britain. Bryder, 1.

p. 266. Nervous system–immune system ties. See, for example, Robert Kanigel, "Where Mind and Body Meet," *Mosaic*, Summer 1986, 52–60.
"*emotional stress associated with separation.*" George W. Comstock et al., "Tuberculosis Morbidity in the U.S. Navy: Its Distribution and Decline," *American Review of Respiratory Disease* 110 (1974): 572–580; interview, George W. Comstock.

p. 267. *wild spasmodic leap.* Bryder, 109–110.
The Danish phenomenon. See Bryder, 109–112.
"*supported by the figures from Denmark.*" Keers, 148.
"*admirable epidemiological detective work.*" Graham A. W. Rook, "The Role of Vitamin D in Tuberculosis," *Annual Review of Respiratory Disease* 138 (1988): 768.

p. 268. Davies paper. *Tubercle* 66 (1985): 301–306.
"*It probably can.*" Rook, 769.
working at night and sleeping by day. Suresh Ram, 40.
Osler. Quoted in Bulstrode, 3.

p. 269. "*this dreadfully cold open room.*" Bollobás, "Ramanujan—A Glimpse," 79.

p. 270. *treatment's rise to popularity*. Smith, 141. Harold Batty Shaw is mentioned in Rankin, "Ramanujan as a Patient," 91.
"*was little influenced by it*." Bryder, 23.
"*towards carrying out the 'open-air' treatment*." Bulstrode, 131.
at least fifty-two British sanatoriums. Keers, 93. But Bryder reports ninety-six "open-air" sanatoriums in England and Wales by 1907.
Mendip Hills was one of them. Bryder, 25.
Matlock, as hydropathic establishment. Rankin, "Ramanujan Manuscripts and Notebooks," 95.
"*bleak and cold*." Dubos, 181.
Architects competed. Bryder, 50.

p. 271. "*the rain blowed through*." Bryder, 53.
"*some quarrel*." P. K. Srinivasan, 71.
"*I haven't got a surname*." Bollobás, "A Glimpse," 79.

p. 272. *slipped into it a brief note*. Interview, Janaki.
Mother-in-law troubles. Interview, S. Sankara Narayanan.
"*It is pitiable for the child-wife*." Compton, 119.

p. 273. "*The tyrannical mother-in-law*." *The Hindu*, 12 May 1899.
who in turn would pass it to him. Carstairs, 45, 66.

p. 274. *more confident and assertive*. Interview, S. Chandrasekhar.
Those close to Janaki. Janaki's litany of abuse based on interviews with T. V. Rangaswami, "Hari" of Madras, Mr. and Mrs. T. S. Bhanumurthy.
out of the question. K. R. Rajagopalan, 41.

p. 275. Ramanujan's father stuck up for Janaki. Interview, Janaki.
Janaki found an excuse to get away. Interview, Janaki.
money for a new sari. K. R. Rajagopalan, 44. Interview, Janaki.
Ramanujan's letters home. Family Record.
he told his friend Chatterji. Suresh Ram, 44.

p. 276. *leaflet advising authors*. Minutes, London Mathematical Society, 14 March 1918.
"*a look of extreme disapproval*." Tresilian Nicholas Remembers," *Trinity Review*, Easter 1969, 13.
"*whose opinions on the war*." Levy, 279.

p. 277. Russell and Trinity. See, for example, Hardy's own *Russell and Trinity*.
"*little book*." *Daily Telegraph*, 30 April 1970.
Hardy kept one mathematician out of the war. Letter, Hardy to Larmor, 28 March 1916. Royal Society. The mathematician was Chapman.
"*perfectly tangible and definite*." Letter, Hardy to Jenkinson, 12 June 1918. Cambridge University.

p. 278. "*very standoffish or possibly shy*." Hodges, 117.
"*multiple layers of reserve*." Ibid., 118.
"*parental solicitude*." Seshu Iyer, 85.
Hardy would choose the former. Hardy, *Apology*, 153.
"*always a little difficult*." Hardy, *Ramanujan*, 1.

p. 279. "*the first to disclaim*." Snow, "The Mathematician," 69.
Ramanujan's tastes in literature and politics. Hardy, *Collected Papers*, 715.
"*I rely for the facts*." Hardy, *Ramanujan*, 2.

p. 279. "*I am to blame.*" Hardy, *Ramanujan*, 11.
p. 280. "*all but supernatural insight.*" Bell, 140.
"*We have no idea.*" Ivars Peterson, "The Formula Man," *Science News*, 25 April 1987, 266.
"*that has barely been drawn.*" Bruce Berndt, "Ramanujan—100 Years Old," 29.
"*a touch of real mystery.*" Littlewood, *Miscellany*, 90.
p. 281. "*An ordinary genius.*" Mark Kac, *Enigmas of Change* (New York: Harper & Row, 1985), xxv.
"*unfold before his eyes.*" S. R. Ranganathan, 87.
"*appeared to him in a dream.*" Ibid., 73.
p. 282. *penchant for interpreting dreams.* P. K. Srinivasan, 97.
a street peddler hawking pills. S. R. Ranganathan, 84.
time to prepare astrological projections. Ibid., 91.
Predicted death. P. K. Srinivasan, 123.
a temple near Trichinopoly. Ibid., 101.
"*space-time-junction point.*" S. R. Ranganathan, 85.
"*the way* maya *works in this world.*" Ibid., 89.
p. 283. "*I do not believe.*" Hardy, *Ramanujan*, 5.
"*equally true.*" Hardy, *Collected Papers*, 715.
"*Ramanujan was no mystic.*" Hardy, *Ramanujan*, 5.
"*I should not trust his insight far.*" Snow, *Apology* foreword, 35.
p. 284. "*thinking vaguely.*" Hardy, *Collected Papers*, 733.
p. 285. "*who gazes at a distant range of mountains.*" Ibid., 598.
p. 286. "*the evidence for all this seems overwhelming.*" For this and what follows, see Ibid., 834–838.
p. 287. "*with some kinds of spiritual beings.*" Hadamard, 40–41.
"*never discarded a particle.*" Bell, 144.
strong mystical bent. Bell, 457.
"*the queerest mixture.*" Bell, 43.
"*by the grace of God.*" "True Genius," by RGK, *Illustrated Weekly of India*, 20 December 1987, 31.
pp. 287–288. "*the Great Architect of the Universe.*" James Jeans, *The Mysterious Universe* (New York: Macmillan), 1930.
p. 288. "*If we may reject divine bounty.*" Bollobás, *Littlewood's Miscellany*, 146.
"*no exception.*" Hardy, *Collected Papers*, 719.
"*Hardy's deep reverence for mathematics.*" E. A. Milne, obituary of Hardy in *Monthly Notices of the Royal Astronomical Society* 108 (1948): 45.
"*if a strict Brahmin.*" Hardy, *Ramanujan*, 4.
"*a quite harmless . . . economy of truth.*" Ibid., 5.
p. 289. "*atheist evangelical.*" Hodges, 118.
"*I never got the feeling they were happy.*" *New York Times*, January 20, 1988.
p. 290. *racism was a factor.* See page 299.
"*cowed down by Dr. Ram.*" Rankin, "Ramanujan as a Patient," 86.
"*make visits difficult.*" Bryder, 200. War-borne slashes in rail services around January 1917 must have left Ramanujan all the more isolated (Marwick, 195).
"*cold weary journey.*" Rankin, "Ramanujan as a Patient," 85.

p. 290. "*comparative solitude.*" P. K. Srinivasan, 76.
he craved the hot dosai. Ibid., 96.

p. 291. "*singularities.*" *Letters from an Indian Clerk.*
On December 6, 1917. Minutes, London Mathematical Society.
Royal Society. *Year-book of the Royal Society of London, 1919* (London: Harrison and Sons, 1919).
"*one honor he longed for.*" Snow, *Variety of Men,* 84–85.

p. 292. *On January 24, 1918. Year-book of the Royal Society 1919,* 212.
"*no time must be lost.*" Rankin, "Ramanujan as a Patient," 91.

p. 293. "*useless to speculate.*" Hardy, "Obituary," 495.
"*a poor and solitary Hindu.*" Hardy, *Ramanujan,* 10.
"*the critical years.*" Ibid., 6.

p. 294. *On February 11.* Family Record.
One day in January or February of 1918. The account of Ramanujan's suicide attempt is derived almost solely from S. Chandrasekhar in notes deposited in the Royal Society and appearing in *Notes and Records of the Royal Society* 30 (1976). Chandrasekhar gives the time of the incident as February 1918. Nandy, 127, says it was "some time in the second half of 1917." But this seems contradicted by the dates surrounding Ramanujan's election to the Royal Society, which are intimately bound up with Chandrasekhar's story. On the other hand, late January, rather than February 1918, cannot be ruled out. By then, Ramanujan had already been put up for membership. On January 25, Hardy wrote Dewsbury and in his reply of March 5 Dewsbury thanked Hardy for "the very clear and explicit statement on the position of Mr. Ramanujan." Further, he promised to treat Hardy's letter "as personal and confidential." Could Hardy have been advising Dewsbury of the suicide attempt? (Dewsbury's letter is at Trinity College.)
"*We in Scotland Yard.*" Chandrasekhar account.

p. 295. *He read the telegram once.* Bollobás, *Littlewood's Miscellany,* 137.
"*the aid and guidance.*" Letter to Hardy at Trinity College.

p. 296. *too sick to travel.* Letter, Ramanujan to Royal Society, 15 May 1918. Duplicated in *Indian Express,* 19 December 1987.
A. S. Ramalingam. Information about him, and most of the account that follows, is drawn from the correspondence appearing in Rankin, "Ramanujan as a Patient," 79–100.

p. 297. *fattening up the patient.* Bryder, 53.
"*warped temperament.*" Bryder, 208.

p. 298. Bovril. *Oxford English Dictionary.*
no such dispensation. Or so my South Indian informants advise me.
Abruptly, he packed his bags. S. R. Ranganathan, 79.
likely the raid of October 19, 1917. There are inconsistencies in the Desmukh account, which records a bombing raid "in early 1918, when the Zeppelin raids were at their worst." But the *Times* of London accounts record no zeppelin raids on the capital then, nor indeed raids of any kind. The most recent substantial raid apt to impress Ramanujan as Desmukh records is that on the date given.

p. 299. "*tasteless and insipid.*" Gandhi, 40.

p. 299. "*novel and unpalatable.*" Neville, "The Late" *(Nature* 106), 662.
"*should Ramanujan be given up?*" Rankin, "Ramanujan as a Patient," 88.
Ramanujan's Trinity fellowship. Bollobás, *Littlewood's Miscellany,* 136–138; Bollobás, "Ramanujan—A Glimpse," 79.

p. 300. *pain no one could trace.* Bollobás, *Littlewood's Miscellany,* 137.
"*My heartfelt thanks.*" Ibid.

p. 301. "*your pains and their encouragement.*" Letter, Ramanujan to Hardy, undated. Trinity College.
"*a brief period of brilliant invention.*" Neville, "Ramanujan" (*Nature* 149), 294.
"*A recent paper.*" Ramanujan, *Collected Papers,* 210–213.

p. 302. "*great practical importance.*" Hardy and Wright, 49.

p. 305. "*practically no reason to doubt its truth.*" Percy Alexander MacMahon, *Combinatory Analysis,* vol. 2 (Cambridge: Cambridge University Press, 1916), 33.
Ramanujan was rummaging. Hardy, *Ramanujan,* 91, puts the year at 1917. But Richard Askey thinks the correspondence surrounding the discovery points more forcibly to 1916.
someone already had. "It is unlikely," George Andrews has written of Rogers's work, "that any important series of papers ever dropped further from sight than these did for the next 20 years."
"*remember very well his surprise.*" Hardy, *Ramanujan,* 91.
"*no position in diplomacy.*" *Nature* 132 (1933), 701.

p. 306. "*very little ambition.*" *Obituary Notices of Fellows of the Royal Society.* 1 (1932–1935), 299.
The correction. *Proceedings of the London Mathematical Society* 14, Series 2, (1915), endpaper of bound volume.
"*they seem to have escaped notice.*" Hardy, in Ramanujan, *Collected Papers,* 344.
yet unproved. MacMahon, *Combinatory Analysis,* vol. 2, 33.
"*regretting that he had overlooked my work.*" Letter, Rogers to F. H. Jackson, 13 February [1918], quoted in George E. Andrews, "L. J. Rogers and the Rogers-Ramanujan Identities," *Mathematical Chronicle* 11 (1982), 1–15. Letter furnished Andrews by Lucy Slater, Cambridge University.

p. 307. "*a very creditable pre-war bonfire.*" *Cambridge Review,* quoted in Howarth, 17–18.
"*as if Time and Eternity were one.*" Howarth, 18.
"*I think it is now time.*" P. K. Srinivasan, 76.

CHAPTER EIGHT

p. 309. "*on the road to a real recovery.*" P. K. Srinivasan, 76.

p. 310. "*and caused such grave anxiety.*" Seshu Iyer, 85.
restored the right to make cakes and pastries. Marwick, 271.

p. 311. Cambridge casualties. Howarth, 16.
"*rocky headlands rejoiced.*" Trevelyan, *An Autobiography & Other Essays* (London: Longmans, Green and Co., 1949), 37.
Colinette House. Interview with its current owners, Deborah and Bryan J. B. Gauld; tour of the house and grounds.
Samuel Mandeville Phillips. Rankin, "Ramanujan as a Patient," 82.

p. 312. *a few weeks after the armistice.* "She was laid to rest on December 6th in the grave where her husband lies in the churchyard." *The Cranleighan* 13 (1919), 76.

p. 312. *1729.* Hardy, *Collected Works*, 719–720; Snow, *Apology* foreword, 37.
would be named University Professor. Memo, Narayana Iyer to Sir Francis Spring, 12 March 1918. Madras Port Trust.

p. 313. "*the help that has been given me.*" P. K. Srinivasan, 46.
Passport photo. Photo, P. K. Srinivasan, opposite 136.

pp. 313–314. Ramanujan brought back to India. *The Hindu*, 19 December 1987; P. K. Srinivasan, 162; Berndt, "Ramanujan—100 Years Old," 26.

p. 314. *the night sky was still aglow.* Hoyt, 101. Shell fragments from the attack can be seen today in the Fort St. George Museum, Madras.
included Ramanujan's parents and wife. Jagjit Singh, "Srinivasa Ramanujan: A Short Biography—Part I," *Mathematical Education*, July–September 1987, 15.
"*When you wear the King-Emperor's uniform.*" Slater, *Southern India*, 291.

p. 315. "*keen interest and high anticipation.*" *Journal of the Indian Mathematical Society* 9, no. 1 (February 1917): 13.
"*some surprising discoveries.*" S. R. Ranganathan, 37.
"*the emergence of Ramanujan into fame.*" Ibid., 64.
claimed five lives already. Journal of the Indian Mathematical Society 10, no. 6 (December 1918): 1.

p. 316. Ramanujan in Bombay and movements after that. Interview, Janaki. Family Record. K. R. Rajagopalan, 46–48.
Why fret over Janaki? Interview, Janaki.
two identical letters. Ibid.
Brother advised her. Ibid.
"*asked to a funeral.*" Neville, "Ramanujan" (*Nature* 149), 294.
boarded the Bombay Mail. P. K. Srinivasan, 92.
"*I foresaw the end.*" Ramachandra Rao, 88.
Komalatammal's excuse. Interview, Janaki.

p. 317. *bundled on to a* jutka. P. K. Srinivasan, 92.
"*does not . . . go far enough.*" Spring memo, March 27, 1919. Madras Port Trust.
when his health improved. Berndt, "Ramanujan—100 Years Old," 26.
Madrasi notables. K. R. Rajagopalan, 47–48.
Luz Church. Muthiah, 137.
on April 6. Family Record.

p. 318. *in monthly installments.* P. K. Srinivasan, 182.
a real relationship. Interview, Janaki.
drank coffee now. Ibid.
"*not a cheerful . . . Ramanujan.*" S. R. Ranganathan, 71.
"*not the original Ramanujan.*" P. K. Srinivasan, 101.
"*only devils.*" Ibid.
he seemed troubled. Letter, S. Chandrasekhar to Hardy, 4 August 1937. Trinity College.
Other accounts. See Nandy, 131.

p. 319. *another irritant.* Letter, A. Ranganathan, 12 January 1981. Royal Society.
"*diamond eardrops.*" Transcript, *Nova* 1508, 16.
Coimbatore vs. Kodumudi. Interview with Janaki. Family Record. But in an interview, T. V. Rangaswami claims Ramanujan was, at least briefly, in

Coimbatore, perhaps later in the year. And K. R. Rajagopalan, 48, also mentions a stay in Coimbatore. But if so, it was brief.

p. 319. Ramanujan's stay in Kodumudi. Interviews with townspeople, including R. Chandrasekhar and K. Elangovan.
registrar's office. Interview, Janaki.

p. 320. *Ramanujan openly rebelled.* The story, its timing, and the trouble leading up to it, are based on interviews with Janaki supplemented by the Family Record.
"*If only you had come.*" Janaki has repeated this many times over the years, as in the *Hindu,* 19 December 1987, and plainly clings to the memory of it.
oil and wash her hair. Interview, "Hari" of Madras, who reports being told the story by Janaki.
send Janaki packing. Nandy, 130.
Every Sunday. Family Record
Fearnside. P. K. Srinivasan, 182.
Thanjavur pun. This appears in many forms. See, for example, S. R. Ranganathan, 93.
Komalatammal decreed. Interview, Janaki.
went on ahead to look for a house. Family Record.

p. 321. "*insisted on seeing my aunt.*" Seshu Iyer, 82.
dried brinjal slices. P. K. Srinivasan, 103.
"*powerful and penetrating eyes.*" S. R. Ranganathan, 75.
Dr. Chandrasekar. K. R. Rajagopalan, 49; P. K. Srinivasan, 108; letters, A. Ranganathan, 12 January and 26 March 1981. Royal Society.
"*this tuberculosis fever.*" K. R. Rajagopalan, 48.
dead within five years. Keers, 144.

p. 322. "*still in very bad health.*" Letter, Dewsbury to Hardy, 22 December 1919. Trinity College.
refused to be further treated. Seshu Iyer and Ramachandra Rao, xviii.
given up the will to live. Letter, A. Ranganathan, 12 January 1981. Royal Society.
Resisted going to Madras. Seshu Iyer and Ramachandra Rao, xviii.
probably a little after the first of the year. Letter, Dewsbury to Hardy, 15 January 1920. Trinity College.
got only as far as the railway station. Interview, Janaki.
known as "Gometra." Interview, Mr. and Mrs. T. C. Krishna, current owners of Gometra.

p. 323. "*which I call 'Mock' theta functions.*" Ramanujan, *Collected Papers*, xxxi.
"*in some ways the very best.*" S. Chandrasekhar, *Letters from an Indian Clerk.*

p. 324. "*we know how beautifully.*" G. N. Watson, "The Final Problem," *Journal of the London Mathematical Society* 11 (1936): 57.

p. 325. "*cold, immortal hands.*" Ibid., 80.
"*It is very difficult.*" Interview, George E. Andrews, in "Spectacular Genius," *The Hindu*, 21 January 1987.
final flurry of creativity. See McMurry for a perceptive study of the myths surrounding tuberculosis.
"*proportionately keener and brighter.*" Seshu Iyer, 85.

p. 326. "*wild eccentric genius of his youth.*" Interview, George E. Andrews, in "Spectacular Genius," *The Hindu*, 21 January 1987.
Crynant inauspicious. Interview, T. V. Rangaswami.
Namberumal Chetty. *Who's Who in Madras*, 1934.
"Gometra?" Tour of the house conducted by Mr. and Mrs. T. C. Krishna.
may briefly have stayed there. Janaki, in an interview, recalls a large "hall" in which Ramanujan stayed, which could have been the large living room.
scant furniture. P. K. Srinivasan, 163.
Visitors moved silently. " 'His Papers Disappeared Mysteriously,' " *The Hindu*, 21 June 1981.
Access jealously guarded. P. K. Srinivasan, 162.

p. 327. Janaki's brother with her. Ibid., 163.
Narayanaswamy Iyer. S. R. Ranganathan, 14.
"*Buoyed up by* spes phthisca." McMurry, 140.

p. 328. *subscribing to new math journals.* Letter, Dewsbury to Hardy, 15 January 1920. Trinity College.
Chetput pun. S. R. Ranganathan, 93.
"*uniformly kind to me.*" S. R. Ranganathan, 91. Janaki confirms that Ramanujan was sure he was going to die.
piece of hot pepper . . . K. R. Rajagopalan, 50.
Bell and stick. Interview, Janaki.
craved the rasam. P. K. Srinivasan, 136.
pounded it all together. Interview, Janaki.

p. 329. "*mentally dead to the world.*" P. K. Srinivasan, 111.
"*skin and bones.*" S. R. Ranganathan, 91. The account of Ramanujan's final days is drawn largely from interviews with Janaki and from various accounts she has given over the years.
relatives stayed away. K. R. Rajagopalan, 51.
Ramachandra Rao arranged the cremation. P. K. Srinivasan, 88.
officially recorded his death. The Hindu, 3 January 1988.
Subrahmanyan Chandrasekhar. See Kameshwar C. Wali, *Chandra* (Chicago: University of Chicago Press, 1990).

p. 330. *"I can still recall the gladness."* Andrews et al., 3.
"*We were proud of Mahatma Gandhi.*" Muthiah, 6.
"*inspired by his example.*" Andrews et al., 5.
neglected his studies. S. R. Ranganathan, 20.

p. 331. "*too deep for tears.*" Letter, A. Ranganathan, 12 January 1981. Royal Society.
"*entirely under the control.*" Letter, Lakshmi Narasimhan to Hardy, 29 April 1920. Trinity College.

p. 332. Death of Ramanujan's father. Interview, V. Viswanathan. I have not learned when Ramanujan's mother died, but a sister of V. Viswanathan (and granddaughter of Narayana Iyer) reports that when Madras was evacuated during World War II, probably in 1942, she saw Komalatammal in Kumbakonam. She would have been seventy-four.
"*became very often sullen.*" Bharathi, 51.

p. 332. Komalatammal's letter. Letter, Komalatammal to Hardy, 25 August 1927. Trinity College.

p. 333. "*console herself by seeing us.*" P. K. Srinivasan, 101.
worried about how she'd be treated. Interview, Janaki.
two days before Ramanujan's death. Ibid.
next half century. In 1931, according to T. V. Rangaswami, she once met A. S. Ramalingam in Madras.
work a sewing machine. Interview, Janaki. See also K. R. Rajagopalan, 53.
"*relatives having swindled her.*" Letter, S. Chandrasekhar to Hardy, 4 August 1937. Trinity College.
Janaki and Narayanan. Interview, Janaki. See also K. R. Rajagopalan, 53.

p. 334. "*And he is no more.*" Ramachandra Rao, 87.
"*inefficient and inelastic educational system.*" Hardy, *Ramanujan*, 7.
"*Indianality.*" Ramachandra Rao, 89.

p. 335. "*raised India in the estimation of the . . . world.*" *Journal of the Indian Mathematical Society* 11 (April 1919), 1.
"*a fluctuation from the norm.*" S. Chandrasekhar, "On Ramanujan," in G. E. Andrews et al., 4.
"*more deeply bound up with the West.*" Nandy, 139.
"*boosted our morale.*" Interview with P. K. Srinivasan.

p. 336. "*by the hand of Srinivasa Ramanujan.*" Neville, Reading Manuscript.

p. 337. "*some permanent memorial.*" P. K. Srinivasan, 79.
Wilson and Watson. "My respect for R[amanujan] has increased considerably in the last three months," Watson wrote Wilson on 28 June 1929. "I have retired from the J.M.B. to gain time for R." Trinity College.
"*no more romantic personality.*" R. D. Carmichael, "Some Recent Researches in the Theory of Numbers," *American Mathematical Monthly* 39, no. 3 (1932): 140.

p. 338. "*so full of human interest.*" Mordell, "Ramanujan," 642.
"*which I immediately read with great interest.*" Paul Erdos, "Ramanujan and I," in *Number Theory, Madras 1987,* ed. K. Alladi, no. 1395 in the series, Lecture Notes in Mathematics, ed. A. Dold and B. Eckmann (Berlin: Springer-Verlag, 1989), 1. The Hardy-Ramanujan paper is no. 35 in Ramanujan, *Collected Papers.*

p. 339. "*unusual and somewhat strange.*" Atle Selberg, "Ramanujan and I," *Science Age,* February 1988, 37.
"*The chapter . . . I loved the most.*" Freeman Dyson, "A Walk Through Ramanujan's Garden," in G. E. Andrews et al., 9.

p. 340. "*destruction of 1917.*" Ibid., 15.
"*originated with Ramanujan.*" Morris Newman, "Congruence Properties of the Partition Function," *Report of the Institute in the Theory of Numbers* (University of Colorado, 1959).
"*tamed a little.*" Hardy, *Collected Papers,* 720.
"*into touch with Euler.*" Littlewood, *Miscellany,* 86.

p. 341. *another ten years.* Letter, R. J. L. Kingsford to Hardy, 23 August 1929. Trinity College.
"*all by myself as a devotee.*" Interview, Freeman Dyson.
"*a stack of browned old paper.*" S. Ramaseshan, "Srinivasa Ramanujan," *Pro-*

ceedings of the Ramanujan Centennial International Conference, ed. R. Balakrishnan et al. (Madras, 1988), 6.

p. 341. "*even more definite picture.*" Littlewood, *Miscellany*, 86.

p. 342. Srinivasan's project. Interview, P. K. Srinivasan. See also P. K. Srinivasan, vi–x.

p. 343. "*On Certain Arithmetical Functions.*" Ramanujan, *Collected Papers*, 136–162.
"*very imperfectly understood.*" Hardy, *Ramanujan*, 161.
"*kept at bay.*" S. Raghavan, "Impact of Ramanujan's Work on Modern Mathematics," *Srinivasan Ramanujan Centenary 1987* [special issue of the *Journal of the Indian Institute of Science*], 46.
An account by Hardy. Hardy, *Ramanujan*, 170.

p. 344. Bach comparison. Berndt, "Ramanujan—100 Years Old," 24–29.
Discovery of the Lost Notebook. Interview, George Andrews. *Letters from an Indian Clerk*. Andrews, *Q-Series*.

p. 346. "*And that bothered me.*" Interview, Bruce Berndt.
"*substance and depth.*" Interview, Freeman Dyson.

p. 347. "*asked to explain.*" Hardy, *Some Famous Problems*, 4.
"*missing pecuniary reward.*" *Journal of the Indian Mathematical Society* 13, no. 3 (June 1921): 100.

p. 348. "*better blast furnaces.*" Bharathi, 93.
Plastics. S. Ramaseshan, "Srinivasa Ramanujan," 11.
Cancer. Meeting held November 10–12, 1988. *The Hindu* article was dated 23 December 1988.

p. 349. "*hard hexagon model.*" R. J. Baxter, "Ramanujan's Identities in Statistical Mechanics," in Andrews et al., 69–84. See also Gurney Williams III, "The Master of Math," *Omni*, December 1987, 58–64.
"*a computer algebra package in his head.*" Saraswathi Menon, "Beautiful Important Work," *The Hindu*, 22 December 1987.
"*he really didn't need them.*" Andrews, *Q-Series*.
"*Hype.*" Interview, Freeman Dyson. Bruce Berndt has written me with "another 'application' of Ramanujan's mathematics that Dyson calls 'hype.' " William Beyer, a physicist at Los Alamos National Laboratory, has shown how one of Ramanujan's formulas can be "very useful in predicting nuclear war." The formula does not, of course, truly "predict" nuclear war, but rather supplies the basis for a theoretical model for estimating its likelihood.

p. 350. "*appreciate Ramanujan early.*" Askey, 72.
"*I was stunned.*" *Letters from an Indian Clerk*.
"*a thrill . . . indistinguishable.*" G. N. Watson, "The Final Problem," *Journal of the London Mathematical Society* 11 (1936): 80.

p. 351. "*a certain character of permanence.*" Hardy, *Some Famous Problems*, 4–5.
framed and garlanded . . . portrait. Noted in a videotape made of the celebration.

p. 352. *the university had noticed.* S. R. Ranganathan, 55. S. Chandrasekhar also played a role in seeing to it that Janaki was better cared for.
Janaki financed Interview, Janaki.
"Where is the statue?" Hindu, 21 June 1981.
Granlund sculpture. Interview, Richard Askey.

p. 353. *Each afternoon.* Interview, T. V. Rangaswami. In the mornings, he reports, Janaki was occupied with devotions or cooking.
broke down and cried. Interview, T. V. Rangaswami.
left to foreigners. Ramanujan, said P. K. Srinivasan in an interview, "was born in India, reborn in the U.K., and now born again in the U.S."
"*could not get even a lectureship.*" J. B. S. Haldane quoted in sidebar to "True Genius," by "RGK," 30.

p. 354. "*languishing in obscurity.*" Bharathi, 36.
"*a new India and a new world?*" Nehru, *Discovery of India* (London: Meridean Books), 1960.
many Ramanujans. Interview, R. Viswanathan, Kumbakonam.

p. 355. "*of defaming Ramanujan's name.*" S. Chandrasekhar, "An Incident in the Life of S. Ramanujan . . . ," notes deposited in Royal Society.
Ramanujan Institute. Interview, K. S. Padmanabhan.

p. 356. "*How many registrars.*" S. Ramaseshan, 3.
"*luminescence.*" "True Genius," by "RGK," 31.
"*our bankruptcy was made manifest.*" Report of Joint Secretary D. D. Kapadia, Second Conference of the Indian Mathematical Society, 11–13 January 1919, reported in the Society's *Journal.*

pp. 357–358. "*to win the confidence.*" Neville, *The Farey Series of Order 1025* (Cambridge: University Press, 1950), xxvii.

p. 358. Hardy's greatest contribution. Quoted in *The Hindu,* 19 December 1987.
svayambhu. "True Genius," by "RGK," 26.

p. 359. "*I did not invent him.*" Hardy, *Ramanujan,* 1.

EPILOGUE

p. 361. "*principal permanent happiness.*" Letter, Hardy to Thomson, 13 December 1919. Cambridge University.
hard feelings left over from the war. "Life in College was through all these years [of the war] . . . definitely unpleasant, and the recollection of them was an important factor in my own decision to try to move to Oxford." Hardy, *Russell and Trinity,* 10.
"*darker for Hardy.*" Snow, *Apology* foreword, 38.
"*felt he must get away.*" Young, 287.

p. 362. "*Cambridge beguiles.*" Rose and Ziman, 44–45.
"*By direction of the Syndicate.*" Letter to Hardy, 29 April 1920. Trinity College.
"*a great shock and surprise.*" P. K. Srinivasan, 78.

p. 363. "*all Germans will be relegated.*" Quoted in Joseph W. Dauben, "Mathematicians of World War I: The International Diplomacy of G. H. Hardy and Gösta Mittag-Leffler as Reflected in Their Personal Correspondence," *Historia Mathematica* 7 (1980): 261–288.
"*trivial changes of sign.*" The story has been told in many places. Young, 280, is one of them.
"*modified my former views.*" Letter, Hardy to Mittag-Leffler, quoted in Dauben, "Mathematicians of World War I."
"*happiest time of his life.*" Snow, *Apology* foreword, 40. "After Hardy's social discomforts at the High Table of Trinity, Hardy expanded and mellowed in

a wonderful way, in what was to him the more benign atmosphere of New College. His popularity there was immediate and assured, and he gave back in the intellectual exhilaration of his conversation what he received in sympathetic friendship." E. A. Milne, obituary of Hardy in *Monthly Notices of the Royal Astronomical Society* 108 (1948): 46.

p. 364. "*customary deference of those days.*" Mary Cartwright, "Moments in a Girl's Life," unpublished manuscript, 5.

"*a delightful time.*" Letter, Hardy to president of Princeton University, 24 April 1929. (The letterhead is that of the Mayflower Hotel, Washington, D.C.)

Ruth as familiar as Hobbs. "He had a similar passion for baseball [as for cricket], which he watched whenever he could when he was in the States." From a journal referee's report commenting on an article about Ramanujan and Hardy. Robert Rankin, Glasgow.

"*a* wonderful *book.*" Postcard, Hardy to R. P. Boas, 10 January 1940.

Oxford tennis photo. Trinity College.

p. 365. *retirement.* Snow, *Apology* foreword, 43.

"*one could often hear the party's laughter.*" Snow, "The Mathematician," 72.

"*a* real *mathematician.*" Snow, *Apology* foreword, 9.

"*suddenly at least the equal.*" Young, 293.

"*deep solicitude.*" A. V. Hill, obituary notice in *The Mathematical Gazette* 22 (May 1948), 51.

"*then I cannot remain in it.*" Letter, Hardy to Mordell. St. John's College.

p. 366. "*more uncharitable conclusion.*" Hardy, *Collected Papers,* 611.

"*he was with the miners.*" J. B. S. Haldane, "A 'Pure' Scientist—and a Great One," *Daily Worker* (London), 29 December 1947.

"*some mysterious terror to exact thought.*" Hardy, *Collected Papers,* 260.

some would grumble later. J. C. Burkill, review of *Collected Papers of G. H. Hardy,* vol. 7, *Bulletin of the London Mathematical Society* 12 (1980): 226.

Congratulations from Soviets. Letter, Soviet Ambassador to Hardy, 20 February 1934. Trinity College.

Honorary degrees. After the ceremony in which he received his degree from Edinburgh, Hardy was strolling with a group of mathematicians when he spotted a mouse. His colleagues, reported *The Scotsman* (3 December 1947), "were granted the privilege of watching one of the world's greatest mathematicians stalking it, on hands and knees, around a tree."

p. 367. *Hardy's New Year's resolutions.* J. C. Burkill, entry on Hardy in *Dictionary of Scientific Biography.*

his madcap bent. Young, 276.

"*old brandy.*" Snow, *Apology* foreword, 48.

"*aged and shriveled replica.*" Wiener, *I Am a Mathematician,* 152.

p. 368. "*youthful quickness and power.*" Hardy, *Collected Papers,* 745.

while dusting his bookcase. Interview, Mary Cartwright.

his most important papers. Titchmarsh, 458–461.

"*like Wanda Landowska playing Bach.*" Letter, Freeman Dyson to C. P. Snow, 22 May 1967. Freeman Dyson.

"*makes him look so old.*" Interview, Mary Cartwright. Written in blue ink, almost certainly by his sister, on the photo in the Trinity archives: "He never looked as old as this!"

p. 368. Hardy's suicide attempt. Snow, *Apology* foreword, 54.

p. 369. "*one conclusion to be drawn.*" Ibid., 57.
"*Hardy is now dying.*" Philip Snow, 95.
"*the passing of a great age.*" Norbert Wiener, *Bulletin of the American Mathematical Society* 55 (1949): 77.
"*book of haunting sadness.*" Snow, *Apology* foreword, 50.
"*It is a melancholy experience.*" Hardy, *Apology*, 61.
"*on something like equal terms.*" Ibid., 148.
"*about which he felt strongly.*" Mary Cartwright, "Moments in a Girl's Life," unpublished manuscript, 7.

p. 370. "*the one romantic incident.*" Hardy, *Ramanujan*, 2.
"*most careful editing.*" Letter, Hardy to Mittag-Leffler, about 1920.
"*a short but characteristic example.*" Hardy, in Ramanujan, *Collected Papers*, 232.
"*the decisive event of my life.*" Hardy, *Apology*, 148.

p. 371. "*poor and solitary Hindu.*" Hardy, *Ramanujan*, 10.
"*among the most remarkable.*" *Manchester Guardian*, 2 December 1947.
"*one of the most fascinating obituary notices.*" Quoted in Andrews, *Q-Series.*
a prime subject of his own book. Nandy, 3.
Hardy's report on Ramanujan. Hardy, *Collected Papers*, 491–503.

p. 372. "*greater than the gain.*" Hardy, *Collected Papers*, 720.
"*no secrecy at all.*" Snow, *Apology* foreword, 30.
"*Hardy was terribly proud.*" Interview, Mary Cartwright.
"*I am going to give some lectures.*" Letter, Hardy to Chandrasekhar, 19 February 1936. Trinity College.
"*labor of love.*" Mordell, "Ramanujan," 642.
Harvard Tercentenary celebration. See, for example, *Harvard Alumni Bulletin*, 30 September 1936. *The Tercentenary of Harvard College* (Cambridge, Mass.: Harvard University Press, 1937), compiled by Jerome D. Greene; *Harvard Magazine*, May–June 1986.
"*who has led the advance.*" Trinity College.

p. 373. "*the rule in Shelley's case.*" Newman, *The World of Mathematics*, vol. 4 (New York: Simon and Schuster, 1956), 2036. According to Newman, the story is told by Supreme Court Justice Felix Frankfurter.
in Hitler's army. "Highbrows at Harvard," *Time*, 14 September 1936.
about nine in the evening. Conference schedule, in *The Tercentenary of Harvard College*, Appendix N, 465.
"*a very great mathematician.*" Hardy, *Ramanujan*, 1.

Selected Bibliography

Alexanderson, G. L. *The Polya Picture Album: Encounters of a Mathematician*. Boston: Birkhäuser, 1987.

Andrews, George E. "An Introduction to Ramanujan's 'Lost' Notebook." *American Mathematical Monthly* 86 (1979): 89–108.

———. *Number Theory*. Philadelphia: W. B. Saunders Company, 1971.

———. *Q-Series: Their Development and Application in Analysis, Number Theory, Combinatorics, Physics, and Computer Algebra*. Providence, R.I.: American Mathematical Society, Regional Conference Series in Mathematics, No. 66, 1986.

———, Richard A. Askey, Bruce C. Berndt, K. G. Ramanathan, and Robert A. Rankin. *Ramanujan Revisited*. San Diego: Academic Press, 1988.

Askey, Richard A. "Ramanujan and Hypergeometric and Basic Hypergeometric Series." In *Ramanujan International Symposium on Analysis*, edited by N. K. Thakare. Macmillan India, 1989.

Association of Mathematics Teachers of India. *Ramanujan Centenary Year Souvenir*. Proceedings of 22d annual conference, 3–6 December 1987.

Atkinson, Thomas Dinham. *Cambridge Described and Illustrated*. London: Macmillan, 1897.

Balakrishnan, R., K. S. Padmanabhan, and V. Thangaraj. *Proceedings of the Ramanujan Centennial International Conference*, 15–18 December 1987, at Annamalai University, Annamalainagar. Ramanujan Mathematical Society, 1988.

Balasubrahmanyan, S. R. *Later Chola Temples*. Mugdala Trust, 1975.

Barham, Jane. *Backstairs Cambridge*. Herts, England: Ellisons' Editions, 1986.

Barlow, Glyn. *The Story of Madras*. London: Humphrey Milford, Oxford University Press, 1921.

Barnes, John. *Ahead of His Age—Bishop Barnes of Birmingham*. London: Collins, 1979.

Bateman, P., and H. Diamond. "John E. Littlewood (1885–1977)," *Mathematical Intelligencer* 1 (1978): 28–33.

Bell, E. T. *Men of Mathematics*. New York: Simon and Schuster, 1986.

Berndt, Bruce C. "A Pilgrimage." *Mathematical Intelligencer* 8, no. 1 (1986): 25–30.

———. "The Quarterly Reports of S. Ramanujan." *American Mathematical Monthly* 90, no. 8 (October 1983): 505–516.

———. "Ramanujan—100 Years Old (Fashioned) or 100 Years New (Fangled)?" *Mathematical Intelligencer* 10, no. 3 (1988): 24–29.

———. *Ramanujan's Notebooks*. 2 parts. New York: Springer-Verlag, 1985–1989.

Bhanumurthy, Kalyanalakshmy. "[Ramanujan:] The Man Behind His Mathematics." *The Hindu*, 20 December 1987.

Bharathi, Ramananda. *Prof. Srinivasa Ramanujan Commemoration Volume*. Madras: Jupiter Press, 1974.

Bhattacharaya, Jogendra Nath. *Hindu Castes and Sects*. Calcutta: Editions Indian, 1896. Reprinted, Calcutta, 1973.

Bishop, T. J. H., in collaboration with Rupert Wilkinson. *Winchester and the Public School Elite*. London: Faber and Faber, 1967.

Bollobás, Béla, ed. *Littlewood's Miscellany*. Cambridge: Cambridge University Press, 1986.

———. "Ramanujan—A Glimpse of His Life and His Mathematics." *Cambridge Review*, June 1988, 76–80.

Borwein, Jonathan M., and Peter B. Borwein. "Ramanujan and Pi." *Scientific American* 258, no. 2 (February 1988): 112–117.

Brandon, Peter. *A History of Surrey*. Darwen County History Series. Phillimore, 1977.

Bromwich, T. J. I'A. *An Introduction to the Theory of Infinite Series*. London: Macmillan, 1908.

Bryder, Linda. *Below the Magic Mountain*. Oxford: Clarendon Press, 1988.

Bulstrode, H. Timbrell. *Report on Sanatoria for Consumption*. 35th Annual Report 1905–06, Supplement in Continuation of the Report of the Medical Officer for 1905–06. London: His Majesty's Stationery Office, 1908.

Burkill, J. C. "John Edensor Littlewood." *Bulletin of the London Mathematical Society* 11 (1979): 59–103.

Butler, J. R. M. *Henry Montagu Butler: Master of Trinity College Cambridge, 1886–1918*. London: Longmans, Green and Co., 1925.

Buxton, John, and Penry Williams, eds. *New College, Oxford 1379–1979*. The Wardens and Fellows of New College, Oxford, 1979.

Campbell, Douglas M., and John C. Higgins. *Mathematics: People, Problems, Results*. 2 vols. Belmont, Calif.: Wadsworth, 1984.

Carr, G. S. *Formulas and Theorems in Pure Mathematics*. 2d ed. New York: Chelsea, 1970. (Originally published as *A Synopsis of Results in Pure and Applied Mathematics*.)

Carstairs, G. Morris. *The Twice-Born*. Bloomington, Ind.: Indiana University Press, 1967.

Cartwright, Mary. "John Edensor Littlewood, FRS, FRAS, Hon FIMA." *Bulletin of the Institute of Mathematics and Its Applications*. April 1978, 87–90.

Centenary Celebration Souvenir. Kumbakonam: Town High School, 1964.

Chaudhuri, Nirad C. *A Passage to England*. London: Macmillan, 1959.

Chopra, P. N., T. K. Ravindran, and N. Subrahmanian, *A History of South India*. 3 vols. New Delhi: S. Chad & Co., 1979.

Clark, Geoffrey, and W. Harding Thompson. *The Surrey Landscape*. London: A. & C. Black, 1934.

Clark, Ronald W. *The Life of Bertrand Russell*. New York: Knopf, 1976.

"Close Up: Professor J. E. Littlewood, F.R.S." *Trinity Review*, Michaelmas 1959.

Collins, Randall. ***The Case of the Philosophers' Ring by Dr. John H. Watson***. New York: Crown Publishers, 1978.

Colonial Students in Britain. London: PEP (Political and Economic Planning), 1955.

Compton, Herbert. ***Indian Life in Town and Country***. New York: G. P. Putnam's, 1904.

Das, R. K. ***Temples of Tamilnad***. Bombay: Bharatiya Vidya Bhavan, 1964.

Davis, Philip J., and Reuben Hersh. ***The Mathematical Experience***. Boston: Birkhäuser, 1981.

Deacon, Richard. ***The Cambridge Apostles***. London: Robert Royce Limited, 1985.

Dubos, Rene and Jean. ***The White Plague: Tuberculosis, Man and Society***. Boston: Little, Brown and Company, 1952.

Dunning, Eric, and Kenneth Sheard. ***Barbarians, Gentlemen and Players: A Sociological Study of the Development of Rugby Football***. New York: New York University Press, 1979.

Ellis, Havelock. ***A Study of British Genius***. London: Hurst and Blackett, 1904.

———, and John Addington Symonds. ***Sexual Inversion***. London: Wilson and Macmillan, 1897.

Eves, Howard. ***An Introduction to the History of Mathematics***. 4th ed. New York: Holt, Rinehart, and Winston, 1976.

Foster, M., and Lewis Shore. ***Physiology for Beginners***. London, 1894.

Fowler, Laurence, and Helen Fowler. ***Cambridge Commemorated***. Cambridge University Press, 1984.

Fuller, Bampfylde. ***The Empire of India***. London: Sir Isaac Pitman & Sons, 1913.

Furbank, P. N. ***E. M. Forster: A Life***. New York: Harcourt Brace Jovanovich, 1978.

Gandhi, Mohandas K. ***Autobiography: The Story of My Experiments with Truth***. Public Affairs Press, 1948. Reprint. New York: Dover, 1983.

Gleick, James. "An Isolated Genius Is Given His Due." ***New York Times***, July 14, 1987, 17.

Hadamard, Jacques. ***The Psychology of Invention in the Mathematical Field***. Princeton: Princeton University Press, 1949.

Halperin, John. ***C. P. Snow: An Oral Biography***. New York: St. Martin's Press, 1983.

Hardy, G. H. ***Bertrand Russell and Trinity***. Cambridge University Press, 1970.

———. ***Collected Papers of G. H. Hardy***. Vol. 7. Oxford: Clarendon Press, 1979.

———. ***A Course of Pure Mathematics***. Cambridge University Press, 1908.

———. ***A Mathematician's Apology***. Cambridge University Press, 1940. Reprint, with a foreword by C. P. Snow, 1967.

———. "Obituary, S. Ramanujan." ***Nature*** 105 (June 17, 1920), 494–495.

———. ***Orders of Infinity: The "Infinitarcalcul" of Paul Du Bois-Reymond***. Cambridge Tracts in Mathematics and Mathematical Physics, 12. Cambridge University Press, 1910.

———. ***Ramanujan***. Cambridge University Press, 1940. 3d ed. New York: Chelsea, 1978.

———. ***Some Famous Problems of the Theory of Numbers***. Oxford: Clarendon Press, 1920.

———, and E. M. Wright. ***An Introduction to the Theory of Numbers***. 3d ed. Oxford: Clarendon Press, 1954.

Hemingway, F. R. ***Tanjore District Gazetteer***. 1915.

Himmelfarb, Gertrude. ***Marriage and Morals Among the Victorians***. New York: Alfred A. Knopf, 1986.

Historical Register of University of Cambridge. Supplement 1, 1911–1920. Cambridge University Press, 1922.

Hodges, Andrew. ***Alan Turing: The Enigma***. New York: Simon and Schuster, 1983.

Howarth, T. E. B. *Cambridge Between Two Wars*. London: Collins, 1978.
Hoyt, Edwin P. *The Last Cruise of the Emden*. New York: Macmillan, 1966.
Hynes, Samuel. *The Edwardian Turn of Mind*. Princeton: Princeton University Press, 1968.
Imperial Gazetteer. Madras Presidency, 1908.
Indian Scientists. Madras: G. A. Natesan & Co., 1929.
Jekyll, Gertrude. *Old West Surrey*. London: Longmans, Green and Co., 1904.
Jordan, M. C. *Cours d'analyse de l'Ecole Polytechnique*. Paris: Gauthier-Villars, 1882.
Kandaswamy Moopanar, R. "My Reminiscences." *Centenary Celebration Souvenir*. Kumbakonam: Town High School, 1964.
Keers, R. Y. *Pulmonary Tuberculosis: A Journey Down the Centuries*. London: Balliere Tindall, 1978.
Keynes, Margaret Elizabeth. *A House by the River*. Cambridge: Darwin College, 1976.
Kitcher, Philip. *The Nature of Mathematical Knowledge*. New York: Oxford University Press, 1983.
Krishnaswami Nayadu, W. S. *Old Madras*, 1965.
Lanchester, H. V. *Town Planning in Madras*. (Based on lectures given in Madras in January 1916.) London: Methuen, 1926.
Levy, Paul. *Moore: G. E. Moore and the Cambridge Apostles*. New York: Holt, Rinehart, and Winston, 1979.
Lewandowski, Susan. *Migration and Ethnicity in Urban India: Kerala Migrants in the City of Madras, 1870–1970*. New Delhi: Manohar, 1980.
Littlewood, J. E. "Adventures in Ballistics, 1915–1918." *Bulletin of the Institute of Mathematics and Its Applications*, September/October 1974, 323–328.
———. "Mathematical Life and Teaching in Cambridge Since 1880." Lecture notes.
———. "The Mathematician's Art of Work." *Rockefeller University Review* 5 (September–October 1967): 1–7.
———. *A Mathematician's Miscellany*. London: Methuen & Co., 1953.
———. "Reminiscences of Cambridge and Elsewhere." Lecture notes.
Marcus, Geoffrey. *Before the Lamps Went Out*. Boston: Little, Brown, 1965.
Marvin, F. S. *India and the West*. London: Longmans, Green and Co., 1927.
Marwick, Arthur. *The Deluge: British Society and the First World War*. New York: W. W. Norton, 1970.
Mayo, Katherine. *Mother India*. New York: Harcourt, Brace, 1927.
McMurry, Nan Marie. "*And I? I Am in a Consumption*": *The Tuberculosis Patient, 1780–1930*. Dissertation. Ann Arbor: University Microfilms. 1985.
Megahey, Alan J. *A History of Cranleigh School*. London: Collins, 1983.
Molony, J. Chartres. *A Book of South India*. London: Methuen, 1926.
Mordell, L. J. "Hardy's *A Mathematician's Apology*." *American Mathematical Monthly* 77 (1970): 831–836.
———. "Ramanujan." *Nature* 148, no. 3761 (November 1941): 642–647.
Muthiah, S. *Madras Discovered*. New Delhi: Affiliated East-West Press, 1987.
Nagarajan, K. R., and T. Soundararajan, eds. *Srinivasa Ramanujan, 1887–1920: A Tribute*. Madras: Macmillan India, 1988.
Nairn, Ian, and Nikolaus Pevsner. *Surrey*. Harmondsworth, England: Penguin Books, 1962.
Nandy, Ashis. *Alternative Sciences: Creativity and Authenticity in Two Indian Scientists*. New Delhi: Allied Publishers, 1980.

Neville, E. H. "The Late Srinivasa Ramanujan," *Nature* 106, no. 2673 (January 1921): 661–662.

———. "Srinivasa Ramanujan," *Nature* 149, no. 3776 (March 1942): 292–295.

———. "Srinivasa Ramanujan." Original text of a talk first broadcast in English not later than 1938 and in Hindustani, April 22, 1941. A substantially longer version of the *Nature* article, above. Referred to in the Notes as the Reading Manuscript.

Newman, M. H. A., et al. "Godfrey Harold Hardy, 1877–1947." *Mathematical Gazette* 32 (1948): 49–51.

Padfield, J. E. *The Hindu at Home*. 1908. 2d facsimile ed. Delhi: B. R. Publishing Corporation, 1975.

Parry, R. St. John. *Henry Jackson, O.M.* Cambridge: Cambridge University Press, 1926.

Polya, George. "Some Mathematicians I Have Known." *American Mathematical Monthly* 76 (1969): 746–753.

The Port of Madras: Past, Present and Future. Madras Port Trust, 1967.

"R.G.K." "True Genius." In *Illustrated Weekly of India*, 20 December 1987, 26–30.

"Ragami" [pseud.]. *Ramanujan, the Mathematical Genius* [Tamil]. Series. *Dinamani Kadir* (Madras weekly newspaper), 1985–1986.

Rajagopalan, K. R. *Srinivasa Ramanujan*. Madras: Sri Aravinda-Bharati, 1988.

Rajagopalan, T. R. *Pilgrim's Guide (South India)*. Madurai: Sri Karthikeiya Publication, 1977.

Ram, Suresh. *Srinivasa Ramanujan*. New Delhi: National Book Trust, 1972.

Ramachandra Rao, R. "In Memoriam: S. Ramanujan." *Journal of the Indian Mathematical Society* 12, no. 3 (June 1920): 87–90.

Ramakrishna, G., N. Gayathri, and D. Chattopadhyaya. *An Encyclopaedia of South Indian Culture*. Calcutta: K. P. Bagchi, 1983.

Ramakrishna, T. *My Visit to the West*. London: T. Fisher Unwin, 1915.

Ramanathan, K. G. "Ramanujan: A Life Sketch." *Science Age*, December 1987, 19–26.

———. "Srinivasa Ramanujan: Mathematician Extraordinary," *Science Today*, December 1974, 13–21.

Ramanujan, Srinivasa. *Collected Papers of Srinivasa Ramanujan*. Ed. G. H. Hardy, P. V. Seshu Iyer, and B. M. Wilson. Cambridge University Press, 1927.

———. *The Lost Notebook and Other Unpublished Papers*. New Delhi: Narosa Publishing House, 1988.

———. *Notebooks*. 2 vols. Bombay: Tata Institute of Fundamental Research, 1957.

Ramaseshan, S. "Srinivasa Ramanujan." In *Proceedings of the Ramanujan Centennial International Conference*, 15–18 December 1987, 1–14.

Ranganathan, S. R. *Ramanujan: The Man and the Mathematician*. Bombay: Asia Publishing House, 1967.

Rankin, R. A. "Ramanujan as a Patient." *Proceedings of the Indian Academy of Sciences* (Mathematical Sciences) 93, nos. 2 & 3 (December 1984): 79–90.

———. "Ramanujan's Manuscripts and Notebooks." *Bulletin of the London Mathematical Society* 14 (1982): 81–97.

———. "Ramanujan's Manuscripts and Notebooks, II." *Bulletin of the London Mathematical Society* 21 (1989): 351–365.

———. "Srinivasa Ramanujan (1887–1920)." *Bulletin of the Institute of Mathematics and Its Applications* 23, nos. 10/11 (October/November 1987): 145–152.

Report of the [Lytton] Committee on Indian Students, 1921–1922. Pt. 1, Report and Appendices; Pt. 2, Evidence. London: His Majesty's Stationery Office for India Office, 1922.

Rose, Jasper, and John Ziman. *Camford Observed.* London: Victor Gollancz, 1964.

Rouse Ball, W. W. *A History of the Study of Mathematics at Cambridge.* Cambridge: Cambridge University Press, 1889.

Russell, Dora. *The Tamarisk Tree.* New York: G. P. Putnam's, 1975.

Sabben-Clare, James. *Winchester College.* Southampton, England: Paul Cave Publications, 1982.

Satthianadhan, S. *Four Years in an English University.* Madras: Lawrence Asylum Press, 1890.

Seshu Iyer, P. V. "The Late Mr. S. Ramanujan, B.A., F.R.S." *Journal of the Indian Mathematical Society* 12, no. 3 (June 1920): 81–86.

———, and R. Ramachandra Rao. "Srinivasa Ramanujan." In *Collected Papers of Srinivasa Ramanujan,* ed. G. H. Hardy, P. V. Seshu Iyer, and B. M. Wilson. Cambridge University Press, 1927: xi–xix.

Seymour, B., and M. Warrington. *Bygone Cranleigh.* Chichester, Sussex: Phillimore, 1984.

Singer, Milton, ed. *Traditional India: Structure and Change.* Philadelphia: American Folklore Society, 1959.

Singh, Amar Kumar. "The Impact of Foreign Study: The Indian Experience," *Minerva,* Autumn 1962.

———. *Indian Students in Britain.* London: Asia Publishing House, 1963.

Singh, Jagjit. "Srinivasa Ramanujan: A Short Biography." 2 pt. *Mathematical Education,* July–September and October–December 1987.

Slater, Gilbert, ed. *Economic Studies.* Vol. 1, *Some South Indian Villages.* London: Oxford University Press, 1918.

———. *Southern India: Its Political and Economic Problems.* London: George Allen Unwin, 1936.

Smith, F. B. *The Retreat of Tuberculosis, 1850–1950.* London: Croom Helm, 1988.

Snow, C. P. "The Mathematician on Cricket." *Saturday Book.* 8th year. 1941–1942.

———. *Variety of Men.* New York: Charles Scribner's Sons, 1966. (Piece on Hardy appeared in slightly different form as foreword to 1967 edition of Hardy's *A Mathematician's Apology.*)

Snow, Philip. *Stranger and Brother.* London: Macmillan, 1982.

Srinivas, M. N. *Social Change in Modern India.* Berkeley: University of California Press, 1971.

Srinivasachari, C. S. *History of the City of Madras.* Madras: P. Varadachary & Co., 1939.

Srinivasan, P. K., ed. *Ramanujan Memorial Number.* Vol. 1, *Letters and Reminiscences.* Madras: Muthialpet High School, 1968.

———. *Ramanujan Memorial Number.* Vol. 2, *An Inspiration.* Madras: Muthialpet High School, 1967.

Steegman, John. *Cambridge: As It Was, and As It Is Today.* 4th rev. ed. London: B. T. Batsford, 1949.

Steevens, G. W. *In India.* Edinburgh: Blackwood and Sons, 1899.

Stein, Burton, ed. *South Indian Temples: An Analytical Reconsideration.* New Delhi: Vikas Publishing House, 1978.

Struik, Dirk J. *A Concise History of Mathematics.* 4th rev. ed. New York: Dover, 1987.
Stubbs, Charles W. *The Story of Cambridge.* London: J. M. Dent & Co., 1905.
Sud, Babu Lal. *How to Become a Barrister and Take a Degree at Oxford or Cambridge.* Kapurthala State, India: self-published, 1917.
Sykes, Christopher, *Letters from an Indian Clerk*, television documentary, Channel 4 International, 1987.
Tennyson, Charles. *Cambridge from Within.* Philadelphia: George W. Jacobs, 1913.
Thomson, J. J. *Recollections and Reflections.* New York: Macmillan, 1937.
Thurston, Edgar. *The Madras Presidency.* Cambridge University Press, 1914.
Titchmarsh, E. C. "Godfrey Harold Hardy, 1877–1947." *Obituary Notices of Fellows of the Royal Society* 6 (1949): 447–458.
Tuckwell, Rev. W. *The Ancient Ways: Winchester Fifty Years Ago.* London: Macmillan & Co., 1893.
Turnbull, H. W. *The Great Mathematicians.* London: Methuen & Co., 1929.
Urwick, W. *India Illustrated.* Rev. and enl. by Edward P. Thwing. New York: Hurst, 1891.
Venkataraman, G. 3 pt. series on Ramanujan in *The Hindu*: "Once in a Few Centuries," 20 December 1987; "A Turning Point for Ramanujan," 27 December 1987; "The Last Years of Ramanujan," 5 January 1988.
Wagh, D. B. "Srinivasa Ramanujan: Number Theorist Extraordinary," *Bhavan's Journal* 33, no. 23 (July 1–15, 1987): 21–29.
"The War." In *Historical Register of the University of Cambridge, Supplement, 1911–1920.* Cambridge University Press, 1922.
Watson, Francis. *A Concise History of India.* New York: Thames and Hudson, 1979.
Weinstein, Jay A. "Madras: An Analysis of Urban Ecological Structure in India." *Sage Research Papers in the Social Sciences.* Social Ecology of the Community Series, No. 90-015. Beverly Hills, 1974.
White, William M. *Visitor's Guide to Cambridge.* 2d ed. Cambridge: Metcalfe & Co., 1890.
Whitehead, Henry. *The Village Gods of South India.* Calcutta: The Association Press (in association with Humphrey Milford, Oxford University Press), 1916.
Wiener, Norbert. *Ex-Prodigy.* New York: Simon and Schuster, 1953.
———. "Godfrey Harold Hardy, 1877–1947." *Bulletin of the American Mathematical Society* 55 (1949): 72–77.
———. *I Am a Mathematician.* Garden City, N.Y.: Doubleday, 1956.
Williams, Mary Elizabeth. *A Bibliography of John Edensor Littlewood.* Elizabeth, N.J.: Pageant-Poseidon Press, 1974.
Woolf, Leonard. *Beginning Again.* New York: Harcourt, Brace & World, 1963.
———. *Sowing.* New York: Harcourt, Brace & Co., 1960.
Worswick, Clark, and Ainslie Embree. *The Last Empire.* Millerton, N.Y., 1976.
Wurmser, Leon. *The Mask of Shame.* Baltimore: Johns Hopkins University Press, 1981.
Young, Laurence. *Mathematicians and Their Times.* Amsterdam: North-Holland Publishing Co., 1981.

Author's Note and Acknowledgments

In writing the life of Ramanujan, I faced the barriers of two foreign cultures, a challenging discipline, and a distant time. As I am expert in none of these, I owe a debt of gratitude to the many persons who have helped me surmount those barriers—who have consented to interviews, spent hours explaining recondite areas of mathematics or Indian cultural life, guided me to out of the way documents in libraries and archives, read and criticized early drafts, befriended me in England and India—and, back in Baltimore, offered a supportive hand or word of advice. I am abashed at how much space I require to simply say thank you, but it is an apt measure of my debt.

First, thanks are due Judy, who has borne more husbandly changes of mood than any person should have to bear, and to whom I owe much in the way of nurturing, encouragement, and support. Those intangibles are easy to take for granted when you have them, but almost impossible to get along with when you don't.

To Judy and my sister Rachele, and their refusal to tell me what I wished to hear at a crucial point, I owe an approach to writing about the mathematics to which otherwise I might not have turned.

Thanks go to Davy, whose Daddy was gone in England and India for three months and then, for the year and a half it took to write this book, was too much in the office and not enough with him. "So how's the Ramanujan book?" he asked me one day when he was five. I hope that one day he'll read it and be able to answer for himself.

To my parents, Bea and Charles Kanigel, for leaving me fascinated equally by words, numbers, people, and ideas.

To Michael, Kevin, and Jonathan, to Harry and Rachele, and to Elise and Liz—all of them, each in their own way, irreplaceable parts of my life.

To Bill Stump for encouraging a young writer twenty years ago.

To V. Viswanathan, of Madras, who made room for a confused American in an already overstuffed auto-rickshaw. To him, his brother V. Meenakshisundaram of the Madras Port Trust, to S. Sankara Narayanan and V. Subramanyam, and to the many other members of his family who befriended me in Madras, I owe a debt of kindness I cannot possibly repay.

To Sambandam, Vijaya, Mahalingam, and all their family in Kumbakonam, for their boundless hospitality.

To the "Gang of Three"—three American mathematicians, admirers and students of Ramanujan, who have helped me to understand his work, and who have read the manuscript along the way, invariably peppering their excellent advice with dollops of needed encouragement: George Andrews, Pennsylvania State University; Richard Askey, University of Wisconsin; and Bruce Berndt, University of Illinois. Without the help of these three men this book could not have been written.

To Robert Rankin of Glasgow, Scotland, whose research into the lives of Ramanujan and Hardy has materially contributed to this book, and who has been unfailingly patient in responding to my trans-Atlantic queries.

To Freeman Dyson, for taking such interest in the book, rounding up old letters, reading the manuscript, and making important suggestions.

To Barbara Grossman, whose idea this book was, and who has placed the full force of her personality behind it. To Joy Smith for her diligence and unfailing good nature. To Erich Hobbing, David Frost, and others at Scribner's for their help and talents. To Zoë English Kharpertian for her fine job of copyediting.

To Vicky Bijur, my agent, for the important role she played in bringing this book into being, and for the almost frightening efficiency with which she has acted in my behalf along the way.

To Jane Alexander, who sent me to India the first time and has been a friend since.

So many have helped, in small ways and large, to make this book possible that it seems scarcely possible to remember them all. I apologize to any I may have inadvertently omitted.

In America: Sudarshan Bhatia and Asha Rijhsinghani. V. Anantharaman and Malini. Ashvin Rajan. Ranjan Roy, Beloit College. Henry S. Tropp, Humboldt State University. S. Chandrasekhar, University of Chicago. Arthur Magida. Gary Leventhal. Ann Finkbeiner and Cal Walker. Ken Gershman. Lee and Phyllis Jaslow. Mildred Foster. Carolyne, Kathy, and everyone at the Red Balloon. William Dyal and Thomas Slakey, St. John's College, Annapolis. Warren Kornberg, *Mosaic* magazine. Jacek Mostwin, Johns Hopkins Hospital. Steve Fisher. Alan Sea. Adrianne Pierce, Johns Hopkins University. John Halperin, Vanderbilt University. Maurice St. Pierre, Morgan State University. Suzanne Holland, *Harvard* magazine. George W. Comstock, Johns Hopkins University School of Hygiene and Public Health. S. Bhargava, University of Illinois, Urbana. Wayne Markert, University of Baltimore.

In Britain: Emma and Jonathan Leigh, Cranleigh School. Kevin Gray, Trinity College. Béla Bollobás, Trinity College. Charles Burkill, Cambridge. Mary Cartwright, Cambridge. Rajiv Krishnan, Christ's College. Roger David Hugh Custance, Winchester College. Guy Newcombe, Trinity College. Vince Darley, Trinity College. John Vickers, St. John's College. R. Robson, Trinity College. Constance Willis, Cambridge. Deborah and Bryan J. B. Gauld, Putney. Theodor Schuchat and Louise Harper, London. Paul and Clare Friedman, London. S. J. Mann, Cranleigh School. Susan M. Oakes, London Mathematical Society. J. D. Webb, Cambridge City Council. Tom Doig, Cambridge Folk Museum. Pat Kattenhorn, India Office Library.

* * *

In India: Janaki Ammal, the widow of Ramanujan, and her son, W. Narayanan, and his family, Triplicane, Madras. T. V. Rangaswami ("Ragami"), Triplicane, Madras. A. P. Victor, Kuala Lumpur, Malaysia. P. K. Srinivasan, Association of Mathematics Teachers of India, Madras. T. M. Srinivasan and T. M. Kasturirangan, Madras. John Herbet Anand, Mylapore, Madras. R. Janarthanan, Kumbakonam. R. Viswanathan, headmaster, and V. Vaidynathan, assistant headmaster, Town Higher Secondary School, Kumbakonam. S. Subbarathinan, Erode. A. Nazimuddin and H. Sharmila, Madras. K. Elangovan, Kodumudi. R. Chandrasekhar, Kodumudi. A. Sanguttuvan, Erode. L. Rajagopolan, S. Elango and D. G. Ramamurthy, Kumbakonam. A. Saranathan and his family, Kumbakonam. S. Govindaraja Battachariar and P. Vasunathan, Uppiliapan Koil, Thirunageswaram. T. U. Bhanumurthy and Kalyanalakshmi, Madras. M. Vinnanasan, Kumbakonam. K. S. Padmanabhan, Ramanujan Institute, Madras. K. Narayanan and A. V. Chandrasekhar, *The Hindu*. A. Ranganathan, Madras. K. Rajamani, Kumbakonam. T. C. Krishna and his family, Madras. Bhama Srinivasan, University of Illinois, Chicago. P. P. Kulkarni, Nagpur. And, of course, the unforgettable Hari, of Madras.

I wish also to express my gratitude to the staffs of the many libraries in the United States, England, and India who have helped me in the quiet, faceless, but unfailingly competent way we all expect of them. That, of course, is the problem: it's so easy to take the library for granted when it has the book or document you need, and to grumble when it doesn't. But the very concept of the library, as a place to store, preserve, and give access to books, was thrown into sharp focus during my five weeks in India. There, libraries cannot always treat their treasures with the expensive care Western libraries can lavish on theirs. In one, I found books and journals set out on the floor, piled this way and that, the pages of even recent books crumbling, dusty and mildewed. And yet never have I seen libraries so intensively used, books so hungrily devoured. In one small library in Erode, I saw every seat at every table taken, and many people standing in the aisles to read.

My appreciation, then, goes to librarians, archivists, and other staff at the following institutions:

Baltimore: The Milton S. Eisenhower Library of Johns Hopkins University. William H. Welch Medical Library. Enoch Pratt Free Library—central library, St. Paul Street branch, and telephone reference service.

Washington, D.C.: Library of Congress.

Cambridge, Mass.: Harvard University Archives.

Princeton, N.J.: Seely G. Mudd Manuscript Library.

Cambridge, England: University Library of Cambridge University, and the libraries of these Cambridge colleges: Trinity, St. John's, Gonville & Caius. Scientific Periodicals Library. Cambridgeshire Collection. Cambridge Folk Museum.

London: Royal Society Library. India Office Library. University of London. National Maritime Museum, Greenwich.

Elsewhere in England: Guildford Public Library (Local Studies Collection). Cranleigh School, library and archives. Cranleigh Library. St. Catherine's School Library, Bramley. New College Library, Oxford. University of Reading.

Erode: Public Library.

Madras: Connemara Library, University of Madras. Archives of *The Hindu*. British Embassy Library. Fort St. George Museum. Madras Museum.

Index

Index